This book is due for return on or before the last date shown below.

Lecture Notes in Mathematics

Edited by J.-M. Morel, F. Takens and B. Teissier

Editorial Policy for Multi-Author Publications: Summer Schools / Intensive Courses

1. Lecture Notes aim to report new developments in all areas of mathematics and their applications – quickly, informally and at a high level. Mathematical texts analysing new developments in modelling and numerical simulation are welcome. Manuscripts should be reasonably self-contained and rounded off. Thus they may, and often will, present not only results of the author but also related work by other people. They should provide sufficient motivation, examples and applications. There should also be an introduction making the text comprehensible to a wider audience. This clearly distinguishes Lecture Notes from journal articles or technical reports which normally are very concise. Articles intended for a journal but too long to be accepted by most journals, usually do not have this "lecture notes" character.

2. In general SUMMER SCHOOLS and other similar INTENSIVE COURSES are held to present mathematical topics that are close to the frontiers of recent research to an audience at the beginning or intermediate graduate level, who may want to continue with this area of work, for a thesis or later. This makes demands on the didactic aspects of the presentation. Because the subjects of such schools are advanced, there often exists no textbook, and so ideally, the publication resulting from such a school could be a first approximation to such a textbook.

   Usually several authors are involved in the writing, so it is not always simple to obtain a unified approach to the presentation.

   For prospective publication in LNM, the resulting manuscript should not be just a collection of course notes, each of which has been developed by an individual author with little or no co-ordination with the others, and with little or no common concept. The subject matter should dictate the structure of the book, and the authorship of each part or chapter should take secondary importance. Of course the choice of authors is crucial to the quality of the material at the school and in the book, and the intention here is not to belittle their impact, but simply to say that the book should be planned to be written by these authors jointly, and not just assembled as a result of what these authors happen to submit.

   This represents considerable preparatory work (as it is imperative to ensure that the authors know these criteria before they invest work on a manuscript), and also considerable editing work afterwards, to get the book into final shape. Still it is the form that holds the most promise of a successful book that will be used by its intended audience, rather than yet another volume of proceedings for the library shelf.

3. Manuscripts should be submitted (preferably in duplicate) either to Springer's mathematics editorial in Heidelberg, or to one of the series editors (with a copy to Springer). Volume editors are expected to arrange for the refereeing, to the usual scientific standards, of the individual contributions. If the resulting reports can be forwarded to us (series editors or Springer) this is very helpful. If no reports are forwarded or if other questions remain unclear in respect of homogeneity etc, the series editors may wish to consult external referees for an overall evaluation of the volume. A final decision to publish can be made only on the basis of the complete manuscript; however a preliminary decision can be based on a pre-final or incomplete manuscript. The strict minimum amount of material that will be considered should include a detailed outline describing the planned contents of each chapter.

   Volume editors and authors should be aware that incomplete or insufficiently close to final manuscripts almost always result in longer evaluation times. They should also be aware that parallel submission of their manuscript to another publisher while under consideration for LNM will in general lead to immediate rejection.

Continued on inside back-cover

Lecture Notes in Mathematics

1945

Editors:
J.-M. Morel, Cachan
F. Takens, Groningen
B. Teissier, Paris

Editors *Mathematical Biosciences Subseries:*
P.K. Maini, Oxford

Fred Brauer · Pauline van den Driessche
Jianhong Wu (Eds.)

Mathematical Epidemiology

With Contributions by:

L.J.S. Allen · C.T. Bauch · C. Castillo-Chavez · D.J.D. Earn
Z. Feng · M.A. Lewis · J. Li · M. Martcheva · M. Nuño
J. Watmough · M.J. Wonham · P. Yan

 Springer

Editors

Fred Brauer
Department of Mathematics
University of British Columbia
Vancouver, B.C. V6T 1Z2, Canada
brauer@math.ubc.ca

Jianhong Wu
Center for Disease Modeling
Department of Mathematics and Statistics
York University
Toronto, Ontario M3J 1P3, Canada
wujh@mathstat.yorku.ca

Pauline van den Driessche
Department of Mathematics and Statistics
University of Victoria
PO BOX 3060 STN CSC
Victoria, B.C. V8W 3R4, Canada
pvdd@math.uvic.ca

ISBN: 978-3-540-78910-9   e-ISBN: 978-3-540-78911-6
DOI: 10.1007/978-3-540-78911-6

Lecture Notes in Mathematics ISSN print edition: 0075-8434
                              ISSN electronic edition: 1617-9692

Library of Congress Control Number: 2008923889

Mathematics Subject Classification (2000): 92D30, 92D25, 34D05, 60J80, 60G05

Cover design: WMXDesign GmbH, Heidelberg
The artwork on the cover page was designed by Ping Yan

Printed on acid-free paper

9 8 7 6 5 4 3 2 1

springer.com

Preface

Mathematical epidemiology has a long history, going back to the smallpox model of Daniel Bernoulli in 1760. Much of the basic theory was developed between 1900 and 1935, and there has been steady progress since that time. More recently, models to evaluate the effect of control measures have been used to assist in the formulation of policy decisions, notably for the foot and mouth disease outbreak in Great Britain in 2001. The SARS (Severe Acute Respiratory Syndrome) epidemic of 2002–2003 aroused great interest in the use of mathematical models to predict the course of an infectious disease and to compare the effects of different control strategies. This revived interest has been reinforced by the current threat of an influenza pandemic.

Mathematical epidemiology differs from most sciences as it does not lend itself to experimental validation of models. Experiments are usually impossible and would probably be unethical. This gives great importance to mathematical models as a possible tool for the comparison of strategies to plan for an anticipated epidemic or pandemic, and to deal with a disease outbreak in real time.

In response to the SARS epidemic, a team was formed by a Canadian center, MITACS (Mathematics for Information Technology and Complex Systems) to work on models for the transmission dynamics of infectious diseases, with a specific goal of evaluating possible management strategies. This team soon recognized that for mathematical modeling to be of assistance in making health policy decisions, it would be necessary to increase the number of mathematical modelers in epidemiology and also to persuade decision makers in the health sciences that mathematical modeling could be useful for them. In pursuit of these goals, a summer school in mathematical epidemiology was developed in 2004 for graduate students from mathematical and biological sciences. This school consisted of a series of lectures on various topics in mathematical epidemiology together with projects done by groups of students, each group containing students from various disciplines and with different levels of experience. In the summer of 2006, another summer school was held, again for a mixed group but this time including a substantial number of people from epidemiology and health sciences. The 2006 summer school also included a few public lectures that covered a wide range of issues and diseases of great interest to public health, illustrating the general framework and abstract mathematical theory in an applied setting. The experience from these courses was that the projects were an essential valuable component of the school, and the mixing of students in the project groups had a very positive effect for communication between disciplines, but also that differences in mathematical backgrounds caused some difficulties.

This book consists of lecture notes intended for such a school. They cover the main aspects of mathematical modeling in epidemiology and contain more than enough material for a concentrated course, giving students additional resource material to pursue the subject further. Our goal is to persuade epidemiologists and public health workers that mathematical modeling can be of use to them. Ideally, it would teach the art of mathematical modeling, but we believe that this art is difficult to teach and is better learnt by doing. For this reason, we have settled on the more modest goal of presenting the main mathematical tools that will be useful in analyzing models and some case studies as examples. We hope that understanding of the case studies will give some insights into the process of mathematical modeling.

We could give a flow chart for the use of this volume but it would not be very interesting, as there is very little interdependence between chapters. Everyone should read the opening chapter as a gentle introduction to the subject, and Chap. 2 on compartmental models is essential for all that follows. Otherwise, the only real chapter dependence is that Chap. 10 requires an understanding of Chap. 3. The first four chapters are basic material, the next six chapters are developments of the basic theory, and the last four chapters are case studies on childhood diseases, influenza, and West Nile virus. The two case studies on influenza deal with different aspects of the disease and do not depend on each other. There are also some suggested projects, taken in part from the recent book "A Course in Mathematical Biology: Quantitative Modeling with Mathematical and Computational Methods" by Gerda de Vries, Thomas Hillen, Mark Lewis, Johannes Müller, and Birgit Schönfisch, *Mathematical Modeling and Computation* 12, SIAM, Philadelphia (2006).

The necessary mathematical background varies from chapter to chapter but a knowledge of basic calculus, ordinary differential equations, and some matrix algebra is essential for understanding this volume. In addition, Chaps. 3, 4, and 10 require some background in probability. Review notes on calculus, matrix algebra, differential equations, and probability have been prepared and may be downloaded at the web site of the Center for Disease Modelling (www.cdm.yorku.ca). Some chapters use more advanced mathematical topics. Some topics in linear algebra beyond elementary matrix theory are needed for Chaps. 6–8. Hopf bifurcations are used in Chaps. 5 and 13. Some knowledge of partial differential equations is needed for Chapters 8, 9, and 13. Preparation of review notes on these topics is in process.

Students from epidemiology and health sciences would probably need some review of basic mathematics, and a course for students with a meagre mathematical background should probably be restricted to the first two chapters and the case studies in Chaps. 12 and 14. A course for students with a strong mathematical background could include any of the chapters depending on the interests of instructors and students. For example, a course emphasizing stochastic ideas could consist of the first four chapters, Chap. 10, plus some of the case studies in the last four chapters. We believe that every course should include some case studies.

We plan to use this volume as text material for future courses of various lengths and with a variety of audiences. One of the main goals of the courses on which this volume is based was to include students from different disciplines. Our experience suggests that a future course aimed at a mixed group should include a variety of non-mathematical case study descriptions and should probably begin with separate tracks for "calculus users" who are comfortable with basic mathematics but have little or no experience with epidemiology, and "calculus victims" who may have studied calculus but in a form that did not persuade them of the value of applications of mathematics to other sciences. The negative experiences that many students in the health sciences may have had in the past are a substantial obstacle that needs to be overcome.

The chapters of this book are independent units and have different levels of difficulty, although there is some overlap. Tremendous efforts were made to ensure that these lectures are coherent and complementary, but no attempt has been made to achieve a unified writing style, or even a unified notation for this book. Because mathematical epidemiology is a rapidly developing field, one goal of any course should be to encourage students to go to the current literature, and experience with different perspectives should be very helpful in being able to assimilate current developments.

These lecture notes would have been impossible without the two summer schools funded by MITACS, as well as the Banff International Research Station for Mathematical Innovation and Discovery (BIRS), The Fields Institute for Research in Mathematical Sciences, the Mathematical Sciences Research Institute and the Pacific Institute for the Mathematical Sciences. We thank these funding agencies for their support, as well as BIRS and York University for supplying physical facilities. We thank Dr. Guojun Gan and Dr. Hongbin Guo for help in assembling the book manuscript. Finally, we thank all the lecturers as well as Dr. Julien Arino and Dr. Lin Wang for technical support during the summer schools, and the summer school students who contributed much useful feedback.

Vancouver, BC   *Fred Brauer*
Victoria, BC   *P. van den Driessche*
Toronto, ON   *Jianhong Wu*
February 2008

Contents

Part II Advanced Modeling and Heterogeneities

Part III   Case Studies

11   The Role of Mathematical Models in Explaining
Recurrent Outbreaks of Infectious Childhood Diseases.... 297

Chris T. Bauch

List of Contributors

Linda J.S. Allen
Department of Mathematics
and Statistics
Texas Tech University
Lubbock, TX 79409-1042
USA
linda.j.allen@ttu.edu

Chris T. Bauch
Department of Mathematics
and Statistics
University of Guelph
Guelph, ON
Canada N1G 2W1
cbauch@uoguelph.ca

Fred Brauer
Department of Mathematics
University of British Columbia
1984, Mathematics Road
Vancouver, BC, Canada V6T 1Z2
brauer@math.ubc.ca

C. Castillo-Chavez
Department of Mathematics
and Statistics
Arizona State University
Temple, AZ 85287-1804
USA
ccchavez@asu.edu

P. van den Driessche
Department of Mathematics
and Statistics
University of Victoria
P.O. BOX 3060, STN CSC
Victoria, BC
Canada V8W 3R4
pvdd@math.uvic.ca

David J.D. Earn
Department of Mathematics
and Statistics
McMaster University
Hamilton, ON, Canada L8S 4K1
earn@math.mcmaster.ca

Z. Feng
Department of Mathematics
Purdue University
West Lafayette
IN 47907, USA
zfeng@math.purdue.edu

Mark A. Lewis
Centre for Mathematical Biology
Department of Mathematical
and Statistical Sciences
University of Alberta
Edmonton, AB
Canada T6G 2G1
mlewis@math.ualberta.ca

Jia Li
Department of Mathematical
Sciences, University of Alabama
in Huntsville, 301, Sparkman Dr
Huntsville, AL 35899
USA
li@math.uah.edu

M. Martcheva
Department of Mathematics
Florida University, Gainesville
FL 32611-8105
USA
maia@math.ufl.edu

M. Nuño
Department of Biostatistics
UCLA School of Public Health
Los Angeles, CA 90095-1772
USA
miriamnuno@ucla.edu

James Watmough
Department of Mathematics
and Statistics
University of New Brunswick
NB, Canada E3B 5A3
watmough@unb.ca

Marjorie J. Wonham
Centre for Mathematical Biology
Department of Mathematical
and Statistical Sciences
University of Alberta
Edmonton, AB
Canada T6G 2G1
mwonham@ualberta.ca

Jianhong Wu
Center for Disease Modeling
Department of Mathematics
and Statistics
York University, Toronto
ON, Canada M3J 1P3
wujh@mathstat.yorku.ca

Ping Yan
Centre for Communicable Diseases
and Infection Control
Infectious Diseases and Emergency
Preparedness Branch
Public Health Agency of Canada
100 Elangtine Drive, AL0602-B
Tunney's Pasture
Ottawa, ON
Canada K1A 0K9
Ping_Yan@phac-aspc.gc.ca

Part I
Introduction and General Framework

Chapter 1
A Light Introduction to Modelling Recurrent Epidemics

David J.D. Earn

Abstract Epidemics of many infectious diseases occur periodically. Why?

1.1 Introduction

There are many excellent books that provide broad and deep introductions to the mathematical theory of infectious disease epidemics, ranging from monographs and textbooks [1–7] to collections of articles from workshops and conferences [8–12]. My goal in this article is to spark an interest in mathematical epidemiology that might inspire you to dig into the existing literature (starting with the rest of this book [13]) and, perhaps, to engage in some research yourself in this fascinating area of science.

I will discuss some famous epidemics that present challenging theoretical questions, and – without getting bogged down in technical details you can find elsewhere – I will try to convince you that you can fairly easily build and analyze simple models that help us understand the complex patterns evident in these data. Not wishing to give you a false impression of the field, I will also briefly mention some epidemic patterns that do not appear to be explicable in terms of simple models (at least, not in terms of simple models we have thought of!).

Several of the following sections are based in part on an even lighter introduction to the subject of mathematical epidemiology that I wrote for a high school mathematics magazine [14]. Here, I do not limit myself to high school mathematics, but I hope the bulk of the article will be easily accessible to you if you have had an elementary course in ordinary differential equations.

Department of Mathematics and Statistics
McMaster University, Hamilton, ON, Canada L8S 4K1
earn@math.mcmaster.ca

1.2 Plague

One of the most famous examples of an epidemic of an infectious disease in a
human population is the Great Plague of London, which took place in 1665–
1666. We know quite a lot about the progression of the Great Plague because
weekly bills of mortality from that time have been retained. A photograph
of such a bill is shown in Fig. 1.1. Note that the report indicates that the
number of deaths from plague (5,533) was more than 37 times the number
of births (146) in the week in question, and that wasn't the worst week! (An
even worse plague occurred in the fourteenth century, but no detailed records
of that epidemic are available.)

Putting together the weekly counts of plague deaths from all the relevant
mortality bills, we can obtain the *epidemic curve* for the Great Plague, which
I've plotted in the top left panel of Fig. 1.2. The characteristic exponential
rise, turnover and decline is precisely the pattern predicted by the classic
susceptible-infective-recovered (SIR) model of Kermack and McKendrick [15]
that I describe below. While this encourages us to think that mathematical
modelling can help us understand epidemics, some detailed features of the
epidemic curve are not predicted by the simple SIR model. For example,
the model does not explain the jagged features in the plotted curve (and
there would be many more small ups and downs if we had a record of daily
rather than weekly deaths). However, with some considerable mathematical
effort, these "fine details" can be accounted for by replacing the differential
equations of Kermack and McKendrick with equations that include stochastic (i.e., random) processes [2]. We can then congratulate ourselves for our
modelling success... until we look at more data.

The bottom left panel of Fig. 1.2 shows weekly mortality from plague in
London over a period of 70 years. The Great Plague is the rightmost (and
highest) peak in the plot. You can see that on a longer timescale, there was a
complex pattern of plague epidemics including extinctions and re-emergences.
This cannot be explained by the basic SIR model (even if we reformulate it
using stochastic processes). The trouble is likely that we have left out a key
biological fact: there is a reservoir of plague in rodents, so it can persist for
years, unnoticed by humans, and then re-emerge suddenly and explosively. By
including the rodents and aspects of spatial spread in a mathematical model,
it has recently been possible to make sense of the pattern of seventeenth
century plague epidemics in London [16]. Nevertheless, some debate continues
as to whether all those plagues were really caused by the same pathogenic
organism.

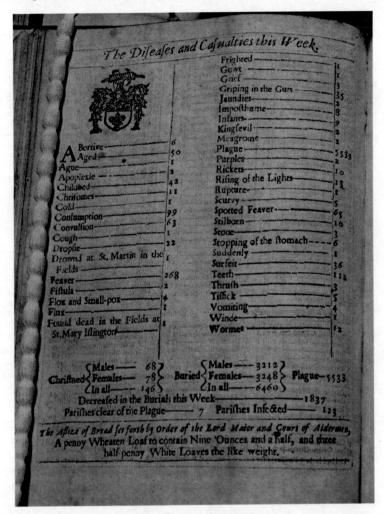

Fig. 1.1 A bill of mortality for the city of London, England, for the week of 26 September to 3 October 1665. This photograph was taken by Claire Lees at the Guildhall in London, England, with the permission of the librarian

1.3 Measles

A less contentious example is given by epidemics of measles, which are definitely caused by a well-known virus that infects the respiratory tract in humans and is transmitted by airborne particles [3]. Measles gives rise to characteristic red spots that are easily identifiable by physicians who have seen many cases, and parents are very likely to take their children to a doctor when such spots are noticed. Consequently, the majority of measles cases in

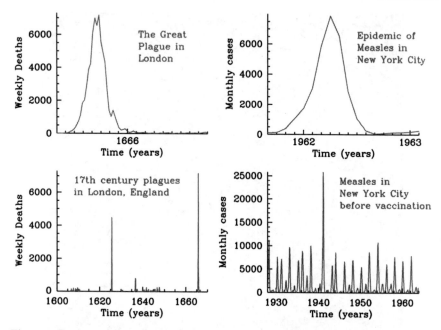

Fig. 1.2 Epidemic curves for plague in London (*left panels*) and measles in New York City (*right panels*). For plague, the curves show the number of deaths reported each week. For measles, the curves show the number of cases reported each month. In the *top panels*, the small ticks on the time axis occur at monthly intervals

developed countries end up in the office of a doctor (who, in many countries, is required to report observed measles cases to a central body). The result is that the quality of reported measles case data is unusually good, and it has therefore stimulated a lot of work in mathematical modelling of epidemics.

An epidemic curve for measles in New York City in 1962 is shown in the top right panel of Fig. 1.2. The period shown is 17 months, exactly the same length of time shown for the Great Plague of London in the top left panel. The 1962 measles epidemic in New York took off more slowly and lasted longer than the Great Plague of 1665. Can mathematical models help us understand what might have caused these differences?

1.4 The SIR Model

Most epidemic models are based on dividing the host population (humans in the case of this article) into a small number of compartments, each containing individuals that are identical in terms of their status with respect to the disease in question. In the SIR model, there are three compartments:

- *Susceptible*: individuals who have no immunity to the infectious agent, so might become infected if exposed
- *Infectious*: individuals who are currently infected and can transmit the infection to susceptible individuals who they contact
- *Removed*: individuals who are immune to the infection, and consequently do not affect the transmission dynamics in any way when they contact other individuals

It is traditional to denote the number of individuals in each of these compartments as S, I and R, respectively. The total host population size is $N = S + I + R$.

Having compartmentalized the host population, we now need a set of equations that specify how the sizes of the compartments change over time. Solutions of these equations will yield, in particular, $I(t)$, the size of the infectious compartment at time t. A plot of $I(t)$ should bear a strong resemblance to observed epidemic curves if this is a reasonable model.

The numbers of individuals in each compartment must be integers, of course, but if the host population size N is sufficiently large we can treat S, I and R as continuous variables and express our model for how they change in terms of a system of differential equations,

$$\frac{dS}{dt} = -\beta SI, \tag{1.1a}$$

$$\frac{dI}{dt} = \beta SI - \gamma I. \tag{1.1b}$$

Here, the *transmission rate* (per capita) is β and the *recovery rate* is γ (so the *mean infectious period* is $1/\gamma$). Note that I have not written a differential equation for the number of removed individuals. The appropriate equation is $dR/dt = \gamma I$ (outflow from the I compartment goes into the R compartment) but since R does not appear in (1.1a) and (1.1b) the equation for dR/dt has no effect on the dynamics of S and I (formalizing the fact that removed individuals cannot affect transmission). This basic SIR model has a long history [15] and is now so standard that you can even find it discussed in some introductory calculus textbooks [17].

If everyone is initially susceptible ($S(0) = N$), then a newly introduced infected individual can be expected to infect other people at the rate βN during the expected infectious period $1/\gamma$. Thus, this first infective individual can be expected to infect $\mathbb{R}_0 = \beta N/\gamma$ individuals. The number $\mathbb{R}_0$ is called the *basic reproduction number* and is unquestionably the most important quantity to consider when analyzing any epidemic model for an infectious disease. In particular, $\mathbb{R}_0$ determines whether an epidemic can occur at all; to see this for the basic SIR model, note in (1.1a) and (1.1b) that I can never increase unless $\mathbb{R}_0 > 1$. This makes intuitive sense, since if each individual transmits the infection to an average of less than one individual then the number of cases must decrease with time.

So, how do we obtain $I(t)$ from the SIR model?

1.5 Solving the Basic SIR Equations

If we take the ratio of (1.1a) and (1.1b) we obtain

$$\frac{dI}{dS} = -1 + \frac{1}{\mathbb{R}_0 S}\,, \tag{1.2}$$

which we can integrate immediately to find

$$I = I(0) + S(0) - S + \frac{1}{\mathbb{R}_0} \ln\left[S/S(0)\right]. \tag{1.3}$$

This is an exact solution, but it gives I as a function of S, not as a function of t. Plots of $I(S)$ for various $\mathbb{R}_0$ show the phase portraits of solutions (Fig. 1.3) but do not give any indication of the time taken to reach any particular points on the curves. While the exact expression for the phase portrait may seem like great progress, it is unfortunately not possible to obtain an exact expression for $I(t)$, even for this extremely simple model.

In their landmark 1927 paper, Kermack and McKendrick [15] found an approximate solution for $I(t)$ in the basic SIR model, but their approximation is valid only at the very beginning of an epidemic (or for all time if $\mathbb{R}_0$ is unrealistically close to unity) so it would not appear to be of much use for understanding measles, which certainly has $\mathbb{R}_0 > 10$.

Computers come to our rescue. Rather than seeking an explicit formula for $I(t)$ we can instead obtain an accurate numerical approximation of the solution. There are many ways to do this [18], but I will briefly mention the simplest approach (Euler's method), which you can implement in a few minutes in any standard programming language, or even a spreadsheet.

Over a sufficiently small time interval Δt, we can make the approximation $dS/dt \simeq \Delta S/\Delta t$, where $\Delta S = S(t+\Delta t) - S(t)$. If we now solve for the number of susceptibles a time Δt in the future, we obtain

$$S(t + \Delta t) = S(t) - \beta S(t)I(t)\Delta t\,. \tag{1.4a}$$

Similarly, we can approximate the number of infectives at time $t + \Delta t$ as

$$I(t + \Delta t) = I(t) + \beta S(t)I(t)\Delta t - \gamma I(t)\Delta t\,. \tag{1.4b}$$

Equations (1.4a) and (1.4b) together provide a scheme for approximating solutions of the basic SIR model. To implement this scheme on a computer, you need to decide on a suitable small time step Δt. If you want to try

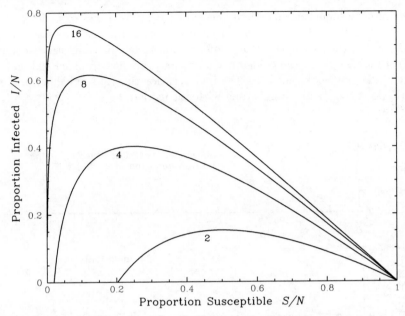

Fig. 1.3 Phase portraits of solutions of the basic SIR model ((1.1a) and (1.1b)) for a newly invading infectious disease. The curves are labelled by the value of the basic reproduction number $\mathbb{R}_0$ (2, 4, 8 or 16). For each curve, the initial time is at the *bottom right corner* of the graph ($I(0) \simeq 0$, $S(0) \simeq N$). All solutions end on the S axis ($I \to 0$ as $t \to \infty$). A simple analytical formula for the phase portrait is easily derived (1.3); from this it is easy to show that $\lim_{t \to \infty} S(t) > 0$ regardless of the value of $\mathbb{R}_0$ (though, as is clear from the phase portraits plotted for $\mathbb{R}_0 = 8$ and 16, nearly everyone is likely to be infected eventually if $\mathbb{R}_0$ is high)

this, I suggest taking Δt to be one tenth of a day. I should point out that I am being extremely cavalier in suggesting the above method. Do try this, but be forewarned that you can easily generate garbage using this simple approach if you're not careful. (To avoid potential confusion, include a line in your program that checks that $S(t) \geq 0$ and $I(t) \geq 0$ at all times. Another important check is to repeat your calculations using a much smaller Δt and make sure your results don't change.)

In order for your computer to carry out the calculations specified by (1.4a) and (1.4b), you need to tell it the parameter values (β and γ, or $\mathbb{R}_0$, N and γ) and initial conditions ($S(0)$ and $I(0)$). For measles, estimates that are independent of the case report data that we're trying to explain indicate that the mean infectious period is $1/\gamma \sim 5$ days and the basic reproduction number is $\mathbb{R}_0 \sim 18$ [3]. The population of New York City in 1960 was $N = 7,781,984$.

If we now assume one infectious individual came to New York before the epidemic of 1962 ($I(0) = 1$), and that everyone in the city was susceptible ($S(0) = N$), then we have enough information to let the computer calculate $I(t)$. Doing so yields the epidemic curve shown in the top panel of Fig. 1.4, which does *not* look much like the real data for the 1962 epidemic in New York. So is there something wrong with our model?

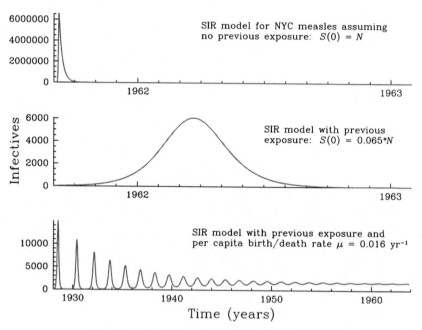

Fig. 1.4 Epidemic curves for measles in New York City, generated by the basic SIR model. The curves show the number of infectives $I(t)$ at time t. In the *top two panels*, the small ticks on the time axis occur at monthly intervals. The parameter values and initial conditions are discussed in the main text, except for the initial proportion susceptible used to generate the *bottom two panels* ($S(0)/N = 0.065$). This initial condition was determined based on the SIR model with vital dynamics, as specified by (1.5a) and (1.5b). The proportion susceptible at equilibrium is $\hat{S} = 1/\mathbb{R}_0 = 1/18 \simeq 0.056$. At the start of each epidemic cycle that occurs as the system approaches the equilibrium, the proportion susceptible must be higher than $\hat{S}$

No, but there is something very wrong with our initial conditions. The bottom right panel of Fig. 1.2 shows reported measles cases in New York City for a 36-year period, the end of which includes the 1962 epidemic. Evidently, measles epidemics had been occurring in New York for decades with no sign of extinction of the virus. In late 1961, most of New York's population had already had measles and was already immune, and the epidemic certainly didn't start because one infectious individual came to the city. The assumptions that $I(0) = 1$ and $S(0) = N$ are ridiculous. If, instead, we take $I(0) = 123 * (5/30)$ (the number of reported cases in September 1961 times the infectious period as a proportion of the length of the month) and $S(0) = 0.065 * N$, then we obtain the epidemic curve plotted in the middle panel of Fig. 1.4, which is much more like the observed epidemic curve of Fig. 1.2 (top right panel).

This is progress – we have a model that can explain a single measles epidemic in New York City – but the model cannot explain the recurrent epidemics observed in the bottom right panel of Fig. 1.2. This is not because we still don't have exactly the right parameter values and initial conditions: no parameter values or initial conditions lead to recurrent epidemics in this simple model. So, it would seem, there must be some essential biological mechanism that we have not included in our model. What might that be?

Let's think about why a second epidemic cannot occur in the model we've discussed so far. The characteristic turnover and decline of an epidemic curve occurs because the pathogen is running out of susceptible individuals to infect. To stimulate a second epidemic, there must be a source of susceptible individuals. For measles, that source cannot be previously infected people, because recovered individuals retain lifelong immunity to the virus. So who is it?

Newborns. They typically acquire immunity from their mothers, but this wanes after a few months. A constant stream of births can provide the source we're looking for.

1.6 SIR with Vital Dynamics

If we expand the SIR model to include B births per unit time and a natural mortality rate μ (per capita) then our equations become

$$\frac{dS}{dt} = B - \beta SI - \mu S, \tag{1.5a}$$

$$\frac{dI}{dt} = \beta SI - \gamma I - \mu I. \tag{1.5b}$$

The timescale for substantial changes in birth rates (decades) is generally much longer than a measles epidemic (a few months), so we'll assume that the population size is constant (thus $B = \mu N$, so there is really only one new parameter in the above equations and we can take it to be B). As before, we can use Euler's trick to convert the equations above into a scheme that enables a computer to generate approximate solutions. An example is shown in the bottom panel of Fig. 1.4, where I have taken the birth rate to be $B = 126,372$ per year (the number of births in New York City in 1928, the

first year for which we have data). The rest of the parameters and initial
conditions are as in the middle panel of the figure.

Again we seem to be making progress. We are now getting recurrent epidemics, but the oscillations in the numbers of cases over time damp out, eventually reaching an equilibrium $(\hat{S}, \hat{I})$. Of course, the bottom plot in Fig. 1.4
shows what happens for only one set of possible initial conditions. Perhaps for
different initial conditions the oscillations don't damp out? If you try a different set of initial conditions – or lots of different sets of initial conditions –
then I guarantee that you will see the same behaviour. The system will always
undergo damped oscillations and converge to $(\hat{S}, \hat{I})$. How can I be so sure,
you might ask?

First of all, by setting the derivatives in (1.5a) and (1.5b) to zero, you can
easily calculate (in terms of the model's parameters) expressions for $\hat{S}$ and $\hat{I}$
that are positive (hence meaningful) provided $\mathbb{R}_0 > 1$. Then, by linearizing
(1.5a) and (1.5b) about the equilibrium and computing the eigenvalues of the
Jacobian matrix for the system, you will find that the equilibrium is locally
stable (the eigenvalues have negative real parts) and approach to the equilibrium is oscillatory (the eigenvalues have non-zero imaginary parts) [3, 7].
But maybe if we are far enough from the equilibrium undamped oscillations
are possible?

No, we can prove rigorously that the equilibrium $(\hat{S}, \hat{I})$ is *globally asymptotically stable*, i.e., all initial conditions with $S(0) > 0$ and $I(0) > 0$ yield
solutions that converge onto this equilibrium. One way to see this is to scale
the variables by population size $(S \rightarrow S/N, I \rightarrow I/N)$ and consider the
function

$$V(S, I) = S - \hat{S} \log S + I - \hat{I} \log I, \qquad S, I \in (0, 1). \qquad (1.6)$$

With a little work you can show that the time derivative of V along solutions
of the model, i.e., $\nabla V \cdot (dS/dt, dI/dt)$ with dS/dt and dI/dt taken from (1.1a)
and (1.1b), is strictly negative for each $S, I \in (0, 1)$. V is therefore a *Lyapunov
function* [19] for the basic SIR model. The existence of such a V ensures the
global asymptotic stability of the equilibrium $(\hat{S}, \hat{I})$ [19].

Finding a Lyapunov function is generally not straightforward, but functions similar the one given in (1.6) have recently been used to prove global stability of equilibria in many epidemic models [20]. The upshot for our present
attempt to understand measles dynamics is that this rigorous argument allows us to rule out the basic SIR model: it cannot explain the real oscillations
in measles incidence in New York City from 1928 to 1964, which showed no
evidence of damping out. Back to the drawing board?

1.7 Demographic Stochasticity

One thing we have glossed over is the presence of noise. While it is true that for sufficiently large population size N it is reasonable to treat S and I as continuous variables, it is not true that the true discreteness of the number of individuals in each compartment has no observable effect. This was recognized by Bartlett [1] who found that a relatively small amount of noise was sufficient to prevent the oscillations of the basic SIR model from damping out. Whether we recast the SIR model as a stochastic process [2,21] or simply add a small noise term to the deterministic equations, we can sustain the oscillations that damp out in the bottom panel of Fig. 1.4.

Again, this is progress that has arisen from an important mechanistic insight. But we are left with another puzzle. If you look carefully at the New York City measles reports in the bottom right panel of Fig. 1.2 you'll see that before about 1945 the epidemics were fairly irregular, whereas after 1945 they followed an almost perfect 2-year cycle. Even with oscillations sustained by noise, the SIR model cannot explain why the measles epidemic pattern in New York City changed in this way. Have we missed another important mechanism?

1.8 Seasonal Forcing

So far, we have been assuming implicitly that the transmission rate β (or, equivalently, the basic reproduction number $\mathbb{R}_0$) is simply a constant and, in particular, that it does not change in time. Let's think about that assumption. The transmission rate is really the product of the rate of contact among individuals and the probability that a susceptible individual who is contacted by an infectious individual will become infected. But the contact rate is *not* constant throughout the year. To see that, consider the fact that in the absence of vaccination the average age at which a person is infected with measles is about 5 years [3], hence most susceptibles are children. Children are in closer contact when school is in session, so the transmission rate must vary seasonally.

A crude approximation of this seasonality is to assume that β varies sinusoidally,

$$\beta(t) = \beta_0(1 + \alpha \cos 2\pi t) \,. \tag{1.7}$$

Here, β_0 is the mean transmission rate, α is the *amplitude* of seasonal variation and time t is assumed to be measured in years. If, as above, β is assumed to be a periodic function (with period 1 year) then the SIR model is said to be *seasonally forced*. We can still use Euler's trick to solve the equations approximately, and I encourage you to do that using a computer for various values of the seasonal amplitude α ($0 \le \alpha \le 1$).

You might think that seasonal forcing is just a minor tweak of the model. In fact, this forcing has an enormous impact on the epidemic dynamics that the model predicts. If you've ever studied the forced pendulum then you might already have some intuition for this. A pendulum with some friction will exhibit damped oscillations and settle down to an equilibrium. But if you tap the pendulum with a hammer periodically then it will never settle down and it can exhibit quite an exotic range of behaviours including chaotic dynamics [22,23] (oscillations that look random). Similarly complex dynamics can occur in the seasonally forced SIR model.

Importantly, with seasonal forcing the SIR model displays *un*damped oscillations (and it does this, incidentally, even in the absence of stochasticity). More than that, for different parameter values, the seasonally forced SIR model can produce all the different types of oscillatory measles patterns I have ever seen in real data. So are we done?

No. As noted in the previous section, the observed measles epidemics in New York City show very clearly that the dynamical pattern changed over time (bottom right panel of Fig. 1.2) and other significant qualitative changes have been observed in measles case series in other places [24]. How can we explain *changes over time* in the *pattern* of measles epidemics?

1.9 Slow Changes in Susceptible Recruitment

Once again, the missing ingredient in the model is a changing parameter value. This time it is the birth rate B, which is not really constant. Birth rates fluctuate seasonally, but to such a small extent that this effect is negligible. What turns out to be more important is the much slower changes that occur in the average birth rate over decades. For example, in New York City the birth rate was much lower during the 1930s (the "Great Depression") than after 1945 (the "baby boom") and this difference accounts for the very different patterns of measles epidemics in New York City during these two time periods [24].

A little more analysis of the SIR model is very useful. Intuitively, reducing the birth rate or increasing the proportion of children vaccinated both affect the rate at which new susceptible individuals are recruited into the population. In fact, it is possible to prove that changes in the birth rate have exactly the same effect on disease dynamics as changes of the same relative magnitude in the transmission rate or the proportion of the population that is vaccinated [24]. This equivalence makes it possible to explain historical case report data for a variety of infectious diseases in many different cities.

Interestingly, it turns out that while most aspects of the dynamics of measles can be explained by employing seasonal forcing without noise, both seasonal forcing and stochasticity are essential to explain the dynamics of other childhood diseases [25].

1.10 Not the Whole Story

I should emphasize that while the seasonally forced SIR model is adequate to explain observed incidence patterns for childhood diseases, it is definitely not adequate for many other diseases that display recurrent epidemics. An important example is influenza (Fig. 1.5). Influenza viruses evolve in ways that evade the human immune system within a few years, making it possible for each of us to be infected by influenza many times. Influenza models must take into account the simultaneous presence in the population of many different strains that interact immunologically and compete for human hosts. The epidemic pattern shown in Fig. 1.5 bears some similarities to the measles pattern shown in Fig. 1.2, and the effects of seasonal forcing help explain this and other influenza patterns to some extent [26]. But we are far from having a simple model that can account for both the annual incidence patterns of influenza in humans and the evolution of the virus [27].

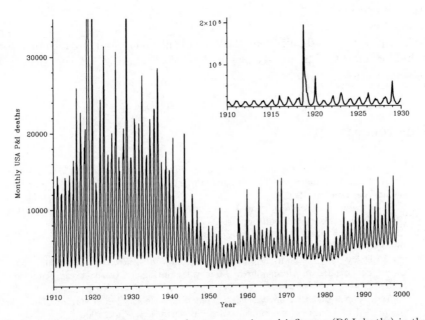

Fig. 1.5 Monthly mortality attributed to pneumonia and influenza (P&I deaths) in the USA in the twentieth century. The inset plot shows the period 1910–1930 on a much larger scale, revealing the magnitude of the three peaks that extend beyond the top of the main panel: 1918–1919, 1919–1920 and 1928–1929. Mortality before 1934 is underestimated. It is traditional to combine pneumonia and influenza because many deaths categorized as having pneumonia as the underlying cause are triggered by an influenza infection

1.11 Take Home Message

One thing that you may have picked up from this article is that successful mathematical modelling of biological systems tends to proceed in steps. We begin with the simplest sensible model and try to discover everything we can about it. If the simplest model cannot explain the phenomenon we're trying to understand then we add more biological detail to the model, and it's best to do this in steps because we are then more likely to be able to determine which biological features have the greatest impact on the behaviour of the model.

In the particular case of mathematical epidemiology, we are lucky that medical and public health personnel have painstakingly conducted surveillance of infectious diseases for centuries. This has created an enormous wealth of valuable data [28] with which to test hypotheses about disease spread using mathematical models, making this a very exciting subject for research.

Acknowledgements

It is a pleasure to thank Sigal Balshine, Will Guest and Fred Brauer for helpful comments. The weekly plague data plotted in Fig. 1.2 were digitized by Seth Earn.

References

1. M. S. Bartlett. *Stochastic population models in ecology and epidemiology*, volume 4 of *Methuen's Monographs on Applied Probability and Statistics*. Spottiswoode, Ballantyne, London, 1960.
2. N. T. J. Bailey. *The Mathematical Theory of Infectious Diseases and Its Applications*. Hafner, New York, second edition, 1975.
3. R. M. Anderson and R. M. May. *Infectious Diseases of Humans: Dynamics and Control*. Oxford University Press, Oxford, 1991.
4. D. J. Daley and J. Gani. *Epidemic modelling, an introduction*, volume 15 of *Cambridge: Studies in Mathematical Biology*. Cambridge university press, Cambridge, 1999.
5. H. Andersson and T. Britton. *Stochastic epidemic models and their statistical analysis*, volume 151 of *Lecture Notes in Statistics*. Springer, New York, 2000.
6. O. Diekmann and J. A. P. Heesterbeek. *Mathematical epidemiology of infectious diseases: model building, analysis and interpretation*. Wiley Series in Mathematical and Computational Biology. Wiley, New York, 2000.
7. F. Brauer and C. Castillo-Chavez. *Mathematical models in population biology and epidemiology*, volume 40 of *Texts in Applied Mathematics*. Springer, New York, 2001.
8. D. Mollison, editor. *Epidemic Models: Their Structure and Relation to Data*. Publications of the Newton Institute. Cambridge University Press, Cambridge, 1995.
9. V. Isham and G. Medley, editors. *Models for Infectious Human Diseases: Their Structure and Relation to Data*. Publications of the Newton Institute. Cambridge University Press, Cambridge, 1996.

10. B. T. Grenfell and A. P. Dobson, editors. *Ecology of Infectious Diseases in Natural Populations*. Publications of the Newton Institute. Cambridge University Press, Cambridge, 1995.

11. C. Castillo-Chavez, with S. Blower, P. van den Driessche, D. Kirschner, and A-A. Yakubu, editors. *Mathematical approaches for emerging and reemerging infectious diseases: an introduction*, volume 125 of *The IMA Volumes in Mathematics and Its Applications*. Springer, New York, 2002.

12. C. Castillo-Chavez, with S. Blower, P. van den Driessche, D. Kirschner, and A-A. Yakubu, editors. *Mathematical approaches for emerging and reemerging infectious diseases: models, methods and theory*, volume 126 of *The IMA Volumes in Mathematics and Its Applications*. Springer, New York, 2002.

13. F. Brauer, P. van den Driessche and J. Wu, editors. *Mathematical Epidemiology* (this volume).

14. D. J. D. Earn. Mathematical modelling of recurrent epidemics. *Pi in the Sky*, 8:14–17, 2004.

15. W. O. Kermack and A. G. McKendrick. A contribution to the mathematical theory of epidemics. *Proceedings of the Royal Society of London Series A*, 115:700–721, 1927.

16. M. J. Keeling and C. A. Gilligan. Metapopulation dynamics of bubonic plague. *Nature*, 407:903–906, 2000.

17. D. Hughes-Hallett, A. M. Gleason, P. F. Lock, D. E. Flath, S. P. Gordon, D. O. Lomen, D. Lovelock, W. G. McCallum, B. G. Osgood, D. Quinney, A. Pasquale, K. Rhea, J. Tecosky-Feldman, J. B. Thrash, and T. W. Tucker. *Applied Calculus*. Wiley, Toronto, second edition, 2002.

18. W. H. Press, S. A. Teukolsky, W. T. Vetterling, and B. P. Flannery. *Numerical Recipes in C*. Cambridge University Press, Cambridge, second edition, 1992.

19. S. Wiggins. *Introduction to applied nonlinear dynamical systems and chaos*, volume 2 of *Texts in Applied Mathematics*. Springer, New York, 2 edition, 2003.

20. A. Korobeinikov and P. K. Maini. A Lyapunov function and global properties for SIR and SEIR epidemiological models with nonlinear incidence. *Mathematical Biosciences and Engineering*, 1(1):57–60, 2004.

21. D. T. Gillespie. A general method for numerically simulating the stochastic time evolution of coupled chemical reactions. *Journal of Computational Physics*, 22:403–434, 1976.

22. J. Gleick. *Chaos*. Abacus, London, 1987.

23. S. H. Strogatz. *Nonlinear Dynamics and Chaos*. Addison Wesley, New York, 1994.

24. D. J. D. Earn, P. Rohani, B. M. Bolker, and B. T. Grenfell. A simple model for complex dynamical transitions in epidemics. *Science*, 287(5453):667–670, 2000.

25. C. T. Bauch and D. J. D. Earn. Transients and attractors in epidemics. *Proceedings of the Royal Society of London Series B-Biological Sciences*, 270(1524):1573–1578, 2003.

26. J. Dushoff, J. B. Plotkin, S. A. Levin, and D. J. D. Earn. Dynamical resonance can account for seasonality of influenza epidemics. *Proceedings of the National Academy of Sciences of the United States of America*, 101(48):16915–16916, 2004.

27. D. J. D. Earn, J. Dushoff, and S. A. Levin. Ecology and evolution of the flu. *Trends in Ecology and Evolution*, 17(7):334–340, 2002.

28. IIDDA. The International Infectious Disease Data Archive, http://iidda.mcmaster.ca.

Chapter 2
Compartmental Models in Epidemiology

Fred Brauer

Abstract We describe and analyze compartmental models for disease transmission. We begin with models for epidemics, showing how to calculate the basic reproduction number and the final size of the epidemic. We also study models with multiple compartments, including treatment or isolation of infectives. We then consider models including births and deaths in which there may be an endemic equilibrium and study the asymptotic stability of equilibria. We conclude by studying age of infection models which give a unifying framework for more complicated compartmental models.

2.1 Introduction

Communicable diseases such as measles, influenza, or tuberculosis, are a fact of modern life. The mechanism of transmission of infections is now known for most diseases. Generally, diseases transmitted by viral agents, such as influenza, measles, rubella (German measles), and chicken pox, confer immunity against reinfection, while diseases transmitted by bacteria, such as tuberculosis, meningitis, and gonorrhea, confer no immunity against reinfection. Other diseases, such as malaria, are transmitted not directly from human to human but by vectors, which are agents (usually insects) who are infected by humans and who then transmit the disease to humans. The West Nile virus involves two vectors, mosquitoes and birds. For sexually transmitted diseases with heterosexual transmission each sex acts as a vector and disease is transmitted back and forth between the sexes.

Department of Mathematics, University of British Columbia, 1984, Mathematics Road, Vancouver BC, Canada V6T 1Z2 brauer@math.ubc.ca

We will be concerned both with epidemics which are sudden outbreaks of a disease, and endemic situations, in which a disease is always present. Epidemics such as the 2002 outbreak of SARS, the Ebola virus and avian flu outbreaks are events of concern and interest to many people. The 1918 Spanish flu epidemic caused millions of deaths, and a recurrence of a major influenza epidemic is a dangerous possibility. An introduction of smallpox is of considerable concern to government officials dealing with terrorism threats.

An endemic situation is one in which a disease is always present. The prevalence and effects of many diseases in less developed countries are probably less well-known but may be of even more importance. Every year millions of people die of measles, respiratory infections, diarrhea and other diseases that are easily treated and not considered dangerous in the Western world. Diseases such as malaria, typhus, cholera, schistosomiasis, and sleeping sickness are endemic in many parts of the world. The effects of high disease mortality on mean life span and of disease debilitation and mortality on the economy in afflicted countries are considerable.

Our goal is to provide an introduction to mathematical epidemiology, including the development of mathematical models for the spread of disease as well as tools for their analysis. Scientific experiments usually are designed to obtain information and to test hypotheses. Experiments in epidemiology with controls are often difficult or impossible to design and even if it is possible to arrange an experiment there are serious ethical questions involved in withholding treatment from a control group. Sometimes data may be obtained after the fact from reports of epidemics or of endemic disease levels, but the data may be incomplete or inaccurate. In addition, data may contain enough irregularities to raise serious questions of interpretation, such as whether there is evidence of chaotic behaviour [12]. Hence, parameter estimation and model fitting are very difficult. These issues raise the question of whether mathematical modeling in epidemiology is of value.

Our response is that mathematical modeling in epidemiology provides understanding of the underlying mechanisms that influence the spread of disease and, in the process, it suggests control strategies. In fact, models often identify behaviours that are unclear in experimental data – often because data are non-reproducible and the number of data points is limited and subject to errors in measurement. For example, one of the fundamental results in mathematical epidemiology is that *most* mathematical epidemic models, including those that include a high degree of heterogeneity, usually exhibit "threshold" behaviour. In epidemiological terms this can be stated as follows: *If the average number of secondary infections caused by an average infective, called the basic reproduction number, is less than one a disease will die out, while if it exceeds one there will be an epidemic.* This broad principle, consistent with observations and quantified via epidemiological models, has been consistently used to estimate the effectiveness of vaccination policies and the likelihood that a disease may be eliminated or eradicated. Hence, even if it is not possible to verify hypotheses accurately, agreement with hypotheses of a qualitative nature is often valuable. Expressions for the basic reproduction number for HIV in various populations have been used to test the possible effectiveness of vaccines that may provide temporary protection by reducing either HIV-infectiousness or susceptibility to HIV. Models are used to estimate how widespread a vaccination plan must be to prevent or reduce the spread of HIV.

In the mathematical modeling of disease transmission, as in most other areas of mathematical modeling, there is always a trade-off between simple models, which omit most details and are designed only to highlight general qualitative behaviour, and detailed models, usually designed for specific situations including short-term quantitative predictions. Detailed models are generally difficult or impossible to solve analytically and hence their usefulness for theoretical purposes is limited, although their strategic value may be high. In these notes we describe simple models in order to establish broad principles. Furthermore, these simple models have additional value as they are the building blocks of models that include more detailed structure.

Many of the early developments in the mathematical modeling of communicable diseases are due to public health physicians. The first known result in mathematical epidemiology is a defense of the practice of inoculation against smallpox in 1760 by Daniel Bernoulli, a member of a famous family of mathematicians (eight spread over three generations) who had been trained as a physician. The first contributions to modern mathematical epidemiology are due to P.D. En'ko between 1873 and 1894 [11], and the foundations of the entire approach to epidemiology based on compartmental models were laid by public health physicians such as Sir Ross, R.A., W.H. Hamer, A.G. McKendrick and W.O. Kermack between 1900 and 1935, along with important contributions from a statistical perspective by J. Brownlee. A particularly instructive example is the work of Ross on malaria. Dr. Ross was awarded the second Nobel Prize in Medicine for his demonstration of the dynamics of the transmission of malaria between mosquitoes and humans. Although his work received immediate acceptance in the medical community, his conclusion that malaria could be controlled by controlling mosquitoes was dismissed on the grounds that it would be impossible to rid a region of mosquitoes completely and that in any case mosquitoes would soon reinvade the region. After Ross formulated a mathematical model that predicted that malaria outbreaks could be avoided if the mosquito population could be reduced below a critical threshold level, field trials supported his conclusions and led to sometimes brilliant successes in malaria control. However, the Garki project provides a dramatic counterexample. This project worked to eradicate malaria from a region temporarily. However, people who have recovered from an attack of malaria have a temporary immunity against reinfection. Thus elimination of malaria from a region leaves the inhabitants of this region without immunity when the campaign ends, and the result can be a serious outbreak of malaria.

We will begin with an introduction to epidemic models. Next, we will incorporate demographic effects into the models to explore endemic states, and finally we will describe models with infectivity depending on the age of infection. Our approach will be *qualitative*. By this we mean that rather than attempting to find explicit solutions of the systems of differential equations which will form our models we will be concerned with the asymptotic behaviour, that is, the behaviour as $t \to \infty$ of solutions.

This material is meant to be an introduction to the study of compartmental models in mathematical epidemiology. More advanced material may be found in many other sources, including Chaps. 5–9 of this volume, the case studies in Chaps. 11–14, and [2, 4–6, 9, 17, 29, 35].

2.1.1 Simple Epidemic Models

An epidemic, which acts on a short temporal scale, may be described as a sudden outbreak of a disease that infects a substantial portion of the population in a region before it disappears. Epidemics usually leave many members untouched. Often these attacks recur with intervals of several years between outbreaks, possibly diminishing in severity as populations develop some immunity. This is an important aspect of the connection between epidemics and disease evolution.

The Book of Exodus describes the plagues that Moses brought down upon Egypt, and there are several other biblical descriptions of epidemic outbreaks. Descriptions of epidemics in ancient and medieval times frequently used the term "plague" because of a general belief that epidemics represented divine retribution for sinful living. More recently some have described AIDS as punishment for sinful activities. Such views have often hampered or delayed attempts to control this modern epidemic .

There are many biblical references to diseases as historical influences, such as the decision of Sennacherib, the king of Assyria, to abandon his attempt to capture Jerusalem about 700 BC because of the illness of his soldiers (Isaiah 37, 36–38). The fall of empires has been attributed directly or indirectly to epidemic diseases. In the second century AD the so-called Antonine plagues (possibly measles and smallpox) invaded the Roman Empire, causing drastic population reductions and economic hardships. These led to disintegration of the empire because of disorganization, which facilitated invasions of barbarians. The Han empire in China collapsed in the third century AD after a very similar sequence of events. The defeat of a population of millions of Aztecs by Cortez and his 600 followers can be explained in part by a smallpox epidemic that devastated the Aztecs but had almost no effect on the invading Spaniards thanks to their built-in immunities. The Aztecs were not only weakened by disease but also confounded by what they interpreted as a divine force favoring the invaders. Smallpox then spread southward to the Incas in Peru and was an important factor in the success of Pizarro's invasion a few years later. Smallpox was followed by other diseases such as measles and diphtheria imported from Europe to North America. In some regions, the indigenous populations were reduced to one tenth of their previous levels by these diseases. Between 1519 and 1530 the Indian population of Mexico was reduced from 30 million to 3 million.

The Black Death spread from Asia throughout Europe in several waves during the fourteenth century, beginning in 1346, and is estimated to have caused the death of as much as one third of the population of Europe between 1346 and 1350. The disease recurred regularly in various parts of Europe for more than 300 years, notably as the Great Plague of London of 1665–1666. It then gradually withdrew from Europe. As the plague struck some regions harshly while avoiding others, it had a profound effect on political and economic developments in medieval times. In the last bubonic plague epidemic in France (1720–1722), half the population of Marseilles, 60% of the population in nearby Toulon, 44% of the population of Arles and 30% of the population of Aix and Avignon died, but the epidemic did not spread beyond Provence.

The historian W.H. McNeill argues, especially in his book [26], that the spread of communicable diseases has frequently been an important influence in history. For example, there was a sharp population increase throughout the world in the eighteenth century; the population of China increased from 150 million in 1760 to 313 million in 1794 and the population of Europe increased from 118 million in 1700 to 187 million in 1800. There were many factors involved in this increase, including changes in marriage age and technological improvements leading to increased food supplies, but these factors are not sufficient to explain the increase. Demographic studies indicate that a satisfactory explanation requires recognition of a decrease in the mortality caused by periodic epidemic infections. This decrease came about partly through improvements in medicine, but a more important influence was probably the fact that more people developed immunities against infection as increased travel intensified the circulation and co-circulation of diseases.

Perhaps the first epidemic to be examined from a modeling point of view was the Great Plague in London (1665–1666). The plague was one of a sequence of attacks beginning in the year 1346 of what came to be known as the Black Death. It is now identified as the bubonic plague, which had actually invaded Europe as early as the sixth century during the reign of the Emperor Justinian of the Roman Empire and continued for more than three centuries after the Black Death. The Great Plague killed about one sixth of the population of London. One of the few "benefits" of the plague was that it caused Cambridge University to be closed for two years. Isaac Newton, who was a student at Cambridge at the time, was sent to his home and while "in exile" he had one of the most productive scientific periods of any human in history. He discovered his law of gravitation, among other things, during this period.

The characteristic features of the Great Plague were that it appeared quite suddenly, grew in intensity, and then disappeared, leaving part of the population untouched. The same features have been observed in many other epidemics, both of fatal diseases and of diseases whose victims recovered with immunity against reinfection.

In the nineteenth century recurrent invasions of cholera killed millions in India. The influenza epidemic of 1918–1919 killed more than 20 million people overall, more than half a million in the United States. One of the questions that first attracted the attention of scientists interested in the study of the spread of communicable diseases was why diseases would suddenly develop in a community and then disappear just as suddenly without infecting the entire community. One of the early triumphs of mathematical epidemiology [21] was the formulation of a simple model that predicted behaviour very similar to the behaviour observed in countless epidemics. The Kermack–McKendrick model is a compartmental model based on relatively simple assumptions on the rates of flow between different classes of members of the population.

There are many questions of interest to public health physicians confronted with a possible epidemic. For example, how severe will an epidemic be? This question may be interpreted in a variety of ways. For example, how many individuals will be affected altogether and thus require treatment? What is the maximum number of people needing care at any particular time? How long will the epidemic last? How much good would quarantine or isolation of victims do in reducing the severity of the epidemic? These are some of the questions we would like to study with the aid of models.

2.1.2 The Kermack–McKendrick Model

We formulate our descriptions as *compartmental models*, with the population under study being divided into compartments and with assumptions about the nature and time rate of transfer from one compartment to another. Diseases that confer immunity have a different compartmental structure from diseases without immunity. We will use the terminology SIR to describe a disease which confers immunity against re-infection, to indicate that the passage of individuals is from the susceptible class S to the infective class I to the removed class R. On the other hand, we will use the terminology SIS to describe a disease with no immunity against re-infection, to indicate that the passage of individuals is from the susceptible class to the infective class and then back to the susceptible class. Other possibilities include $SEIR$ and $SEIS$ models, with an exposed period between being infected and becoming infective, and $SIRS$ models, with temporary immunity on recovery from infection.

The independent variable in our compartmental models is the time t and the rates of transfer between compartments are expressed mathematically as derivatives with respect to time of the sizes of the compartments, and as a result our models are formulated initially as *differential equations*. Possible generalizations, which we shall not explore in these notes, include models in which the rates of transfer depend on the sizes of compartments over the past as well as at the instant of transfer, leading to more general types of functional equations, such as differential-difference equations, integral equations, or integro-differential equations.

In order to model such an epidemic we divide the population being studied into three classes labeled S, I, and R. We let $S(t)$ denote the number of individuals who are susceptible to the disease, that is, who are not (yet) infected at time t. $I(t)$ denotes the number of infected individuals, assumed infectious and able to spread the disease by contact with susceptibles. $R(t)$ denotes the number of individuals who have been infected and then removed from the possibility of being infected again or of spreading infection. Removal is carried out either through isolation from the rest of the population or through immunization against infection or through recovery from the disease with full immunity against reinfection or through death caused by the disease. These characterizations of removed members are different from an epidemiological perspective but are often equivalent from a modeling point of view which takes into account only the state of an individual with respect to the disease.

In formulating models in terms of the derivatives of the sizes of each compartment we are assuming that the number of members in a compartment is a differentiable function of time. This may be a reasonable approximation if there are many members in a compartment, but it is certainly suspect otherwise. In formulating models as differential equations, we are assuming that the epidemic process is *deterministic*, that is, that the behaviour of a population is determined completely by its history and by the rules which describe the model. In other chapters of this volume Linda Allen and Ping Yan describe the study of *stochastic* models in which probabilistic concepts are used and in which there is a distribution of possible behaviours. The developing study of network science, introduced in Chap. 4 of this volume and described in [28, 30, 33], is another approach.

The basic compartmental models to describe the transmission of communicable diseases are contained in a sequence of three papers by W.O. Kermack and A.G. McKendrick in 1927, 1932, and 1933 [21–23]. The first of these papers described epidemic models. What is often called the Kermack–McKendrick epidemic model is actually a special case of the general model introduced in this paper. The general model included dependence on age of infection, that is, the time since becoming infected. Curiously, Kermack and McKendrick did not explore this situation further in their later models which included demographic effects. Age of infection models have become important in the study of HIV/AIDS, and we will return to them in the last section of this chapter.

The special case of the model proposed by Kermack and McKendrick in 1927 which is the starting point for our study of epidemic models is

$$S' = -\beta SI$$
$$I' = \beta SI - \alpha I$$
$$R' = \alpha I .$$

A flow chart is shown in Fig. 2.1. It is based on the following assumptions:

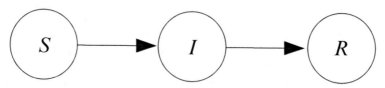

Fig. 2.1 Flow chart for the SIR model

(1) An average member of the population makes contact sufficient to transmit infection with βN others per unit time, where N represents total population size (mass action incidence) .
(2) Infectives leave the infective class at rate αI per unit time.
(3) There is no entry into or departure from the population, except possibly through death from the disease.

According to (1), since the probability that a random contact by an infective is with a susceptible, who can then transmit infection, is S/N, the number of new infections in unit time per infective is $(\beta N)(S/N)$, giving a rate of new infections $(\beta N)(S/N)I = \beta SI$. Alternately, we may argue that for a contact by a susceptible the probability that this contact is with an infective is I/N and thus the rate of new infections per susceptible is $(\beta N)(I/N)$, giving a rate of new infections $(\beta N)(I/N)S = \beta SI$. Note that both approaches give the same rate of new infections; there are situations which we shall encounter where one is more appropriate than the other. We need not give an algebraic expression for N since it cancels out of the final model, but we should note that for a disease that is fatal to all who are infected $N = S+I$; while, for a disease from which all infected members recover with immunity, $N = S+I+R$. Later, we will allow the possibility that some infectives recover while others die of the disease. The hypothesis (3) really says that the time scale of the disease is much faster than the time scale of births and deaths so that demographic effects on the population may be ignored. An alternative view is that we are only interested in studying the dynamics of a single epidemic outbreak. In later sections we shall consider models that are the same as those considered in this first section except for the incorporation of demographic effects (births and deaths) along with the corresponding epidemiological assumptions.

The assumption (2) requires a fuller mathematical explanation, since the assumption of a recovery rate proportional to the number of infectives has no clear epidemiological meaning. We consider the "cohort" of members who were all infected at one time and let $u(s)$ denote the number of these who are still infective s time units after having been infected. If a fraction α of these leave the infective class in unit time then

$$u' = -\alpha u \ ,$$

and the solution of this elementary differential equation is

$$u(s) = u(0)\, e^{-\alpha s} \ .$$

Thus, the fraction of infectives remaining infective s time units after having become infective is $e^{-\alpha s}$, so that the length of the infective period is distributed exponentially with mean $\int_0^\infty e^{-\alpha s} ds = 1/\alpha$, and this is what (2) really assumes.

The assumptions of a rate of contacts proportional to population size N with constant of proportionality β, and of an exponentially distributed recovery rate are unrealistically simple. More general models can be constructed and analyzed, but our goal here is to show what may be deduced from extremely simple models. It will turn out that many more realistic models exhibit very similar qualitative behaviours.

In our model R is determined once S and I are known, and we can drop the R equation from our model, leaving the system of two equations

$$S' = -\beta SI \tag{2.1}$$
$$I' = (\beta S - \alpha)I \ .$$

We are unable to solve this system analytically but we learn a great deal about the behaviour of its solutions by the following qualitative approach. To begin, we remark that the model makes sense only so long as $S(t)$ and $I(t)$ remain non-negative. Thus if either $S(t)$ or $I(t)$ reaches zero we consider the system to have terminated. We observe that $S' < 0$ for all t and $I' > 0$ if and only if $S > \alpha/\beta$. Thus I increases so long as $S > \alpha/\beta$ but since S decreases for all t, I ultimately decreases and approaches zero. If $S(0) < \alpha/\beta$, I decreases to zero (no epidemic), while if $S(0) > \alpha/\beta$, I first increases to a maximum attained when $S = \alpha/\beta$ and then decreases to zero (epidemic). We think of introducing a small number of infectives into a population of susceptibles and ask whether there will be an epidemic. The quantity $\beta S(0)/\alpha$ is a threshold quantity, called the *basic reproduction number* and denoted by $\mathcal{R}_0$, which determines whether there is an epidemic or not. If $\mathcal{R}_0 < 1$ the infection dies out, while if $\mathcal{R}_0 > 1$ there is an epidemic.

The definition of the basic reproduction number $\mathcal{R}_0$ is that the basic reproduction number is the number of secondary infections caused by a single infective introduced into a wholly susceptible population of size $K \approx S(0)$

over the course of the infection of this single infective. In this situation, an infective makes βK contacts in unit time, all of which are with susceptibles and thus produce new infections, and the mean infective period is $1/\alpha$; thus the basic reproduction number is actually $\beta K/\alpha$ rather than $\beta S(0)/\alpha$.

Instead of trying to solve for S and I as functions of t, we divide the two equations of the model to give

$$\frac{I'}{S'} = \frac{dI}{dS} = \frac{(\beta S - \alpha)I}{-\beta S I} = -1 + \frac{\alpha}{\beta S} ,$$

and integrate to find the orbits (curves in the (S, I)-plane, or phase plane)

$$I = -S + \frac{\alpha}{\beta} \log S + c , \tag{2.2}$$

with c an arbitrary constant of integration. Here, we are using log to denote the natural logarithm. Another way to describe the orbits is to define the function

$$V(S, I) = S + I - \frac{\alpha}{\beta} \log S$$

and note that each orbit is a curve given implicitly by the equation $V(S, I) = c$ for some choice of the constant c. The constant c is determined by the initial values $S(0)$, $I(0)$ of S and I, respectively, because $c = V(S(0), I(0)) = S(0) + I(0) - \alpha \log S(0)/\beta$. Note that the maximum value of I on each of these orbits is attained when $S = \alpha/\beta$. Note also that since none of these orbits reaches the I - axis, $S > 0$ for all times. In particular, $S_\infty = \lim_{t\to\infty} S(t) > 0$, which implies that part of the population escapes infection.

Let us think of a population of size K into which a small number of infectives is introduced, so that $S_0 \approx K$, $I_0 \approx 0$, and $\mathcal{R}_0 = \beta K/\alpha$. If we use the fact that $\lim_{t\to\infty} I(t) = 0$, and let $S_\infty = \lim_{t\to\infty} S(t)$, then the relation $V(S_0, I_0) = V(S_\infty, 0)$ gives

$$K - \frac{\alpha}{\beta} \log S_0 = S_\infty - \frac{\alpha}{\beta} \log S_\infty ,$$

from which we obtain an expression for β/α in terms of the measurable quantities S_0 and S_∞, namely

$$\frac{\beta}{\alpha} = \frac{(\log S_0 - \log S_\infty)}{K - S_\infty} .$$

We may rewrite this in terms of $\mathcal{R}_0$ as the *final size relation*

$$\log S_0 - \log S_\infty = \mathcal{R}_0 \left[1 - \frac{S_\infty}{K} \right] . \tag{2.3}$$

In particular, since the right side of (2.3) is finite, the left side is also finite, and this shows that $S_\infty > 0$.

It is generally difficult to estimate the contact rate β which depends on the particular disease being studied but may also depend on social and behavioural factors. The quantities S_0 and S_∞ may be estimated by serological studies (measurements of immune responses in blood samples) before and after an epidemic, and from these data the basic reproduction number $\mathcal{R}_0$ may be estimated by using (2.3). This estimate, however, is a retrospective one which can be determined only after the epidemic has run its course.

Initially, the number of infectives grows exponentially because the equation for I may be approximated by

$$I' = (\beta K - \alpha)I$$

and the initial growth rate is

$$r = \beta K - \alpha = \alpha(\mathcal{R}_0 - 1) .$$

This initial growth rate r may be determined experimentally when an epidemic begins. Then since K and α may be measured β may be calculated as

$$\beta = \frac{r + \alpha}{K} .$$

However, because of incomplete data and under-reporting of cases this estimate may not be very accurate. This inaccuracy is even more pronounced for an outbreak of a previously unknown disease, where early cases are likely to be misdiagnosed.

The maximum number of infectives at any time is the number of infectives when the derivative of I is zero, that is, when $S = \alpha/\beta$. This maximum is given by

$$I_{max} = S_0 + I_0 - \frac{\alpha}{\beta} \log S_0 - \frac{\alpha}{\beta} + \frac{\alpha}{\beta} \log \frac{\alpha}{\beta} , \qquad (2.4)$$

obtained by substituting $S = \alpha/\beta$, $I = I_{max}$ into (2.2).

Example. (The Great Plague in Eyam) The village of Eyam near Sheffield, England suffered an outbreak of bubonic plague in 1665–1666 the source of which is generally believed to be the Great Plague of London. The Eyam plague was survived by only 83 of an initial population of 350 persons. As detailed records were preserved and as the community was persuaded to quarantine itself to try to prevent the spread of disease to other communities, the disease in Eyam has been used as a case study for modeling [31]. Detailed examination of the data indicates that there were actually two outbreaks of which the first was relatively mild. Thus we shall try to fit the model (2.1) over the period from mid-May to mid-October 1666, measuring time in months with an initial population of seven infectives and 254 susceptibles, and a final population of 83. Values of susceptibles and infectives in Eyam are given in [31] for various dates, beginning with $S(0) = 254, I(0) = 7$, shown in Table 2.1.

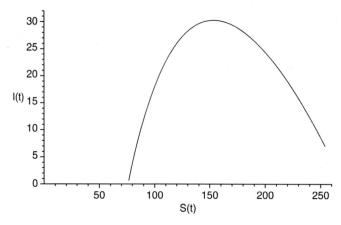

Fig. 2.2 The S–I plane

Table 2.1 Eyam Plague data

Date (1666)	Susceptibles	Infectives
July 3/4	235	14.5
July 19	201	22
August 3/4	153.5	29
August 19	121	21
September 3/4	108	8
September 19	97	8
October 4/5	Unknown	Unknown
October 20	83	0

The relation (2.3) with $S_0 = 254$, $I_0 = 7$, $S_\infty = 83$ gives $\beta/\alpha = 6.54 \times 10^{-3}$, $\alpha/\beta = 153$. The infective period was 11 days, or 0.3667 month, so that $\alpha = 2.73$. Then $\beta = 0.0178$. The relation (2.4) gives an estimate of 30.4 for the maximum number of infectives. We use the values obtained here for the parameters β and α in the model (2.1) for simulations of both the phase plane, the (S, I)-plane, and for graphs of S and I as functions of t (Figs. 2.2, 2.3, and 2.4). Figure 2.5 plots these data points together with the phase portrait given in Fig. 2.2 for the model (2.1).

The actual data for the Eyam epidemic are remarkably close to the predictions of this very simple model. However, the model is really too good to be true. Our model assumes that infection is transmitted directly between people. While this is possible, bubonic plague is transmitted mainly by rat fleas. When an infected rat is bitten by a flea, the flea becomes extremely hungry and bites the host rat repeatedly, spreading the infection in the rat. When the host rat dies its fleas move on to other rats, spreading the disease further. As the number of available rats decreases the fleas move to human hosts, and this is how plague starts in a human population (although the second phase of the epidemic may have been the pneumonic form of bubonic plague, which

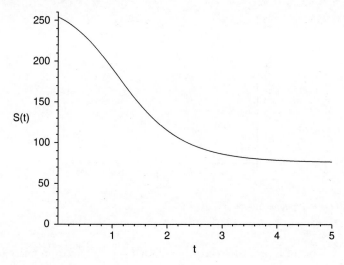

Fig. 2.3 S as a function of t

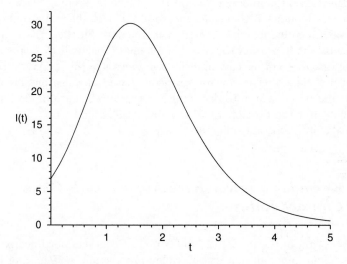

Fig. 2.4 I as a function of t

can be spread from person to person). One of the main reasons for the spread of plague from Asia into Europe was the passage of many trading ships; in medieval times ships were invariably infested with rats. An accurate model of plague transmission would have to include flea and rat populations, as well as movement in space. Such a model would be extremely complicated and its predictions might well not be any closer to observations than our simple unrealistic model. In [31] a stochastic model was also used to fit the data, but the fit was rather poorer than the fit for the simple deterministic model (2.1).

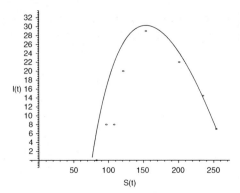

Fig. 2.5 The *S–I* plane, model and data

In the village of Eyam the rector persuaded the entire community to quarantine itself to prevent the spread of disease to other communities. This policy actually increased the infection rate in the village by keeping fleas, rats, and people in close contact with one another, and the mortality rate from bubonic plague was much higher in Eyam than in London. Further, the quarantine could do nothing to prevent the travel of rats and thus did little to prevent the spread of disease to other communities. One message this suggests to mathematical modelers is that control strategies based on false models may be harmful, and it is essential to distinguish between assumptions that simplify but do not alter the predicted effects substantially, and wrong assumptions which make an important difference.

2.1.3 Kermack–McKendrick Models with General Contact Rates

The assumption in the model (2.1) of a rate of contacts per infective which is proportional to population size N, called *mass action incidence* or bilinear incidence, was used in all the early epidemic models. However, it is quite unrealistic, except possibly in the early stages of an epidemic in a population of moderate size. It is more realistic to assume a contact rate which is a non-increasing function of total population size. For example, a situation in which the number of contacts per infective in unit time is constant, called *standard incidence*, is a more accurate description for sexually transmitted diseases.

We generalize the model (2.1) by replacing the assumption (1) by the assumption that an average member of the population makes $C(N)$ contacts in unit time with $C'(N) \geq 0$ [7, 10], and we define

$$\beta(N) = \frac{C(N)}{N} .$$

It is reasonable to assume $\beta'(N) \leq 0$ to express the idea of saturation in the number of contacts. Then mass action incidence corresponds to the choice $C(N) = \beta N$, and standard incidence corresponds to the choice $C(N) = \lambda$. The assumptions $C(N) = N\beta(N), C'(N) \geq 0$ imply that

$$\beta(N) + N\beta'(N) \geq 0 . \tag{2.5}$$

Some epidemic models [10] have used a Michaelis–Menten type of interaction of the form

$$C(N) = \frac{aN}{1 + bN} .$$

Another form based on a mechanistic derivation for pair formation [14] leads to an expression of the form

$$C(N) = \frac{aN}{1 + bN + \sqrt{1 + 2bN}} .$$

Data for diseases transmitted by contact in cities of moderate size [25] suggests that data fits the assumption of a form

$$C(N) = \lambda N^a$$

with $a = 0.05$ quite well. All of these forms satisfy the conditions $C'(N) \geq 0, \beta'(N) \leq 0$.

Because the total population size is now present in the model we must include an equation for total population size in the model. This forces us to make a distinction between members of the population who die of the disease and members of the population who recover with immunity against reinfection. We assume that a fraction f of the αI members leaving the infective class at time t recover and the remaining fraction $(1 - f)$ die of disease. We use S, I, and N as variables, with $N = S + I + R$. We now obtain a three-dimensional model

$$\begin{aligned}
S' &= -\beta(N)SI \\
I' &= \beta(N)SI - \alpha I \\
N' &= -(1 - f)\alpha I .
\end{aligned} \tag{2.6}$$

We also have the equation $R' = -f\alpha I$, but we need not include it in the model since R is determined when S, I, and N are known. We should note that if $f = 1$ the total population size remains equal to the constant K, and the model (2.6) reduces to the simpler model (2.1) with β replaced by the constant $\beta(K)$.

We wish to show that the model (2.6) has the same qualitative behaviour as the model (2.1), namely that there is a basic reproduction number which distinguishes between disappearance of the disease and an epidemic outbreak, and that some members of the population are left untouched when the epidemic passes. These two properties are the central features of all epidemic models.

For the model (2.6) the basic reproduction number is given by

$$\mathcal{R}_0 = \frac{K\beta(K)}{\alpha}$$

because a single infective introduced into a wholly susceptible population makes $C(K) = K\beta(K)$ contacts in unit time, all of which are with susceptibles and thus produce new infections, and the mean infective period is $1/\alpha$. In addition to the basic reproduction number $\mathcal{R}_0$ there is also a time-dependent running reproduction number which we call $\mathcal{R}^*$, representing the number of secondary infections caused by a single individual in the population who becomes infective at time t. In this situation, an infective makes $C(N) = N\beta(N)$ contacts in unit time and a fraction S/N are with susceptibles and thus produce new infections. Thus it is easy to see that for the model (2.6) the running reproduction number is given by

$$\mathcal{R}^* = \frac{S\beta(N)}{\alpha}.$$

If $\mathcal{R}^* < 1$ for all large t, the epidemic will pass. We may calculate the rate of change of the running reproduction number with respect to time, using (2.6) and (2.5) to find that

$$\frac{d}{dt}\mathcal{R}^* = \frac{S'(t)\beta(N) + S(t)\beta'(N)N'(t)}{\alpha} = \frac{(-\beta(N))^2 SI - S\alpha(1-f)\beta'(N)}{\alpha}.$$
$$\leq \frac{\beta(N)SI}{\alpha} \cdot \left[\beta(N) - \frac{(1-f)\alpha}{N}\right].$$

Thus $\frac{d}{dt}\mathcal{R}^* < 0$ if $N\beta(N) > \alpha(1-f)$, or $\mathcal{R}^* > (1-f)S/N$. This means that $\mathcal{R}^*$ decreases whenever $\mathcal{R}^* > 1$. Thus if $\mathcal{R}^* < 1$ for $t = T$ then $\mathcal{R}^* < 1$ for $t > T$. If $\mathcal{R}_0 > 1$ then $I'(0) = \alpha(\mathcal{R}_0 - 1)I(0) > 0$, and an epidemic begins. However, $\mathcal{R}^*$ decreases until it is less than 1 and then remains less than 1. Thus the epidemic will pass. If $\mathcal{R}_0 < 1$ then $I'(0) = \alpha(\mathcal{R}_0 - 1)I(0) < 0, \mathcal{R}^* < 1$ for all t, and there is no epidemic.

From (2.6) we obtain

$$S' + I' = -\alpha I$$
$$N' = -\alpha(1 - f)I .$$

Integration of these equations from 0 to t gives

$$S(t) + I(t) - S(0) - I(0) = -\alpha \int_0^t I(s)ds \qquad (2.7)$$

$$N(t) - N(0) = -\alpha(1 - f) \int_0^t I(s)ds .$$

When we combine these two equations, eliminating the integral expression, and use $N(0) = S(0) + I(0) = K$, we obtain

$$K - N(t) = (1 - f)[K - S(t) - I(t)] .$$

If we let $t \to \infty$, $S(t)$ and $N(t)$ decrease monotonically to limits S_∞ and N_∞ respectively and $I(t) \to 0$. This gives the relation

$$K - N_\infty = (1 - f)[K - S_\infty] . \qquad (2.8)$$

In this equation, $K - N_\infty$ is the change in population size, which is the number of disease deaths over the course of the epidemic, while $K - S_\infty$ is the change in the number of susceptibles, which is the number of disease cases over the course of the epidemic. In this model, (2.8) is obvious, but we shall see in a more general setting how to derive an analogous equation from which we can calculate an average disease mortality. Equation (2.8) generalizes to the infection age epidemic model of Kermack and McKendrick.

If we use the same approach as was used for (2.1) to show that $S_\infty > 0$, we obtain

$$\frac{dI}{dS} = -1 + \frac{\alpha}{S\beta(N)}$$

and we are unable to proceed because of the dependence on N. However, we may use a different approach to obtain the desired result. We assume that $\beta(0)$ is finite, thus ruling out standard incidence. If we let $t \to \infty$ in the second equation of (2.7) we obtain

$$\alpha \int_0^\infty I(s)ds = S(0) + I(0) - S_\infty = K - S_\infty.$$

The first equation of (2.6) may be written as

$$-\frac{S'(t)}{S(t)} = \beta(N(t))I(t).$$

Since

$$\beta(N) \leq \beta(0),$$

integration from 0 to ∞ gives

$$\log \frac{S(0)}{S_\infty} = \int_0^\infty \beta(N(t))I(t)dt$$

$$\leq \beta(0) \int_0^\infty I(t)dt$$

$$= \frac{\beta(0)(K - S_\infty)}{\alpha K}.$$

Since the right side of this inequality is finite, the left side is also finite and this establishes that $S_\infty > 0$.

In addition, if we use the same integration together with the inequality

$$\beta(N) \geq \beta(K),$$

we obtain a final size inequality

$$\log \frac{S(0)}{S_\infty} = \int_0^\infty \beta(N(t))I(t)dt$$

$$\geq \beta(K) \int_0^\infty I(t)dt = \mathcal{R}_0 \left[1 - \frac{S_\infty}{K} \right].$$

If $\beta(N) \to \infty$ as $N \to 0$ we must use a different approach to analyze the limiting behaviour. It is possible to show that $S_\infty = 0$ is possible only if $N \to 0$ and $\int_0^K \beta(N)dN$ diverges, and this is possible only if $f = 0$, that is, only if all infectives ide of disease. The assumption that $\beta(N)$ is unbounded as $N \to 0$ is biologically unreasonable. In particular, standard incidence is not realistic for small population sizes. A more realistic assumption would be that the number of contacts per infective in unit time is linear for small population size and saturates for larger population sizes, which rules out the possibility that the epidemic sweeps through the entire population.

2.1.4 Exposed Periods

In many infectious diseases there is an exposed period after the transmission of infection from susceptibles to potentially infective members but before these potential infectives can transmit infection. If the exposed period is short it is often neglected in modeling. A longer exposed period could perhaps lead to significantly different model predictions, and we need to show that this is not the case. To incorporate an exponentially distributed exposed period with mean exposed period $1/\kappa$ we add an exposed class E and use compartments S, E, I, R and total population size $N = S + E + I + R$ to give a generalization of the epidemic model (2.6).

$$S' = -\beta(N)SI$$
$$E' = \beta(N)SI - \kappa E \qquad (2.9)$$
$$I' = \kappa E - \alpha I$$
$$N' = -(1-f)\alpha I \ .$$

We also have the equation $R' = -f\alpha I$, but we need not include it in the model since R is determined when S, I, and N are known. A flow chart is shown in Fig. 2.6.

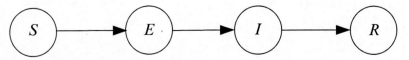

Fig. 2.6 Flow chart for the $SEIR$ model

The analysis of this model is the same as the analysis of (2.6), but with I replaced by $E + I$. That is, instead of using the number of infectives as one of the variables we use the total number of infected members, whether or not they are capable of transmitting infection.

Some diseases have an asymptomatic stage in which there is some infectivity rather than an exposed period. This may be modeled by assuming infectivity reduced by a factor ε_E during an exposed period. A calculation of the rate of new infections per susceptible leads to a model

$$S' = -\beta(N)S(I + \varepsilon_E E)$$
$$E' = \beta(N)S(I + \varepsilon_E E) - \kappa E \qquad (2.10)$$
$$I' = \kappa E - \alpha I \ .$$

For this model

$$\mathcal{R}_0 = \frac{K\beta(K)}{\alpha} + \varepsilon_E \frac{K\beta(K)}{\kappa} \ .$$

There is a final size relation like (2.3) for the model (2.9). Integration of the sum of the first two equations of (2.9) from 0 to ∞ gives

$$K - S_\infty = \kappa \int_0^\infty E(s)ds$$

and division of the first equation of (2.9) by S followed by integration from 0 to ∞ gives

$$\log S_0 - \log S_\infty = \int_0^\infty \beta(N(s))[I(s) + \epsilon_E E(s)ds$$

$$\geq \beta(K) \int_0^\infty [I(s) + \epsilon_E E(s)ds$$

$$= \beta(K)[\epsilon_E + \frac{\kappa}{\alpha}] \int_0^\infty E(s)ds$$

$$= \mathcal{R}_0 \left[1 - \frac{S_\infty}{K}\right].$$

The same integration using $\beta(N) \leq \beta(0) < \infty$ shows as in the previous section that $S_\infty > 0$.

2.1.5 Treatment Models

One form of treatment that is possible for some diseases is vaccination to protect against infection before the beginning of an epidemic. For example, this approach is commonly used for protection against annual influenza outbreaks. A simple way to model this would be to reduce the total population size by the fraction of the population protected against infection. However, in reality such inoculations are only partly effective, decreasing the rate of infection and also decreasing infectivity if a vaccinated person does become infected. To model this, it would be necessary to divide the population into two groups with different model parameters and to make some assumptions about the mixing between the two groups. We will not explore such more complicated models here.

If there is a treatment for infection once a person has been infected, we model this by supposing that a fraction γ per unit time of infectives is selected for treatment, and that treatment reduces infectivity by a fraction δ. Suppose that the rate of removal from the treated class is η. The $SITR$ model, where T is the treatment class, is given by

$$\begin{aligned}
S' &= -\beta(N)S[I + \delta T] \\
I' &= \beta(N)S[I + \delta T] - (\alpha + \gamma)I \\
T' &= \gamma I - \eta T \\
N' &= -(1-f)\alpha I - (1 - f_T)\eta T.
\end{aligned} \qquad (2.11)$$

A flow chart is shown in Fig. 2.7.

It is not difficult to prove, much as was done for the model (2.1) that

$$S_\infty = \lim_{t \to \infty} S(t) > 0, \quad \lim_{t \to \infty} I(t) = \lim_{t \to \infty} T(t) = 0.$$

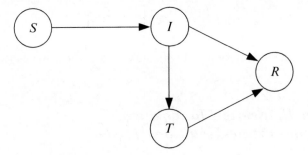

Fig. 2.7 Flow chart for the $SITR$ model

In order to calculate the basic reproduction number, we may argue that an infective in a totally susceptible population causes βK new infections in unit time, and the mean time spent in the infective compartment is $1/(\alpha + \gamma)$. In addition, a fraction $\gamma/(\alpha+\gamma)$ of infectives are treated. While in the treatment stage the number of new infections caused in unit time is $\delta\beta K$, and the mean time in the treatment class is $1/\eta$. Thus $\mathcal{R}_0$ is

$$\mathcal{R}_0 = \frac{\beta K}{\alpha + \gamma} + \frac{\gamma}{\alpha + \gamma} \frac{\delta\beta K}{\eta}. \tag{2.12}$$

It is also possible to establish the final size relation (2.3) by means similar to those used for the simple model (2.1). We integrate the first equation of (2.11) to obtain

$$\log \frac{S(0)}{S_\infty} = \int_0^\infty \beta(N(t))[I(t) + \delta T(t)]dt$$

$$\geq \beta(K) \int_0^\infty [I(t) + \delta T(t)]dt.$$

Integration of the third equation of (2.11) gives

$$\gamma \int_0^\infty I(t)dt = \eta \int_0^\infty T(t)dt.$$

Integration of the sum of the first two equations of (2.11) gives

$$K - S_\infty = (\alpha + \gamma) \int_0^\infty I(t)dt.$$

Combination of these three equations and (2.12) gives

$$\log \frac{S(0)}{S_\infty} \geq \mathcal{R}_0 \left[\frac{K - S_\infty}{K} \right].$$

If β is constant, this relation is an equality, and is the same as (2.3).

2.1.6 An Epidemic Management (Quarantine-Isolation) Model

An actual epidemic differs considerably from the idealized models (2.1) or (2.6), as was shown by the SARS epidemic of 2002–3. Some notable differences are:

1. As we have seen in the preceding section, at the beginning of an epidemic the number of infectives is small and a deterministic model, which presupposes enough infectives to allow homogeneous mixing, is inappropriate.
2. When it is realized that an epidemic has begun, individuals are likely to modify their behaviour by avoiding crowds to reduce their contacts and by being more careful about hygiene to reduce the risk that a contact will produce infection.
3. If a vaccine is available for the disease which has broken out, public health measures will include vaccination of part of the population. Various vaccination strategies are possible, including vaccination of health care workers and other first line responders to the epidemic, vaccination of members of the population who have been in contact with diagnosed infectives, or vaccination of members of the population who live in close proximity to diagnosed infectives.
4. Diagnosed infectives may be hospitalized, both for treatment and to isolate them from the rest of the population.
5. Contact tracing of diagnosed infectives may identify people at risk of becoming infective, who may be quarantined (instructed to remain at home and avoid contacts) and monitored so that they may be isolated immediately if and when they become infective.
6. In some diseases, exposed members who have not yet developed symptoms may already be infective, and this would require inclusion in the model of new infections caused by contacts between susceptibles and asymptomatic infectives from the exposed class.
7. Isolation may be imperfect; in-hospital transmission of infection was a major problem in the SARS epidemic.

In the SARS epidemic of 2002–2003 in-hospital transmission of disease from patients to health care workers or visitors because of imperfect isolation accounted for many of the cases. This points to an essential heterogeneity in disease transmission which must be included whenever there is any risk of such transmission.

All these generalizations have been considered in studies of the SARS epidemic of 2002–3. While the ideas were suggested in SARS modelling, they are in fact relevant to any epidemic. One beneficial effect of the SARS epidemic has been to draw attention to epidemic modelling which may be of great value in coping with future epidemics.

If a vaccine is available for a disease which threatens an epidemic outbreak, a vaccinated class which is protected at least partially against infection should be included in a model. While this is not relevant for an outbreak of a new disease, it would be an important aspect to be considered in modelling an influenza epidemic or a bioterrorist outbreak of smallpox.

For an outbreak of a new disease, where no vaccine is available, isolation and quarantine are the only control measures available. Let us formulate a model for an epidemic once control measures have been started. Thus, we assume that an epidemic has started, but that the number of infectives is small and almost all members of the population are still susceptible.

We formulate a model to describe the course of an epidemic when control measures are begun under the assumptions:

1. Exposed members may be infective with infectivity reduced by a factor $\varepsilon_E, 0 \leq \varepsilon_E < 1$.
2. Exposed members who are not isolated become infective at rate κ_1.
3. We introduce a class Q of quarantined members and a class J of isolated members.
4. Exposed members are quarantined at a proportional rate γ_1 in unit time (in practice, a quarantine will also be applied to many susceptibles, but we ignore this in the model). Quarantine is not perfect, but reduces the contact rate by a factor ε_Q. The effect of this assumption is that some susceptibles make fewer contacts than the model assumes.
5. There may be transmission of disease by isolated members, with an infectivity factor of ε_J.
6. Infectives are diagnosed at a proportional rate γ_2 per unit time and isolated. In addition, quarantined members are monitored and when they develop symptoms at rate κ_2 they are isolated immediately.
7. Infectives leave the infective class at rate α_1 and a fraction f_1 of these recover, and isolated members leave the isolated class at rate α_2 with a fraction f_2 recovering.

These assumptions lead to the $SEQIJR$ model [13]

$$S' = -\beta(N)S[\varepsilon_E E + \varepsilon_E \varepsilon_Q Q + I + \varepsilon_J J]$$
$$E' = \beta(N)S[\varepsilon_E E + \varepsilon_E \varepsilon_Q Q + I + \varepsilon_J J] - (\kappa_1 + \gamma_1)E$$
$$Q' = \gamma_1 E - \kappa_2 Q \qquad\qquad (2.13)$$
$$I' = \kappa_1 E - (\alpha_1 + \gamma_2)I$$
$$J' = \kappa_2 Q + \gamma_2 I - \alpha_2 J$$
$$N' = -(1 - f_1)\alpha_1 I - (1 - f_2)\alpha_2 J \ .$$

Here, we have used an equation for N to replace the equation

$$R' = f_1 \alpha_1 I + f_2 \alpha_2 J \ .$$

The model before control measures are begun is the special case

$$\gamma_1 = \gamma_2 = \kappa_2 = \alpha_2 = f_2 = 0, \ \ Q = J = 0$$

of (2.13). It is the same as (2.10).

We define the *control reproduction number* $\mathcal{R}_c$ to be the number of secondary infections caused by a single infective in a population consisting essentially only of susceptibles with the control measures in place. It is analogous to the basic reproduction number but instead of describing the very beginning of the disease outbreak it describes the beginning of the recognition of the epidemic. The basic reproduction number is the value of the control reproduction number with

$$\gamma_1 = \gamma_2 = \kappa_2 = \alpha_2 = f_2 = 0 \ .$$

In addition, there is a time-dependent *effective reproduction number* $\mathcal{R}^*$ which continues to track the number of secondary infections caused by a single infective as the epidemic continues with control measures (quarantine of asymptomatics and isolation of symptomatics) in place. It is not difficult to show that if the inflow into the population from travellers and new births is small (i.e., if the epidemiological time scale is much faster than the demographic time scale), our model implies that $\mathcal{R}^*$ will become and remain less than unity, so that the epidemic will always pass. Even if $\mathcal{R}_c > 1$, the epidemic will abate eventually when the effective reproduction number becomes less than unity. The effective reproduction number $\mathcal{R}^*$ is essentially $\mathcal{R}_c$ multiplied by a factor S/N, but allows time-dependent parameter values as well.

However, it should be remembered that if the epidemic takes so long to pass that there are enough new births and travellers to keep $\mathcal{R}^* > 1$, there will be an endemic equilibrium meaning that the disease will establish itself and remain in the population.

We have already calculated $\mathcal{R}_0$ for (2.10) and we may calculate $\mathcal{R}_c$ in the same way but using the full model with quarantined and isolated classes. We obtain

$$\mathcal{R}_c = \frac{\varepsilon_E K\beta(K)}{D_1} + \frac{K\beta(K)\kappa_1}{D_1 D_2} + \frac{\varepsilon_Q \varepsilon_E K\beta(K)\gamma_1}{D_1 \kappa_2} + \frac{\varepsilon_J K\beta(K)\kappa_1 \gamma_2}{\alpha_2 D_1 D_2} + \frac{\varepsilon_J K\beta(K)\gamma_1}{\alpha_2 D_1}$$

$$\mathcal{R}^* = \mathcal{R}_c \frac{S}{N},$$

where $D_1 = \gamma_1 + \kappa_1$, $\quad D_2 = \gamma_2 + \alpha_1$.

Each term of $\mathcal{R}_c$ has an epidemiological interpretation. The mean duration in E is $1/D_1$ with contact rate $\varepsilon_E \beta$, giving a contribution to $\mathcal{R}_c$ of $\varepsilon_E K\beta(K)/D_1$. A fraction κ_1/D_1 goes from E to I, with contact rate β and mean duration $1/D_2$, giving a contribution of $K\beta(K)\kappa_1/D_1 D_2$. A fraction γ_1/D_1 goes from E to Q, with contact rate $\varepsilon_E \varepsilon_Q \beta$ and mean duration $1/\kappa_2$, giving a contribution of $\varepsilon_E \varepsilon_Q K\beta(K)\gamma_1/D_1 \kappa_2$. A fraction $\kappa_1 \gamma_2/D_1 D_2$ goes from E to I to J, with a contact rate of $\varepsilon_J \beta$ and a mean duration of $1/\alpha_2$, giving a contribution of $\varepsilon_J K\beta(K)\kappa_1 \gamma_2/\alpha_2 D_1 D_2$. Finally, a fraction γ_1/D_1 goes from E to Q to J with a contact rate of $\varepsilon_J \beta$ and a mean duration of $1/\alpha_2$ giving a contribution of $\varepsilon_J K\beta(K)\gamma_1/D_1 \alpha_2$. The sum of these individual contributions gives $\mathcal{R}_c$.

In the model (2.13) the parameters γ_1 and γ_2 are *control* parameters which may be varied in the attempt to manage the epidemic. The parameters ϵ_Q and ϵ_J depend on the strictness of the quarantine and isolation processes and are thus also control measures in a sense. The other parameters of the model are specific to the disease being studied. While they are not variable, their measurements are subject to experimental error.

The linearization of (2.13) at the disease-free equilibrium $(K, 0, 0, 0, 0, K)$ has matrix

$$\begin{bmatrix} \epsilon_E K\beta(K) - (\kappa_1 + \gamma_1) & \varepsilon_E \varepsilon_Q \beta(K) & K\beta(K) & \varepsilon_J K\beta(K) \\ \gamma_1 & -\kappa_2 & 0 & 0 \\ \kappa_1 & 0 & -(\alpha_1 + \gamma_2) & 0 \\ 0 & \kappa_2 & \gamma_2 & -\alpha_2 \end{bmatrix}.$$

The corresponding characteristic equation is a fourth degree polynomial equation whose leading coefficient is 1 and whose constant term is a positive constant multiple of $1 - \mathcal{R}_c$, thus positive if $\mathcal{R}_c < 1$ and negative if $\mathcal{R}_c > 1$. If $\mathcal{R}_c > 1$ there is a positive eigenvalue, corresponding to an initial exponential growth rate of solutions of (2.13). If $\mathcal{R}_c < 1$ it is possible to show that all eigenvalues of the coefficient matrix have negative real part, and thus solutions of (2.13) die out exponentially [38].

Next, we wish to show that analogues of the relation (2.8) and $S_\infty > 0$ derived for the model (2.6) are valid for the management model (2.13). We begin by integrating the equations for $S + E, Q, I, J$, and N of (2.13) with respect to t from $t = 0$ to $t = \infty$, using the initial conditions

$$S(0) + E(0) = N(0) = K, \qquad Q(0) = I(0) = J(0) = 0.$$

We obtain, since E, Q, I, and J all approach zero at $t \to \infty$,

$$K - S_\infty = (\kappa_1 + \gamma_1) \int_0^\infty E(s)ds$$

$$\gamma_1 \int_0^\infty E(s)ds = \kappa_2 \int_0^\infty Q(s)ds$$

$$\kappa_1 \int_0^\infty E(s)ds = (\alpha_1 + \gamma_2) \int_0^\infty I(s)ds$$

$$\kappa_2 \int_0^\infty Q(s)ds = \alpha_2 \int_0^\infty J(s)ds - \gamma_2 \int_0^\infty I(s)ds$$

$$K - N_\infty = (1 - f_1)\alpha_1 \int_0^\infty I(s)ds + (1 - f_2)\alpha_2 \int_0^\infty J(s)ds .$$

In order to relate $(K - S_\infty)$ to $(K - N_\infty)$, we need to express $\int_0^\infty I(s)ds$ and $\int_0^\infty J(s)ds$ in terms of $\int_0^\infty E(s)ds$.

From the three above relations for integrals we obtain

$$(\alpha_1 + \gamma_2) \int_0^\infty I(s)ds = \kappa_1 \int_0^\infty E(s)ds$$

$$\alpha_2 \int_0^\infty J(s)ds = \frac{\gamma_1\alpha_1 + \gamma_1\gamma_2 + \kappa_1\gamma_2}{\alpha_1 + \gamma_2} \int_0^\infty E(s)ds .$$

Thus we have

$$K - N_\infty = \frac{(1 - f_1)\alpha_1\kappa_1 + (1 - f_2)(\gamma_1\alpha_1 + \gamma_1\gamma_2 + \kappa_1\gamma_2)}{\alpha_1 + \gamma_2} \int_0^\infty E(s)ds$$

$$= \frac{(1 - f_1)\alpha_1\kappa_1 + (1 - f_2)(\gamma_1\alpha_1 + \gamma_1\gamma_2 + \kappa_1\gamma_2)}{(\kappa_1 + \gamma_1)(\alpha_1 + \gamma_2)}[K - S_\infty] .$$

This has the form, analogous to (2.8),

$$K - N_\infty = c[K - S_\infty] \tag{2.14}$$

with c, the disease death rate, given by

$$c = \frac{(1 - f_1)\alpha_1\kappa_1 + (1 - f_2)(\gamma_1\alpha_1 + \gamma_1\gamma_2 + \kappa_1\gamma_2)}{(\kappa_1 + \gamma_1)(\alpha_1 + \gamma_2)} .$$

The mean disease death rate may be measured and this expression gives information about some of the parameters in the model which can not be measured directly. It is easy to see that $0 \le c \le 1$ with $c = 0$ if and only if $f_1 = f_2 = 1$, that is, if and only if there are no disease deaths, and $c = 1$ if and only if $f_1 = f_2 = 0$, that is, if and only if the disease is universally fatal.

An argument similar to the one used for (2.6) but technically more complicated may be used to show that $S_\infty > 0$ for the treatment model (2.13). Thus the asymptotic behaviour of the management model (2.13) is the same as that of the simpler model (2.6). If the control reproduction number $\mathcal{R}_c$ is less than 1 the disease dies out and if $\mathcal{R}_c > 1$ there is an epidemic which will pass leaving some members of the population untouched.

2.1.7 Stochastic Models for Disease Outbreaks

The underlying assumptions of the models of Kermack–McKendrick type studied in this chapter are that the sizes of the compartments are large enough that the mixing of members is homogeneous. While these assumptions are probably reasonable once an epidemic is well underway, at the beginning of a disease outbreak the situation may be quite different. At the beginning of an epidemic most members of the population are susceptible, that is, not (yet) infected, and the number of infectives (members of the population who are infected and may transmit infection) is small. The transmission of infection depends strongly on the pattern of contacts between members of the population, and a description should involve this pattern. Since the number of infectives is small a description involving an assumption of mass action should be replaced by a model which incorporates stochastic effects.

One approach would be a complete description of stochastic epidemic models, for which we refer the reader to the chapter on stochastic models in this volume by Linda Allen. Another approach would be to consider a stochastic model for an outbreak of a communicable disease to be applied so long as the number of infectives remains small, distinguishing a (minor) disease outbreak confined to this initial stage from a (major) epidemic which occurs if the number of infectives begins to grow at an exponential rate. Once an epidemic has started we may switch to a deterministic compartmental model. This approach is described in Chap. 4 on network models in this volume. There is an important difference between the behaviour of network models and the behaviour of models of Kermack–McKendrick type, namely that for a stochastic disease outbreak model if $\mathcal{R}_0 < 1$ the probability that the infection will die out is 1, while if $\mathcal{R}_0 > 1$ there is a positive probability that the infection will persist, and will lead to an epidemic and a positive probability that the infection will increase initially but will produce only a minor outbreak and will die out before triggering a major epidemic.

2.2 Models with Demographic Effects

2.2.1 The SIR Model

Epidemics which sweep through a population attract much attention and arouse a great deal of concern. As we have mentioned in the introduction, the prevalence and effects of many diseases in less developed countries are probably less well-known but may be of even more importance. There are diseases which are endemic in many parts of the world and which cause millions of deaths each year. We have omitted births and deaths in our description of models because the time scale of an epidemic is generally much shorter than the demographic time scale. In effect, we have used a time scale on which the number of births and deaths in unit time is negligible. To model a disease which may be endemic we need to think on a longer time scale and include births and deaths.

For diseases that are endemic in some region public health physicians need to be able to estimate the number of infectives at a given time as well as the rate at which new infections arise. The effects of quarantine or vaccine in reducing the number of victims are of importance, just as in the treatment of epidemics. In addition, the possibility of defeating the endemic nature of the disease and thus controlling or even eradicating the disease in a population is worthy of study.

Measles is a disease for which endemic equilibria have been observed in many places, frequently with sustained oscillations about the equilibrium. The epidemic model of the first section assumes that the epidemic time scale is so short relative to the demographic time scale that demographic effects may be ignored. For measles, however, the reason for the endemic nature of the disease is that there is a flow of new susceptible members into the population, and in order to try to model this we must include births and deaths in the model. The simplest way to incorporate births and deaths in an infectious disease model is to assume a constant number of births and an equal number of deaths per unit time so that the total population size remains constant. This is, of course, feasible only if there are no deaths due to the disease. In developed countries such an assumption is plausible because there are few deaths from measles. In less developed countries there is often a very high mortality rate for measles and therefore other assumptions are necessary.

The first attempt to formulate an SIR model with births and deaths to describe measles was given in 1929 by H.E. Soper [32], who assumed a constant birth rate μK in the susceptible class and a constant death rate μK in the removed class. His model is

$$S' = -\beta SI + \mu K$$
$$I' = \beta SI - \gamma I$$
$$R' = \gamma I - \mu K \ .$$

This model is unsatisfactory biologically because the linkage of births of susceptibles to deaths of removed members is unreasonable. It is also an improper model mathematically because if $R(0)$ and $I(0)$ are sufficiently small then $R(t)$ will become negative. For any disease model to be plausible it is essential that the problem be properly posed in the sense that the number of members in each class must remain non-negative. A model that does not satisfy this requirement cannot be a proper description of a disease model and therefore must contain some assumption that is biologically unreasonable. A full analysis of a model should include verification of this property.

A model of Kermack and McKendrick [22] includes births in the susceptible class proportional to total population size and a death rate in each class proportional to the number of members in the class. This model allows the total population size to grow exponentially or die out exponentially if the birth and death rates are unequal. It is applicable to such questions as whether a disease will control the size of a population that would otherwise grow exponentially. We shall return to this topic, which is important in the study of many diseases in less developed countries with high birth rates. To formulate a model in which total population size remains bounded we could follow the approach suggested by [15] in which the total population size is held constant by making birth and death rates equal. Such a model is

$$S' = -\beta SI + \mu(K - S)$$
$$I' = \beta SI - \gamma I - \mu I$$
$$R' = \gamma I - \mu R \,.$$

Because $S + I + R = K$, we can view R as determined when S and I are known and consider the two-dimensional system

$$S' = -\beta SI + \mu(K - S)$$
$$I' = \beta SI - \gamma I - \mu I \,.$$

We shall examine a slightly more general SIR model with births and deaths for a disease that may be fatal to some infectives. For such a disease the class R of removed members should contain only recovered members, not members removed by death from the disease. It is not possible to assume that the total population size remain constant if there are deaths due to disease; a plausible model for a disease that may be fatal to some infectives must allow the total population to vary in time. The simplest assumption to allow this is a constant birth rate Λ, but in fact the analysis is quite similar if the birth rate is a function $\Lambda(N)$ of total population size N.

Let us analyze the model

$$S' = \Lambda - \beta SI - \mu S$$
$$I' = \beta SI - \mu I - \alpha I \qquad (2.15)$$
$$N' = \Lambda - (1 - f)\alpha I - \mu N \,,$$

where $N = S + I + R$, with a mass action contact rate, a constant number of births Λ per unit time, a proportional natural death rate μ in each class, and a rate of recovery or disease death α of infectives with a fraction f of infectives recovering with immunity against reinfection. In this model if $f = 1$

the total population size approaches a limit $K = \Lambda/\mu$. Then K is the carrying capacity of the population. If $f < 1$ the total population size is not constant and K represents a carrying capacity or maximum possible population size, rather than a population size. We view the first two equations as determining S and I, and then consider the third equation as determining N once S and I are known. This is possible because N does not enter into the first two equations. Instead of using N as the third variable in this model we could have used R, and the same reduction would have been possible.

If the birth or recruitment rate $\Lambda(N)$ is a function of total population size then in the absence of disease the total population size N satisfies the differential equation

$$N' = \Lambda(N) - \mu N .$$

The *carrying capacity* of population size is the limiting population size K, satisfying

$$\Lambda(K) = \mu K, \qquad \Lambda'(K) < \mu .$$

The condition $\Lambda'(K) < \mu$ assures the asymptotic stability of the equilibrium population size K. It is reasonable to assume that K is the only positive equilibrium, so that

$$\Lambda(N) > \mu N$$

for $0 \leq N \leq K$. For most population models,

$$\Lambda(0) = 0, \qquad \Lambda''(N) \leq 0 .$$

However, if $\Lambda(N)$ represents recruitment into a behavioural class, as would be natural for models of sexually transmitted diseases, it would be plausible to have $\Lambda(0) > 0$, or even to consider $\Lambda(N)$ to be a constant function. If $\Lambda(0) = 0$, we require $\Lambda'(0) > \mu$ because if this requirement is not satisfied there is no positive equilibrium and the population would die out even in the absence of disease.

We have used a mass action contact rate for simplicity, even though a more general contact rate would give a more accurate model, just as in the epidemics considered in the preceding section. With a general contact rate and a density-dependent birth rate we would have a model

$$\begin{aligned}
S' &= \Lambda(N) - \beta(N)SI - \mu S \\
I' &= \beta(N)SI - \mu I - \alpha I \\
N' &= \Lambda(N) - (1-f)\alpha I - \mu N.
\end{aligned} \qquad (2.16)$$

If $f = 1$, so that there are no disease deaths, the equation for N is

$$N' = \Lambda(N) - \mu N ,$$

so that $N(t)$ approaches a limiting population size K. The theory of *asymptotically autonomous systems* [8, 24, 34, 37] implies that if N has a constant limit then the system is equivalent to the system in which N is replaced by this limit. Then the system (2.16) is the same as the system (2.15) with β replaced by the constant $\beta(K)$ and N by K, and $\Lambda(N)$ replaced by the constant $\Lambda(K) = \mu K$.

We shall analyze the model (2.15) qualitatively. In view of the remark above, our analysis will also apply to the more general model (2.16) if there are no disease deaths. Analysis of the system (2.16) with $f < 1$ is much more difficult. We will confine our study of (2.16) to a description without details.

The first stage of the analysis is to note that the model (2.15) is a properly posed problem. That is, since $S' \geq 0$ if $S = 0$ and $I' \geq 0$ if $I = 0$, we have $S \geq 0, I \geq 0$ for $t \geq 0$ and since $N' \leq 0$ if $N = K$ we have $N \leq K$ for $t \geq 0$. Thus the solution always remains in the biologically realistic region $S \geq 0, I \geq 0, 0 \leq N \leq K$ if it starts in this region. By rights, we should verify such conditions whenever we analyze a mathematical model, but in practice this step is frequently overlooked.

Our approach will be to identify equilibria (constant solutions) and then to determine the asymptotic stability of each equilibrium. Asymptotic stability of an equilibrium means that a solution starting sufficiently close to the equilibrium remains close to the equilibrium and approaches the equilibrium as $t \to \infty >$, while instability of the equilibrium means that there are solutions starting arbitrarily close to the equilibrium which do not approach it. To find equilibria (S_∞, I_∞) we set the right side of each of the two equations equal to zero. The second of the resulting algebraic equations factors, giving two alternatives. The first alternative is $I_\infty = 0$, which will give a disease-free equilibrium, and the second alternative is $\beta S_\infty = \mu + \alpha$, which will give an endemic equilibrium, provided $\beta S_\infty = \mu + \alpha < \beta K$. If $I_\infty = 0$ the other equation gives $S_\infty = K = \Lambda/\mu$. For the endemic equilibrium the first equation gives

$$I_\infty = \frac{\Lambda}{\mu + \alpha} - \frac{\mu}{\beta} . \tag{2.17}$$

We linearize about an equilibrium (S_∞, I_∞) by letting $y = S - S_\infty$, $z = I - I_\infty$, writing the system in terms of the new variables y and z and retaining only the linear terms in a Taylor expansion. We obtain a system of two linear differential equations,

$$y' = -(\beta I_\infty + \mu)y - \beta S_\infty z$$
$$z' = \beta I_\infty y + (\beta S_\infty - \mu - \alpha)z .$$

The coefficient matrix of this linear system is

$$\begin{bmatrix} -\beta I_\infty - \mu & -\beta S_\infty \\ \beta I_\infty & \beta S_\infty - \mu - \alpha \end{bmatrix} .$$

We then look for solutions whose components are constant multiples of $e^{\lambda t}$; this means that λ must be an eigenvalue of the coefficient matrix. The condition that all solutions of the linearization at an equilibrium tend to zero as $t \to \infty$ is that the real part of every eigenvalue of this coefficient matrix is negative. At the disease-free equilibrium the matrix is

$$\begin{bmatrix} -\mu & -\beta K \\ 0 & \beta K - \mu - \alpha \end{bmatrix},$$

which has eigenvalues $-\mu$ and $\beta K - \mu - \alpha$. Thus, the disease-free equilibrium is asymptotically stable if $\beta K < \mu + \alpha$ and unstable if $\beta K > \mu + \alpha$. Note that this condition for instability of the disease-free equilibrium is the same as the condition for the existence of an endemic equilibrium.

In general, the condition that the eigenvalues of a 2×2 matrix have negative real part is that the determinant be positive and the trace (the sum of the diagonal elements) be negative. Since $\beta S_\infty = \mu + \alpha$ at an endemic equilibrium, the matrix of the linearization at an endemic equilibrium is

$$\begin{bmatrix} -\beta I_\infty - \mu & -\beta S_\infty \\ \beta I_\infty & 0 \end{bmatrix} \tag{2.18}$$

and this matrix has positive determinant and negative trace. Thus, the endemic equilibrium, if there is one, is always asymptotically stable. If the quantity

$$R_0 = \frac{\beta K}{\mu + \alpha} = \frac{K}{S_\infty} \tag{2.19}$$

is less than one, the system has only the disease-free equilibrium and this equilibrium is asymptotically stable. In fact, it is not difficult to prove that this asymptotic stability is *global*, that is, that every solution approaches the disease-free equilibrium. If the quantity $\mathcal{R}_0$ is greater than one then the disease-free equilibrium is unstable, but there is an endemic equilibrium that is asymptotically stable. Again, the quantity $\mathcal{R}_0$ is the basic reproduction number. It depends on the particular disease (determining the parameter α) and on the rate of contacts, which may depend on the population density in the community being studied. The disease model exhibits a *threshold* behaviour: If the basic reproduction number is less than one the disease will die out, but if the basic reproduction number is greater than one the disease will be endemic. Just as for the epidemic models of the preceding section, the basic reproduction number is the number of secondary infections caused by a single infective introduced into a wholly susceptible population because the number of contacts per infective in unit time is βK, and the mean infective period (corrected for natural mortality) is $1/(\mu + \alpha)$.

There are two aspects of the analysis of the model (2.16) which are more complicated than the analysis of (2.15). The first is in the study of equilibria. Because of the dependence of $\Lambda(N)$ and $\beta(N)$ on N, it is necessary to use two of the equilibrium conditions to solve for S and I in terms of N and then substitute into the third condition to obtain an equation for N. Then by comparing the two sides of this equation for $N = 0$ and $N = K$ it is possible to show that there must be an endemic equilibrium value of N between 0 and K.

The second complication is in the stability analysis. Since (2.16) is a three-dimensional system which can not be reduced to a two-dimensional system, the coefficient matrix of its linearization at an equilibrium is a 3×3 matrix and the resulting characteristic equation is a cubic polynomial equation of the form

$$\lambda^3 + a_1\lambda^2 + a_2\lambda + a_3 = 0 .$$

The *Routh–Hurwitz* conditions

$$a_1 > 0, \qquad a_1 a_2 > a_3 > 0$$

are necessary and sufficient conditions for all roots of the characteristic equation to have negative real part. A technically complicated calculation is needed to verify that these conditions are satisfied at an endemic equilibrium for the model (2.16).

The asymptotic stability of the endemic equilibrium means that the compartment sizes approach a steady state. If the equilibrium had been unstable, there would have been a possibility of sustained oscillations. Oscillations in a disease model mean fluctuations in the number of cases to be expected, and if the oscillations have long period could also mean that experimental data for a short period would be quite unreliable as a predictor of the future. Epidemiological models which incorporate additional factors may exhibit oscillations. A variety of such situations is described in [18, 19].

The epidemic models of the first section also exhibited a threshold behaviour but of a slightly different kind. For these models, which were SIR models without births or natural deaths, the threshold distinguished between a dying out of the disease and an epidemic, or short term spread of disease.

From the third equation of (2.15) we obtain

$$N' = \Lambda - \mu N - (1 - f)\alpha I ,$$

where $N = S + I + R$. From this we see that at the endemic equilibrium $N = K - (1 - f)\alpha I/\mu$, and the reduction in the population size from the carrying capacity K is

$$(1 - f)\frac{\alpha}{\mu}I_\infty = (1 - f)[\frac{\alpha K}{\mu + \alpha} - \frac{\alpha}{\beta}] .$$

The parameter α in the SIR model may be considered as describing the pathogenicity of the disease. If α is large it is less likely that $\mathcal{R}_0 > 1$. If α is small then the total population size at the endemic equilibrium is close to the carrying capacity K of the population. Thus, the maximum population decrease caused by disease will be for diseases of intermediate pathogenicity.

2.2.2 The SIS Model

In order to describe a model for a disease from which infectives recover with immunity against reinfection and that includes births and deaths as in the model (2.16), we may modify the model (2.16) by removing the equation for R and moving the term $f\alpha I$ describing the rate of recovery from infection to the equation for S. This gives the model

$$S' = \Lambda(N) - \beta(N)SI - \mu S + f\alpha I \qquad (2.20)$$
$$I' = \beta(N)SI - \alpha I - \mu I$$

describing a population with a density-dependent birth rate $\Lambda(N)$ per unit time, a proportional death rate μ in each class, and with a rate α of departure from the infective class through recovery or disease death and with a fraction f of infectives recovering with no immunity against reinfection. In this model, if $f < 1$ the total population size is not constant and K represents a *carrying capacity*, or maximum possible population size, rather than a constant population size.

It is easy to verify that

$$\mathcal{R}_0 = \frac{K\beta(K)}{\mu + \alpha} \; .$$

If we add the two equations of (2.20), and use $N = S + I$ we obtain

$$N' = \Lambda(N) - \mu N - (1 - f)\alpha I \; .$$

For the SIS model we are able to carry out the analysis with a general contact rate. If $f = 1$ the equation for N is

$$N' = \Lambda(N) - \mu N$$

and N approaches the limit K. The system (2.20) is asymptotically autonomous and its asymptotic behaviour is the same as that of the single differential equation

$$I' = \beta(K)I(K - I) - (\alpha + \mu)I \; , \qquad (2.21)$$

where S has been replaced by $K - I$. But (2.21) is a logistic equation which is easily solved analytically by separation of variables or qualitatively by an equilibrium analysis. We find that $I \to 0$ if $K\beta(K) < (\mu + \alpha)$, or $\mathcal{R}_0 < 1$ and $I \to I_\infty > 0$ with

$$I_\infty = K - \frac{\mu + \alpha}{\beta(K)} = K(1 - \frac{1}{\mathcal{R}_0})$$

if $K\beta(K) > (\mu + \alpha)$ or $\mathcal{R}_0 > 1$.

To analyze the SIS model if $f < 1$, it is convenient to use I and N as variables instead of S and I, with S replaced by $N - I$. This gives the model

$$I' = \beta(N)I(N - I) - (\mu + \alpha)I \qquad (2.22)$$
$$N' = \Lambda(N) - \mu N - (1 - f)\alpha I \ .$$

Equilibria are found by setting the right sides of the two differential equations equal to zero. The first of the resulting algebraic equations factors, giving two alternatives. The first alternative is $I = 0$, which will give a disease-free equilibrium $I = 0, N = K$, and the second alternative is $\beta(N)(N - I) = \mu + \alpha$, which may give an endemic equilibrium. For an endemic equilibrium (I_∞, N_∞) the first equation gives

$$I_\infty \beta(N_\infty) = N_\infty \beta(N_\infty) - (\mu + \alpha) \ .$$

Substitution into the other equilibrium condition gives

$$\Lambda(N_\infty) = \mu N_\infty + (1 - f)\alpha[N_\infty - \frac{\mu + \alpha}{\beta(N_\infty)}] \ ,$$

which can be simplified to

$$\beta(N_\infty)\Lambda(N_\infty) = \mu N_\infty \beta(N_\infty) + (1 - f)\alpha\left[N_\infty \beta(N_\infty) - (\mu + \alpha)\right] \ . \qquad (2.23)$$

At $N = 0$ the left side of (2.23) is $\beta(0)\Lambda(0) \geq 0$, while the right side is $-(1 - f)\alpha(\mu + \alpha)$, which is negative since $f < 1$. At $N = K$ the left side of (2.23) is

$$\beta(K)\Lambda(K) = \mu K \beta(K)$$

while the right side of (2.23) is

$$\mu K \beta(K) + (1 - f)\alpha[K\beta(K) - (\mu + \alpha)] \ .$$

Since

$$\mathcal{R}_0 = \frac{K\beta(K)}{\mu + \alpha} \ ,$$

if $\mathcal{R}_0 > 1$ the left side of (2.23) is less than the right side of (2.23), and this implies that (2.23) has a solution for $N, 0 < N < K$. Thus there is an endemic equilibrium if $\mathcal{R}_0 > 1$. If $\mathcal{R}_0 < 1$ this reasoning may be used to show that there is no endemic equilibrium.

The linearization of (2.22) at an equilibrium (I_∞, N_∞) has coefficient matrix

$$\begin{bmatrix} \beta(N_\infty)(N_\infty - 2I_\infty) - (\mu + \alpha) & \beta(N_\infty)I_\infty + \beta'(N_\infty)I_\infty(N_\infty - I_\infty) \\ -(1 - f)\alpha & \Lambda'(N_\infty) - \mu. \end{bmatrix}$$

At the disease-free equilibrium the matrix is

$$\begin{bmatrix} K\beta(K) - (\mu + \alpha) & 0 \\ -(1-f)\alpha & \Lambda'(K) - \mu \end{bmatrix},$$

which has eigenvalues $\Lambda'(K) - \mu$ and $K\beta K - (\mu + \alpha)$. Thus, the disease-free equilibrium is asymptotically stable if $K\beta(K) < \mu + \alpha$, or $\mathcal{R}_0 < 1$, and unstable if $K\beta(K) > \mu + \alpha$, or $\mathcal{R}_0 > 1$. Note that the condition for instability of the disease-free equilibrium is the same as the condition for the existence of an endemic equilibrium.

At an endemic equilibrium, since $\beta(N_\infty)(N_\infty - I_\infty) = \mu + \alpha$, the matrix is

$$\begin{bmatrix} -I\beta(N_\infty) & I_\infty\beta(N_\infty) + I_\infty(N_\infty - I_\infty)\beta'(N_\infty) \\ -(1-f)\alpha & \Lambda'(N_\infty) - \mu \end{bmatrix}.$$

Since $\beta'(N_\infty) \leq 0$

$$\beta(N_\infty) + (N_\infty - I_\infty)\beta'(N_\infty) \geq \beta(N_\infty) + N_\infty\beta'(N_\infty) \geq 0 \,.$$

Thus if $\Lambda'(N_\infty) < \mu$ the coefficient matrix has sign structure

$$\begin{bmatrix} - & + \\ - & - \end{bmatrix}.$$

It is clear that the coefficient matrix has negative trace and positive determinant if $\Lambda'(N) < \mu$ and this implies that the endemic equilibrium is asymptotically stable. Thus, the endemic equilibrium, which exists if $\mathcal{R}_0 > 1$, is always asymptotically stable. If $\mathcal{R}_0 < 1$ the system has only the disease-free equilibrium and this equilibrium is asymptotically stable. In the case $f = 1$ the verification of these properties remains valid if there are no births and deaths. This suggests that a requirement for the existence of an endemic equilibrium is a flow of new susceptibles either through births, as in the SIR model or through recovery without immunity against reinfection, as in the SIS model with or without births and deaths.

If the epidemiological and demographic time scales are very different, for the SIR model we observed that the approach to endemic equilibrium is like a rapid and severe epidemic. The same happens in the SIS model, especially if there is a significant number of deaths due to disease. If there are few disease deaths the number of infectives at endemic equilibrium may be substantial, and there may be damped oscillations of large amplitude about the endemic equilibrium.

For both the SIR and SIS models we may write the differential equation for I as

$$I' = I[\beta(N)S - (\mu + \alpha)] = \beta(N)I[S - S_\infty] \,,$$

which implies that whenever S exceeds its endemic equilibrium value S_∞, I is increasing and epidemic-like behaviour is possible. If $\mathcal{R}_0 < 1$ and $S < K$ it follows that $I' < 0$, and thus I is decreasing. Thus, if $\mathcal{R}_0 < 1$, I cannot increase and no epidemic can occur.

Next, we will turn to some applications of SIR and SIS models, taken mainly from [3].

2.3 Some Applications

2.3.1 Herd Immunity

In order to prevent a disease from becoming endemic it is necessary to reduce the basic reproduction number $\mathcal{R}_0$ below one. This may sometimes be achieved by immunization. If a fraction p of the Λ newborn members per unit time of the population is successfully immunized, the effect is to replace K by $K(1-p)$, and thus to reduce the basic reproduction number to $\mathcal{R}_0(1-p)$. The requirement $\mathcal{R}_0(1-p) < 1$ gives $1 - p < 1/\mathcal{R}_0$, or

$$ p > 1 - \frac{1}{\mathcal{R}_0} \, . $$

A population is said to have *herd immunity* if a large enough fraction has been immunized to assure that the disease cannot become endemic. The only disease for which this has actually been achieved worldwide is smallpox for which $\mathcal{R}_0$ is approximately 5, so that 80% immunization does provide herd immunity.

For measles, epidemiological data in the United States indicate that $\mathcal{R}_0$ for rural populations ranges from 5.4 to 6.3, requiring vaccination of 81.5–84.1% of the population. In urban areas $\mathcal{R}_0$ ranges from 8.3 to 13.0, requiring vaccination of 88.0–92.3% of the population. In Great Britain, $\mathcal{R}_0$ ranges from 12.5 to 16.3, requiring vaccination of 92–94% of the population. The measles vaccine is not always effective, and vaccination campaigns are never able to reach everyone. As a result, herd immunity against measles has not been achieved (and probably never can be). Since smallpox is viewed as more serious and requires a lower percentage of the population be immunized, herd immunity was attainable for smallpox. In fact, smallpox has been eliminated; the last known case was in Somalia in 1977, and the virus is maintained now only in laboratories (although there is currently some concern that it may be reintroduced as a bioterrorism attack). The eradication of smallpox was actually more difficult than expected because high vaccination rates were achieved in some countries but not everywhere, and the disease persisted in some countries. The eradication of smallpox was possible only after an intensive campaign for worldwide vaccination [16].

2.3.2 Age at Infection

In order to calculate the basic reproduction number $\mathcal{R}_0$ for a disease, we need to know the values of the contact rate β and the parameters μ, K, and α. The parameters μ, K, and α can usually be measured experimentally but the contact rate β is difficult to determine directly. There is an indirect means of estimating $\mathcal{R}_0$ in terms of the life expectancy and the mean age at infection which enables us to avoid having to estimate the contact rate. In this calculation, we will assume that β is constant, but we will also indicate the modifications needed when β is a function of total population size N. The calculation assumes exponentially distributed life spans and infective periods. In fact, the result is valid so long as the life span is exponentially distributed.

Consider the "age cohort" of members of a population born at some time t_0 and let a be the age of members of this cohort. If $y(a)$ represents the fraction of members of the cohort who survive to age (at least) a, then the assumption that a fraction μ of the population dies per unit time means that $y'(a) = -\mu y(a)$. Since $y(0) = 1$, we may solve this first order initial value problem to obtain $y(a) = e^{-\mu a}$. The fraction dying at (exactly) age a is $-y'(a) = \mu y(a)$. The mean life span is the average age at death, which is $\int_0^\infty a[-y'(a)]da$, and if we integrate by parts we find that this life expectancy is

$$\int_0^\infty [-ay'(a)]\, da = [-ay(a)]_0^\infty + \int_0^\infty y(a)\, da = \int_0^\infty y(a)\, da \ .$$

Since $y(a) = e^{-\mu a}$, this reduces to $1/\mu$. The life expectancy is often denoted by L, so that we may write

$$L = \frac{1}{\mu} \ .$$

The rate at which surviving susceptible members of the population become infected at age a and time $t_0 + a$, is $\beta I(t_0 + a)$. Thus, if $z(a)$ is the fraction of the age cohort alive and still susceptible at age a, $z'(a) = -[\mu + \beta I(t_0 + a)]z(a)$. Solution of this first linear order differential equation gives

$$z(a) = e^{-[\mu a + \int_0^a \beta I(t_0 + b)\, db]} = y(a)e^{-\int_0^a \beta I(t_0 + b)\, db} \ .$$

The mean length of time in the susceptible class for members who may become infected, as opposed to dying while still susceptible, is

$$\int_0^\infty e^{-\int_0^a \beta I(t_0 + b)\, db}\, da \ ,$$

and this is the mean age at which members become infected. If the system is at an equilibrium I_∞, this integral may be evaluated, and the mean age at infection, denoted by A, is given by

$$A = \int_0^\infty e^{-\beta I_\infty a}\, da = \frac{1}{\beta I_\infty}.$$

For our model the endemic equilibrium is

$$I_\infty = \frac{\mu K}{\mu + \alpha} - \frac{\mu}{\beta},$$

and this implies

$$\frac{L}{A} = \frac{\beta I_\infty}{\mu} = \mathcal{R}_0 - 1. \tag{2.24}$$

This relation is very useful in estimating basic reproduction numbers. For example, in some urban communities in England and Wales between 1956 and 1969 the average age of contracting measles was 4.8 years. If life expectancy is assumed to be 70 years, this indicates $\mathcal{R}_0 = 15.6$.

If β is a function $\beta(N)$ of total population size the relation (2.24) becomes

$$\mathcal{R}_0 = \frac{\beta(K)}{\beta(N)} \left[1 + \frac{L}{A} \right].$$

If disease mortality does not have a large effect on total population size, in particular if there is no disease mortality, this relation is very close to (2.24).

The relation between age at infection and basic reproduction number indicates that measures such as inoculations, which reduce $\mathcal{R}_0$, will increase the average age at infection. For diseases such as rubella (German measles), whose effects may be much more serious in adults than in children, this indicates a danger that must be taken into account: While inoculation of children will decrease the number of cases of illness, it will tend to increase the danger to those who are not inoculated or for whom the inoculation is not successful. Nevertheless, the number of infections in older people will be reduced, although the fraction of cases which are in older people will increase.

2.3.3 The Interepidemic Period

Many common childhood diseases, such as measles, whooping cough, chicken pox, diphtheria, and rubella, exhibit variations from year to year in the number of cases. These fluctuations are frequently regular oscillations, suggesting that the solutions of a model might be periodic. This does not agree with the predictions of the model we have been using here; however, it would not be inconsistent with solutions of the characteristic equation, which are complex conjugate with small negative real part corresponding to lightly damped oscillations approaching the endemic equilibrium. Such behaviour would look like recurring epidemics. If the eigenvalues of the matrix of the linearization at an endemic equilibrium are $-u \pm iv$, where $i^2 = -1$, then the solutions of the linearization are of the form $Be^{-ut}\cos(vt+c)$, with decreasing "amplitude" Be^{-ut} and "period" $\frac{2\pi}{v}$.

For the model (2.15) we recall from (2.17) that at the endemic equilibrium we have

$$\beta I_\infty + \mu = \mu\mathcal{R}_0, \qquad \beta S_\infty = \mu + \alpha,$$

and from (2.18) the matrix of the linearization is

$$\begin{bmatrix} -\mu\mathcal{R}_0 & -(\mu+\alpha) \\ \mu(\mathcal{R}_0-1) & 0 \end{bmatrix}$$

The eigenvalues are the roots of the quadratic equation

$$\lambda^2 + \mu\mathcal{R}_0\lambda + \mu(\mathcal{R}_0-1)(\mu+\alpha) = 0,$$

which are

$$\lambda = \frac{-\mu\mathcal{R}_0 \pm \sqrt{\mu^2\mathcal{R}_0{}^2 - 4\mu(\mathcal{R}_0-1)(\mu+\alpha)}}{2}.$$

If the mean infective period $1/\alpha$ is much shorter than the mean life span $1/\mu$, we may neglect the terms that are quadratic in μ. Thus, the eigenvalues are approximately

$$\frac{-\mu\mathcal{R}_0 \pm \sqrt{-4\mu(\mathcal{R}_0-1)\alpha}}{2},$$

and these are complex with imaginary part $\sqrt{\mu(\mathcal{R}_0-1)\alpha}$. This indicates oscillations with period approximately

$$\frac{2\pi}{\sqrt{\mu(\mathcal{R}_0-1)\alpha}}.$$

We use the relation $\mu(\mathcal{R}_0-1) = \mu L/A$ and the mean infective period $\tau = 1/\alpha$ to see that the interepidemic period T is approximately $2\pi\sqrt{A\tau}$. Thus, for example, for recurring outbreaks of measles with an infective period of 2 weeks or $1/26$ year in a population with a life expectancy of 70 years with $\mathcal{R}_0$ estimated as 15, we would expect outbreaks spaced 2.76 years apart. Also, as the "amplitude" at time t is $e^{-\mu\mathcal{R}_0 t/2}$, the maximum displacement from equilibrium is multiplied by a factor $e^{-(15)(2.76)/140} = 0.744$ over each cycle. In fact, many observations of measles outbreaks indicate less damping of the oscillations, suggesting that there may be additional influences that are not included in our simple model. To explain oscillations about the endemic equilibrium a more complicated model is needed. One possible generalization would be to assume seasonal variations in the contact rate. This is a reasonable supposition for a childhood disease most commonly transmitted through school contacts, especially in winter in cold climates. Note, however, that data from observations are never as smooth as model predictions and models are inevitably gross simplifications of reality which cannot account for random variations in the variables. It may be difficult to judge from experimental data whether an oscillation is damped or persistent.

2.3.4 "Epidemic" Approach to the Endemic Equilibrium

In the model (2.15) the demographic time scale described by the birth and natural death rates Λ and μ and the epidemiological time scale described by the rate α of departure from the infective class may differ substantially. Think, for example, of a natural death rate $\mu = 1/75$, corresponding to a human life expectancy of 75 years, and epidemiological parameters $\alpha = 25, f = 1$, describing a disease from which all infectives recover after a mean infective period of $1/25$ year, or two weeks. Suppose we consider a carrying capacity $K = 1,000$ and take $\beta = 0.1$, indicating that an average infective makes $(0.1)(1,000) = 100$ contacts per year. Then $\mathcal{R}_0 = 4.00$, and at the endemic equilibrium we have $S_\infty = 250.13$, $I_\infty = 0.40$, $R_\infty = 749.47$. This equilibrium is globally asymptotically stable and is approached from every initial state.

However, if we take $S(0) = 999$, $I(0) = 1$, $R(0) = 0$, simulating the introduction of a single infective into a susceptible population and solve the system numerically we find that the number of infectives rises sharply to a maximum of 400 and then decreases to almost zero in a period of 0.4 year, or about 5 months. In this time interval the susceptible population decreases to 22 and then begins to increase, while the removed (recovered and immune against reinfection) population increases to almost 1,000 and then begins a gradual decrease. The size of this initial "epidemic" could not have been predicted from our qualitative analysis of the system (2.15). On the other hand, since μ is so small compared to the other parameters of the model, we might consider neglecting μ, replacing it by zero in the model. If we do this, the model reduces to the simple Kermack–McKendrick epidemic model (without births and deaths) of the first section.

If we follow the model (2.15) over a longer time interval we find that the susceptible population grows to 450 after 46 years, then drops to 120 during a small epidemic with a maximum of 18 infectives, and exhibits widely spaced epidemics decreasing in size. It takes a very long time before the system comes close to the endemic equilibrium and remains close to it. The large initial epidemic conforms to what has often been observed in practice when an infection is introduced into a population with no immunity, such as the smallpox inflicted on the Aztecs by the invasion of Cortez.

If we use the model (2.15) with the same values of β, K and μ, but take $\alpha = 25, f = 0$ to describe a disease fatal to all infectives, we obtain very similar results. Now the total population is $S + I$, which decreases from an initial size of 1,000 to a minimum of 22 and then gradually increases and eventually approaches its equilibrium size of 250.53. Thus, the disease reduces the total population size to one-fourth of its original value, suggesting that infectious diseases may have large effects on population size. This is true even for populations which would grow rapidly in the absence of infection, as we shall see later.

2.3.5 Disease as Population Control

Many parts of the world experienced very rapid population growth in the eighteenth century. The population of Europe increased from 118 million in 1700 to 187 million in 1800. In the same time period the population of Great Britain increased from 5.8 million to 9.15 million, and the population of China increased from 150 million to 313 million [27]. The population of English colonies in North America grew much more rapidly than this, aided by substantial immigration from England, but the native population, which had been reduced to one tenth of their previous size by disease following the early encounters with Europeans and European diseases, grew even more rapidly. While some of these population increases may be explained by improvements in agriculture and food production, it appears that an even more important factor was the decrease in the death rate due to diseases. Disease death rates dropped sharply in the eighteenth century, partly from better understanding of the links between illness and sanitation and partly because the recurring invasions of bubonic plague subsided, perhaps due to reduced susceptibility. One plausible explanation for these population increases is that the bubonic plague invasions served to control the population size, and when this control was removed the population size increased rapidly.

In developing countries it is quite common to have high birth rates and high disease death rates. In fact, when disease death rates are reduced by improvements in health care and sanitation it is common for birth rates to decline as well, as families no longer need to have as many children to ensure that enough children survive to take care of the older generations. Again, it is plausible to assume that population size would grow exponentially in the absence of disease but is controlled by disease mortality.

The SIR model with births and deaths of Kermack and McKendrick [22] includes births in the susceptible class proportional to population size and a natural death rate in each class proportional to the size of the class. Let us analyze a model of this type with birth rate r and a natural death rate $\mu < r$. For simplicity we assume the disease is fatal to all infectives with disease death rate α, so that there is no removed class and the total population size is $N = S + I$. Our model is

$$S' = r(S + I) - \beta SI - \mu S \qquad (2.25)$$
$$I' = \beta SI - (\mu + \alpha)I .$$

From the second equation we see that equilibria are given by either $I = 0$ or $\beta S = \mu + \alpha$. If $I = 0$ the first equilibrium equation is $rS = \mu S$, which implies $S = 0$ since $r > \mu$. It is easy to see that the equilibrium (0,0) is unstable. What actually would happen if $I = 0$ is that the susceptible population would grow exponentially with exponent $r - \mu > 0$. If $\beta S = \mu + \alpha$ the first equilibrium condition gives

$$r\frac{\mu + \alpha}{\beta} + rI - (\mu + \alpha)I - \frac{\mu(\mu + \alpha)}{\beta} = 0 ,$$

which leads to

$$(\alpha + \mu - r)I = \frac{(r - \mu)(\mu + \alpha)}{\beta} .$$

Thus, there is an endemic equilibrium provided $r < \alpha + \mu$, and it is possible to show by linearizing about this equilibrium that it is asymptotically stable. On the other hand, if $r > \alpha + \mu$ there is no positive equilibrium value for I. In this case we may add the two differential equations of the model to give

$$N' = (r - \mu)N - \alpha I \geq (r - \mu)N - \alpha N = (r - \mu - \alpha)N$$

and from this we may deduce that N grows exponentially. For this model either we have an asymptotically stable endemic equilibrium or population size grows exponentially. In the case of exponential population growth we may have either vanishing of the infection or an exponentially growing number of infectives.

If only susceptibles contribute to the birth rate, as may be expected if the disease is sufficiently debilitating, the behaviour of the model is quite different. Let us consider the model

$$S' = rS - \beta SI - \mu S = S(r - \mu - \beta I) \tag{2.26}$$
$$I' = \beta SI - (\mu + \alpha)I = I(\beta S - \mu - \alpha)$$

which has the same form as the celebrated Lotka–Volterra predator–prey model of population dynamics. This system has two equilibria, obtained by setting the right sides of each of the equations equal to zero, namely (0,0) and an endemic equilibrium $((\mu + \alpha)/\beta, (r - \mu)/\beta)$. It turns out that the qualitative analysis approach we have been using is not helpful as the equilibrium (0,0) is unstable and the eigenvalues of the coefficient matrix at the endemic equilibrium have real part zero. In this case the behaviour of the linearization does not necessarily carry over to the full system. However, we can obtain information about the behaviour of the system by a method that begins with the elementary approach of separation of variables for first order differential equations. We begin by taking the quotient of the two differential equations and using the relation

$$\frac{I'}{S'} = \frac{dI}{dS}$$

to obtain the separable first order differential equation

$$\frac{dI}{dS} = \frac{I(\beta S - \mu - \alpha)}{S(r - \beta I)} .$$

Separation of variables gives

$$\int \left(\frac{r}{I} - \beta \right) dI = \int \left(\beta - \frac{\mu + \alpha}{S} \right) dS .$$

Integration gives the relation

$$\beta(S + I) - r \log I - (\mu + \alpha) \log S = c$$

where c is a constant of integration. This relation shows that the quantity

$$V(S, I) = \beta(S + I) - r \log I - (\mu + \alpha) \log S$$

is constant on each orbit (path of a solution in the $(S, I-$ plane). Each of these orbits is a closed curve corresponding to a periodic solution.

This model is the same as the simple epidemic model of the first section except for the birth and death terms, and in many examples the time scale of the disease is much faster than the time scale of the demographic process. We may view the model as describing an epidemic initially, leaving a susceptible population small enough that infection cannot establish itself. Then there is a steady population growth until the number of susceptibles is large enough for an epidemic to recur. During this growth stage the infective population is very small and random effects may wipe out the infection, but the immigration of a small number of infectives will eventually restart the process. As a result, we would expect recurrent epidemics. In fact, bubonic plague epidemics did recur in Europe for several hundred years. If we modify the demographic part of the model to assume limited population growth rather than exponential growth in the absence of disease, the effect would be to give behaviour like that of the model studied in the previous section, with an endemic equilibrium that is approached slowly in an oscillatory manner if $\mathcal{R}_0 > 1$.

Example. (Fox rabies) Rabies is a viral infection to which many animals, especially foxes, coyotes, wolves, and rats, are highly susceptible. While dogs are only moderately susceptible, they are the main source of rabies in humans. Although deaths of humans from rabies are few, the disease is still of concern because it is invariably fatal. However, the disease is endemic in animals in many parts of the world. A European epidemic of fox rabies thought to have begun in Poland in 1939 and spread through much of Europe has been modeled. We present here a simplified version of a model due to R.M. Anderson and coworkers [1].

We begin with the demographic assumptions that foxes have a birth rate proportional to population size but that infected foxes do not produce offspring (because the disease is highly debilitating), and that there is a natural death rate proportional to population size. Experimental data indicate a birth rate of approximately 1 per capita per year and a death rate of approximately 0.5 per capita per year, corresponding to a life expectancy of 2 years. The fox population is divided into susceptibles and infectives, and the epidemiological assumptions are that the rate of acquisition of infection is proportional to the number of encounters between susceptibles and infectives. We will assume a contact parameter $\beta = 80$, in rough agreement with observations of frequency of contact in regions where the fox density is approximately 1 fox/km^2, and we assume that all infected foxes die with a mean infective period of approximately 5 days or 1/73 year. These assumptions lead to the model

$$S' = -\beta SI + rS - \mu S$$
$$I' = \beta SI - (\mu + \alpha)I$$

with $\beta = 80$, $r = 1.0$, $\mu = 0.5$, $\alpha = 73$. As this is of the form (2.26), we know that the orbits are closed curves in the (S, I) plane, and that both S and I are periodic functions of t. We illustrate with some simulations obtained using Maple (Figs. 2.8, 2.9, and 2.10). It should be noted from the graphs of I in terms of t that the period of the oscillation depends on the amplitude, and thus on the initial conditions, with larger amplitudes corresponding to longer periods.

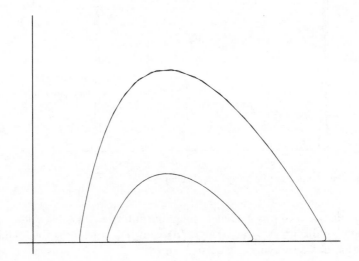

Fig. 2.8 The (S, I) plane

A warning is in order here. The model predicts that for long time intervals the number of infected foxes is extremely small. With such small numbers, the continuous deterministic models we have been using (which assume that population sizes are differentiable functions) are quite inappropriate. If the density of foxes is extremely small an encounter between foxes is a random event, and the number of contacts cannot be described properly by a function of population densities. To describe disease transmission properly when population sizes are very small we would need to use a stochastic model.

Now let us modify the demographic assumptions by assuming that the birth rate decreases as population size increases. We replace the birth rate of r per susceptible per year by a birth rate of re^{-aS} per susceptible per year, with a a positive constant. Then, in the absence of infection, the fox population is given by the first order differential equation

$$N' = N\left(re^{-aN} - \mu\right)$$

and equilibria of this equation are given by $N = 0$ and $re^{-aN} = \mu$, which reduces to $e^{aN} = r/\mu$, or

$$N = \frac{1}{a}\log\frac{r}{\mu}\,.$$

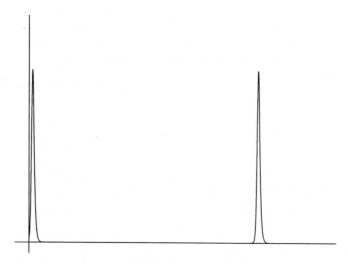

Fig. 2.9 I as a function of t (larger amplitude)

We omit the verification that the equilibrium $N = 0$ is unstable while the positive equilibrium $N = (1/a)\log(r/\mu)$ is asymptotically stable. Thus, the population has a carrying capacity given by

$$K = \frac{1}{a}\log\frac{r}{\mu}\,.$$

The model now becomes

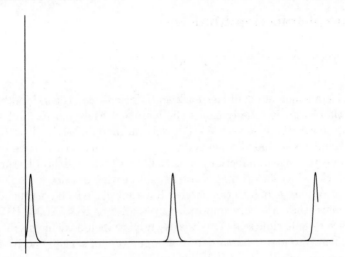

Fig. 2.10 I as a function of t (smaller amplitude)

$$S' = rSe^{-aS} - \beta SI - \mu S$$
$$I' = \beta SI - (\mu + \alpha)I \ .$$

We examine this by looking for equilibria and analyzing their stability. From the second equation, equilibria satisfy either $I = 0$ or $\beta S = \mu + \alpha$. If $I = 0$ the first equilibrium condition reduces to the same equation that determined the carrying capacity, and we have a disease-free equilibrium $S = K$, $I = 0$. If $\beta S = \mu + \alpha$ there is an endemic equilibrium with $\beta I + \mu = re^{-aS}$. A straightforward computation, which we shall not carry out here shows, that the disease-free equilibrium is asymptotically stable if $\mathcal{R}_0 = \beta K/(\mu + \alpha) < 1$ and unstable if $\mathcal{R}_0 > 1$, while the endemic equilibrium, which exists if and only if $\mathcal{R}_0 > 1$, is always asymptotically stable. Another way to express the condition for an endemic equilibrium is to say that the fox population density must exceed a threshold level K_T given by

$$K_T = \frac{\mu + \alpha}{\beta} \ .$$

With the parameter values we have been using, this gives a threshold fox density of $0.92\,\text{fox/km}^2$. If the fox density is below this threshold value, the fox population will approach its carrying capacity and the disease will die out. Above the threshold density, rabies will persist and will regulate the fox population to a level below its carrying capacity. This level may be approached in an oscillatory manner for large $\mathcal{R}_0$.

2.4 Age of Infection Models

2.4.1 The Basic SI*R Model

The 1927 epidemic model of Kermack and McKendrick is considerably more general than what is usually called the Kermack–McKendrick model, which was analyzed in the first section. The general model described by Kermack and McKendrick included a dependence of infectivity on the time since becoming infected (age of infection). The 1932 and 1933 models of Kermack and McKendrick, which incorporated births and deaths, did not include this dependence. While age of infection models have not played a large role in studies of epidemics, they are very important in studies of HIV/AIDS. HIV/AIDS acts on a very long time scale and it is essential to include demographic effects (recruitment into and departure from a population of sexually active individuals). Also, the infectivity of HIV-positive people is high for a relatively short time after becoming infected, then very low for a long period, possibly several years, and then high shortly before developing into full-blown AIDS. Thus, the age of infection models described by Kermack and McKendrick for epidemics but not for endemic situations, have become important in endemic situations.

We will describe a general age of infection model and carry out a partial analysis; there are many unsolved problems in the analysis. We continue to let $S(t)$ denote the number of susceptibles at time t and $R(t)$ the number of members recovered with immunity, but now we let $I^*(t)$ denote the number of infected (but not necessarily infective) members.

We make the following assumptions:

1. The population has a birth rate $\Lambda(N)$, and a natural death rate μ giving a carrying capacity K such that $\Lambda(K) = \mu K, \Lambda'(K) < \mu$.
2. An average infected member makes $C(N)$ contacts in unit time of which S/N are with susceptibles. We define $\beta(N) = C(N)/N$ and it is reasonable to assume that $\beta'(N) \leq 0, C'(N) \geq 0$.
3. $B(\tau)$ is the fraction of infecteds remaining infective if alive when infection age is τ and $B_\mu(\tau) = e^{-\mu\tau}B(\tau)$ is the fraction of infecteds remaining alive and infected when infection age is τ. Let $\hat{B}_\mu(0) = \int_0^\infty B_\mu(\tau)d\tau$.
4. A fraction f of infected members recovers with immunity and a fraction $(1 - f)$ dies of disease.
5. $\pi(\tau)$ with $0 \leq \pi(\tau) \leq 1$ is the infectivity at infection age τ; let $A(\tau) = \pi(\tau)B(\tau), A_\mu(\tau) = \pi(\tau)B_\mu(\tau), \hat{A}_\mu(0) = \int_0^\infty A_\mu(\tau)d\tau$.

In previous sections we have used $B(\tau) = e^{-\alpha\tau}$, which would give $B_\mu(\tau) = e^{-(\mu+\alpha)\tau}$. We let $i_0(t)$ be the number of new infecteds at time t, $i(t, \tau)$ be the number of infecteds at time t with infection age τ, and let $\phi(t)$ be the total infectivity at time t. Then

$$i(t, \tau) = i_0(t - \tau)B_\mu(\tau), \quad 0 \le \tau \le t$$
$$i_0(t) = S\beta(N)\phi(t)$$

and

$$S' = \Lambda(N) - \mu S - \beta(N)S\phi$$

$$I^*(t) = \int_0^\infty i(t, \tau)d\tau$$

$$= \int_0^\infty i_0(t - \tau)B_\mu(\tau)d\tau$$

$$= \int_0^\infty \beta(N(t - \tau))S(t - \tau)\phi(t - \tau)B_\mu(\tau)d\tau$$

$$\phi(t) = \int_0^\infty i_0(t - \tau)A_\mu(\tau)d\tau$$

$$= \int_0^\infty \beta(N(t - \tau))S(t - \tau)\phi(t - \tau)A_\mu(\tau)d\tau .$$

Differentiation of the equation for I^* gives three terms, including the rate of new infections and the rate of natural deaths. The third term gives the rate of recovery plus the rate of disease death as

$$-\int_0^\infty \beta(N(t - \tau))S(t - \tau)\phi(t - \tau)e^{-\mu\tau}B'(\tau)d\tau .$$

Thus the SI^*R model is

$$S' = \Lambda(N) - \mu S - \beta(N)S\phi$$

$$\phi(t) = \int_0^\infty \beta(N(t - \tau))S(t - \tau)\phi(t - \tau)A_\mu(\tau)d\tau \qquad (2.27)$$

$$N'(t) = \Lambda(N) - \mu N$$

$$+ (1 - f)\int_0^\infty \beta(N(t - \tau))S(t - \tau)\phi(t - \tau)e^{-\mu\tau}B'(\tau)d\tau .$$

Since I^* is determined when S, ϕ, N are known we have dropped the equation for I^* from the model, but it will be convenient to recall

$$I^*(t) = \int_0^\infty \beta(N(t - \tau))S(t - \tau)\phi(t - \tau)B_\mu(\tau)d\tau .$$

If $f = 1$ then $N(t)$ approaches the limit K, the model is asymptotically autonomous and its dimension may be reduced to two, replacing N by the constant K. We note, for future use, that

$$\hat{B}_\mu(0) = \int_0^\infty e^{-\mu\tau}B(\tau)d\tau \le \int_0^\infty e^{-\mu\tau}d\tau = 1/\mu \ ,$$

so that

$$0 \le 1 - \mu\hat{B}_\mu(0) \le 1 \ .$$

We define $M = (1-f)(1-\mu\hat{B}_\mu(0)$, and $0 \le M \le 1$. We note, however, that if $f = 1$ then $M = 0$. We also have, using integration by parts,

$$-\int_0^\infty e^{-\mu\tau}B'(\tau)d\tau = 1 - \mu\hat{B}_\mu(0) \ge 0 \ .$$

If a single infective is introduced into a wholly susceptible population, making $K\beta(K)$ contacts in unit time, the fraction still infective at infection age τ is $B_\mu(\tau)$ and the infectivity at infection age τ is $A_\mu(\tau)$. Thus $\mathcal{R}_0$, the total number of secondary infections caused, is

$$\int_0^\infty K\beta(K)A_\mu(\tau)d\tau = K\beta(K)\hat{A}_\mu(0) \ .$$

Example. (Exposed periods) One common example of an age of infection model is a model with an exposed period, during which individuals have been infected but are not yet infected. Thus we may think of infected susceptibles going into an exposed class (E), proceeding from the exposed class to the infective class (I) at rate κE and out of the infective class at rate αI. Exposed members have infectivity 0 and infective members have infectivity 1. Thus $I^* = E + I$ and $\phi = I$.

We let $u(\tau)$ be the fraction of infected members with infection age τ who are not yet infective if alive and $v(\tau)$ the fraction of infected members who are infective if alive. Then the fraction becoming infective at infection age τ if alive is $\kappa u(\tau)$, and we have

$$\begin{aligned} u'(\tau) &= -\kappa u(\tau), \qquad u(0) = 1 \\ v'(\tau) &= \kappa u(\tau) - \alpha v(\tau) \quad v(0) = 0 \ . \end{aligned} \tag{2.28}$$

The solution of the first of the equations of (2.28) is

$$u(\tau) = e^{-\kappa\tau}$$

and substitution of this into the second equation gives

$$v'(\tau) = \kappa e^{-\kappa\tau} - \alpha v(\tau) \ .$$

When we multiply this equation by the integrating factor $e^{\alpha\tau}$ and integrate, we obtain the solution

$$v(\tau) = \frac{\kappa}{\kappa - \alpha}[e^{-\alpha\tau} - e^{-\kappa\tau}],$$

and this is the term $A_\mu(\tau)$ in the general model. The term $B(\tau)$ is $u(\tau)+v(\tau)$. Thus we have

$$A(\tau) = \frac{\kappa}{\kappa - \alpha}[e^{-\alpha\tau} - e^{-\kappa\tau}]$$

$$B(\tau) = \frac{\kappa}{\kappa - \alpha}e^{-\alpha\tau} - \frac{\alpha}{\kappa - \alpha}e^{-\kappa\tau}$$

$$e^{-\mu\tau}B'(\tau) = -\frac{\alpha\kappa}{\kappa - \alpha}[e^{-(\mu+\alpha)\tau} - e^{-(\mu+\kappa)\tau}].$$

With these choices and the identifications

$$I = \phi, \qquad E = I^* - \phi$$

we may verify that the system (2.27) reduces to

$$S' = \Lambda(N) - \beta(N)SI - \mu S$$
$$E' = \beta(N)SI - \kappa E$$
$$I' = \kappa E - (\mu + \alpha)I$$
$$N' = \Lambda(N) - (1 - f)\alpha I - \mu N,$$

which is a standard $SEIR$ model.

For some diseases there is an asymptomatic period during which individuals have some infectivity rather than an exposed period. If the infectivity during this period is reduced by a factor ε, then the model can be described by the system

$$S' = \Lambda(N) - \beta(N)S(I + \varepsilon E) - \mu S$$
$$E' = \beta(N)S(I + \varepsilon E) - \kappa E$$
$$I' = \kappa E - (\mu + \alpha)I$$
$$N' = \Lambda(N) - (1 - f)\alpha I - \mu N.$$

This may be considered as an age of infection model with the same identifications of the variables and the same choice of $u(\tau), v(\tau)$ but with $A(\tau) = \varepsilon u(\tau) + v(\tau)$.

2.4.2 Equilibria

There is a disease-free equilibrium $S = N = K, \phi = 0$ of (2.27). Endemic equilibria (S, N, ϕ) are given by

$$\Lambda(N) = \mu S + S\phi\beta(N)$$
$$S\beta(N)\hat{A}_\mu(0) = 1$$
$$\Lambda(N) = \mu N + (1 - f)(1 - \mu\hat{B}_\mu(0))S\beta(N)\phi \,.$$

If $f = 1$ the third condition gives $\Lambda(N) = \mu N$, which implies $N = K$. Then the second condition may be solved for S, after which the first condition may be solved for ϕ. Thus, there is always an endemic equilibrium.

If $f < 1$ the second of the equilibrium conditions gives

$$\phi = \frac{\hat{A}_\mu(0)}{M}[\Lambda(N) - \mu N].$$

Now substitution of the first two equilibrium conditions into the third gives an equilibrium condition for N, namely

$$(1 - M)\Lambda(N) = \mu N - \frac{\mu M}{\beta(N)\hat{A}_\mu(0)} \tag{2.29}$$

$$= \mu N \left[1 - \frac{M}{C(N)\hat{A}_\mu(0)}\right].$$

If $\mathcal{R}_0 < 1$,

$$C(N)\hat{A}_\mu(0) \leq C(K)\hat{A}_\mu(0) = \mathcal{R}_0 < 1$$

so that

$$1 - \frac{M}{C(N)\hat{A}_\mu(0)} < 1 - M \,.$$

Then we must have $\Lambda(N) < \mu N$. However, this would contradict the demographic condition $\Lambda(N) > \mu N, 0 < N < K$ imposed earlier. This shows that if $\mathcal{R}_0 < 1$ there is no endemic equilibrium.

If $\mathcal{R}_0 > 1$ for $N = 0$ the left side of (2.29) is non-negative while the right side is negative. For $N = K$ the left side of (2.29) is $\mu K(1 - M)$ while the right side is

$$\mu K - \frac{M\mu K}{\mathcal{R}_0} > \mu K(1 - M) \,.$$

This shows that there is an endemic equilibrium solution for N.

2.4.3 The Characteristic Equation

The linearization of (2.27) at an equilibrium (S, N, ϕ) is

$$x' = -[\mu + \phi\beta(N)]x + [\Lambda'(N) - S\phi\beta'(N)]y - S\beta(N)z$$

$$y' = [\Lambda'(N) - \mu]y$$

$$+ (1-f)\int_0^\infty e^{-\mu\tau}B'(\tau)[\phi\beta(N)x(t-\tau) + S\phi\beta'(N)y(t-\tau) + S\beta(N)z(t-\tau)]d\tau$$

$$z(t) = \int_0^\infty A_\mu(\tau)[\phi\beta(N)x(t-\tau) + S\phi\beta'(N)y(t-\tau) + S\beta(N)z(t-\tau)]d\tau \ .$$

The condition that this linearization has solutions which are constant multiples of $e^{-\lambda\tau}$ is that λ satisfies a characteristic equation. The characteristic equation at an equilibrium (S, N, ϕ) is

$$\det \begin{bmatrix} -[\lambda + \mu + \phi\beta(N)] & [\Lambda'(N) - S\phi\beta'(N)] & -S\beta(N) \\ -\phi\beta(N)Q(\lambda) & -[\lambda - \Lambda'(N) + \mu] - S\phi\beta'(N)Q(\lambda) & -S\phi\beta(N)Q(\lambda) \\ \phi\beta(N)\hat{A}_\mu(\lambda) & S\phi\beta'(N)\hat{A}_\mu(\lambda) & S\beta(N)\hat{A}_\mu(\lambda) - 1 \end{bmatrix} = 0$$

with

$$\hat{A}_\mu(\lambda) = \int_0^\infty e^{-\lambda\tau}A_\mu(\tau)d\tau$$

$$\hat{B}_\mu(\lambda) = \int_0^\infty e^{-\lambda\tau}B_\mu(\tau)d\tau$$

$$Q(\lambda) = (1-f)[1 - (\lambda + \mu)\hat{B}_\mu(\lambda)] \ .$$

Here, the choice of $Q(\lambda)$ is motivated by the integration by parts formula

$$\int_0^\infty e^{-(\lambda+\mu)\tau}B'(\tau)d\tau = -1 + \hat{B}_\mu(\lambda) \ .$$

The characteristic equation then reduces to

$$S\beta(N)\hat{A}_\mu(\lambda) + (1-f)\phi S\beta'(N)\hat{B}_\mu(\lambda)$$

$$= 1 + \frac{f\phi\beta(N)}{\lambda + \mu} + \frac{(1-f)\phi P}{\lambda + \mu - \Lambda'(N)}[1 - \Lambda'(N)\hat{B}_\mu(\lambda)] \ , \qquad (2.30)$$

where $P = \beta(N) + S\beta'(N) \geq 0$.

The characteristic equation for a model consisting of a system of ordinary differential equations is a polynomial equation. Now we have a transcendental characteristic equation, but there is a basic theorem that if all roots of the characteristic equation at an equilibrium have negative real part then the equilibrium is asymptotically stable [39, Chap. 4].

At the disease-free equilibrium $S = N = K, \phi = 0$ the characteristic equation is

$$K\beta(K)\hat{A}_\mu(\lambda) = 1 \ .$$

Since the absolute value of the left side of this equation is no greater than $K\beta(K)\hat{A}_\mu(0)$ if $\Re\lambda \geq 0$ the disease-free equilibrium is asymptotically stable if and only if

$$\mathcal{R}_0 = K\beta(K)\hat{A}_\mu(0) < 1 \; .$$

2.4.4 The Endemic Equilibrium

In the analysis of the characteristic equation (2.30) it is helpful to make use of the following elementary result:

If $|P(\lambda)| \le 1, \Re g(\lambda) > 0$ for $\Re\lambda \ge 0$, then all roots of the characteristic equation

$$P(\lambda) = 1 + g(\lambda)$$

satisfy $\Re\lambda < 0$.

To prove this result, we observe that if $\Re\lambda \ge 0$ the left side of the characteristic equation has absolute value at most 1 while the right side has absolute value greater than 1.

If $f = 1$, the characteristic equation reduces to

$$S\beta(N)\hat{A}_\mu(\lambda) = 1 + \frac{\phi\beta(N)}{\lambda + \mu} \; .$$

We have

$$|S\beta(N)\hat{A}_\mu(\lambda)| \le S_\beta(N)\hat{A}_\mu(0) = 1$$

The term

$$\frac{\phi\beta(N)}{\lambda + \mu}$$

in (2.30) has positive real part if $\Re\lambda \ge 0$. It follows from the above elementary result that all roots satisfy $\Re\lambda < 0$, so that the endemic equilibrium is asymptotically stable. Thus all roots of the characteristic equation (2.30) have negative real part if $f = 1$.

The analysis if $f < 1$ is more difficult. The roots of the characteristic equation depend continuously on the parameters of the equation. In order to have a root with $\Re\lambda \ge 0$ there must be parameter values for which either there is a root at "infinity", or there is a root $\lambda = 0$ or there is a pair of pure imaginary roots $\lambda = \pm iy, y > 0$. Since the left side of (2.30) approaches 0 while the right side approaches 1 as $\lambda \to \infty, \Re\lambda \ge 0$, it is not possible for a root to appear at "infinity". For $\lambda = 0$, since $S\beta(N)\hat{A}_\mu(0) = 1$ and $\beta'(N) \le 0$ the left side of (2.30) is less than 1 at $\lambda = 0$, while the right side is greater than 1 since

$$1 - \Lambda'(N)\hat{B}_\mu(0) > 1 - \Lambda'(N)/\mu > 0$$

if $\Lambda'(N) < \mu$. This shows that $\lambda = 0$ is not a root of (2.30), and therefore that all roots satisfy $\Re\lambda < 0$ unless there is a pair of roots $\lambda = \pm iy, y > 0$. According to the Hopf bifurcation theorem [20] a pair of roots $\lambda = \pm iy, y > 0$

indicates that the system (2.27) has an asymptotically stable periodic solution and there are sustained oscillations of the system.

A somewhat complicated calculation using the fact that since $B_\mu(\tau)$ is monotone non-increasing,

$$\int_0^\infty B_\mu(\tau) \sin y\tau dy \geq 0$$

for $0 \leq y < \infty$ shows that the term

$$\frac{(1-f)\phi P}{\lambda + \mu - \Lambda'(N)} \cdot [1 - \Lambda'(N)\hat{B}_\mu(\lambda)]$$

in (2.30) has positive real part at least if

$$-\mu \leq \Lambda'(N) \leq \mu \, .$$

Thus if $-\mu \leq \Lambda'(N) \leq \mu$, instability of the endemic equilibrium is possible only if the term

$$(1-f)\phi S\beta'(N)\hat{B}_\mu(iy)$$

in (2.30) has negative real part for some $y > 0$. This is not possible with mass action incidence, since $\beta'(N) = 0$; thus with mass action incidence the endemic equilibrium of (2.27) is always asymptotically stable. Since $\beta'(N) \leq 0$, instability requires

$$\Re\hat{B}_\mu(iy) = \int_0^\infty B_\mu(\tau) \cos y\tau d\tau < 0$$

for some $y > 0$. If the function $B(\tau)$ is non-increasing and convex, that is, if $B'(\tau) \leq 0, B''(\tau) \geq 0$, then it is easy to show using integration by parts that

$$\int_0^\infty B_\mu(\tau) \cos y\tau d\tau \geq 0$$

for $0 < y < \infty$. Thus if $B(\tau)$ is convex, which is satisfied, for example, by the choice

$$B(\tau) = e^{-\alpha\tau}$$

the endemic equilibrium of (2.22) is asymptotically stable if $-\mu \leq \Lambda'(N) \leq \mu$.

There are certainly less restrictive conditions which guarantee asymptotic stability. However, examples have been given [36,37] of instability, even with $f = 0, \Lambda'(N) = 0$, where constant infectivity would have produced asymptotic stability. Their results indicate that concentration of infectivity early in the infected period is conducive to such instability. In these examples, the instability arises because a root of the characteristic equation crosses the imaginary axis as parameters of the model change, giving a pure imaginary root of the characteristic equation. This translates into oscillatory solutions of the model. Thus infectivity which depends on infection age can cause instability and sustained oscillations.

2.4.5 An SI^*S Model

In order to formulate an SI^*S age of infection model we need only take the SI^*R age of infection model (2.22) and move the recovery term from the equation for R (which was not listed explicitly in the model) to the equation for S. We obtain the model

$$S' = \Lambda(N) - \mu S - \beta(N)S\phi \tag{2.31}$$
$$- f \int_0^\infty \beta(N(t-\tau))S(t-\tau)\phi(t-\tau)e^{-\mu\tau}B'(\tau)d\tau$$

$$\phi(t) = \int_0^\infty \beta(N(t-\tau))S(t-\tau)\phi(t-\tau)A_\mu(\tau)d\tau \tag{2.32}$$

$$N'(t) = \Lambda(N) - \mu N$$
$$+ (1-f)\int_0^\infty \beta(N(t-\tau))S(t-\tau)\phi(t-\tau)e^{-\mu\tau}B'(\tau)d\tau .$$

Although we will not carry out any analysis of this model, it may be attacked using the same approach as that used for (2.27). It may be shown that if $\mathcal{R}_0 = K\beta(K)\hat{A}_\mu(0) < 1$ the disease-free equilibrium is asymptotically stable. If $\mathcal{R}_0 > 1$ there is an endemic equilibrium and the characteristic equation at this equilibrium is

$$S\beta(N)\hat{A}_\mu(\lambda) + (1-f)\phi S\beta'(N)\hat{B}_\mu(\lambda)$$
$$= 1 + f\phi\beta(N)\hat{B}_\mu(\lambda) + \frac{(1-f)\phi P}{\lambda + \mu - \Lambda'(N)} \cdot [1 - \Lambda'(N)\hat{B}_\mu(\lambda)] \tag{2.33}$$

where $P = \beta(N) + S\beta'(N) \geq 0$.

Many diseases, including most strains of influenza, impart only temporary immunity against reinfection on recovery. Such disease may be described by SIS age of infection models, thinking of the infected class I^* as comprised of the infective class I together with the recovered and immune class R. In this way, members of R neither spread or acquire infection. We assume that immunity is lost at a proportional rate κ.

We let $u(\tau)$ be the fraction of infected members with infection age τ who are infective if alive and $v(\tau)$ the fraction of infected members who are not recovered and still immune if alive. Then the fraction becoming immune at infection age τ if alive is $\alpha u(\tau)$, and we have

$$u'(\tau) = -\alpha u(\tau), \qquad u(0) = 1 \tag{2.34}$$
$$v'(\tau) = \alpha u(\tau) - \kappa v(\tau) \quad v(0) = 0 \, .$$

These equations are the same as (2.28) obtained in formulating the $SEIR$ model with α and κ interchanged. Thus we may solve to obtain

$$u(\tau) = e^{-\alpha\tau}$$
$$v(\tau) = \frac{\alpha}{\kappa - \alpha}[e^{-\alpha\tau} - e^{-\kappa\tau}] \, .$$

We take $B(\tau) = u(\tau) + v(\tau)$, $A(\tau) = u(\tau)$. Then if we define $I = \phi, R = I^* - \phi$, the model (2.31) is equivalent to the system

$$S' = \Lambda(N) - \beta(N)SI - \mu S + \kappa R$$
$$I' = \beta(N)SI - (\mu + \alpha)I$$
$$R' = f\alpha E - (\mu + \kappa)R$$
$$N' = \Lambda(N) - (1 - f)\alpha I - \mu N,$$

which is a standard $SIRS$ model.

If we assume that, instead of an exponentially distributed immune period, that there is an immune period of fixed length ω we would again obtain $u(\tau) = e^{-\alpha\tau}$, but now we may calculate that

$$v(\tau) = 1 - e^{-\alpha\tau}, (\tau \leq \omega), \quad v(\tau) = e^{-\alpha\tau}(e^{\alpha\omega} - 1), (\tau > \omega) \, .$$

To obtain this, we note that

$$v'(\tau) = \alpha u(\tau), (\tau \leq \omega), \quad v'(\tau) = \alpha u(\tau) - \alpha u(\tau - \omega), (\tau > \omega) \, .$$

From these we may calculate $A(\tau), B(\tau)$ for an SI^*S model. Since it is known that the endemic equilibrium for an $SIRS$ model with a fixed removed period can be unstable [19], this shows that (2.33) may have roots with non-negative real part and the endemic equilibrium of an SI^*S age of infection model is not necessarily asymptotically stable.

The SI^*R age of infection model is actually a special case of the SI^*S age of infection model. We could view the class R as still infected but having no infectivity, so that $v(\tau) = 0$. The underlying idea is that in infection age models we divide the population into members who may become infected and members who can not become infected, either because they are already infected or because they are immune.

2.4.6 An Age of Infection Epidemic Model

We conclude by returning to the beginning, namely an infection age epidemic model closely related to the original Kermack–McKendrick epidemic model [21]. We simply remove the birth and natural death terms from the SI^*R model (2.27). The result is

$$S' = -\beta(N)S\phi$$

$$\phi(t) = \int_0^\infty \beta(N(t-\tau))S(t-\tau)\phi(t-\tau)A(\tau)d\tau$$

$$N'(t) = (1-f)\int_0^\infty \beta(N(t-\tau))S(t-\tau)\phi(t-\tau)B'(\tau)d\tau$$

which we may rewrite as

$$S' = -\beta(N)S\phi$$

$$\phi(t) = \int_0^\infty [-S'(t-\tau)]A(\tau)d\tau \qquad (2.35)$$

$$N'(t) = (1-f)\int_0^\infty [-S'(t-\tau]B'(\tau)d\tau \ .$$

Then integration of the equation for N with respect to t from 0 to ∞ gives

$$K - N_\infty = (1-f)\int_0^\infty [\int_0^\infty [-S'(t-\tau)]B'(\tau)d\tau dt$$

$$= (1-f)\int_0^\infty [\int_0^\infty [-S'(t-\tau)]dt B'(\tau)d\tau$$

$$= (1-f)\int_0^\infty [S(-\tau) - S_\infty]B'(\tau)d\tau$$

$$= (1-f)(K - S_\infty) \ ,$$

which is the same relation (2.8) obtained for the model (2.6). In this calculation we use the initial data to give $S(-\tau) = K$ and

$$\int_0^\infty B'(\tau)d\tau = B(\infty) - B(0) = -1 \ .$$

The argument that $S_\infty > 0$ for the model (2.35) is analogous to the argument for (2.10). From (2.35) we have

$$-\frac{S'(t)}{S(t)} = \beta(N(t))\int_0^\infty [-S'(t-\tau)]A(\tau)d\tau$$

and integration with respect to t from 0 to ∞ gives

$$\log \frac{S(0)}{S_\infty} = \int_0^\infty \beta(N(t)) \int_0^\infty [-S'(t-\tau)] A(\tau) d\tau dt$$

$$= \int_0^\infty A(\tau) \int_0^\infty \beta(N(t))[-S'(t-\tau)]dt d\tau$$

$$\leq \beta(0) \int_0^\infty A(\tau) \int_0^\infty [-S'(t-\tau)]dt d\tau$$

$$= \beta(0) \int_0^\infty A(\tau)[S(-\tau) - S_\infty]ds d\tau$$

$$= \beta(0)(K - S_\infty) \int_0^\infty A(\tau)d\tau$$

and this shows that $S_\infty > 0$. We recall that we are assuming here that $\beta(0)$ is finite; in other words we are ruling out standard incidence. It is possible to show that S_∞ can be zero only if $N \to 0$ and $\int_0^K \beta(N)dN$ diverges. However, from (2.8) we see that this is possible only if $f = 0$. If there are no disease deaths, so that the total population size N is constant, or if β is constant (mass action incidence), the above integration gives the final size relation

$$\log \frac{S(0)}{S_\infty} = \mathcal{R}_0 \left[1 - \frac{S_\infty}{K} \right].$$

We may view the epidemic management model (2.13) as an age of infection model. We define $I^* = E + Q + I + J$, and we need only calculate the kernels $A(\tau), B(\tau)$. We let $u(\tau)$ denote the number of members of infection age τ in E, $v(\tau)$ the number of members of infection age τ in Q, $w(\tau)$ the number of members of infection age τ in I, and $z(\tau)$ the number of members of infection age τ in J. Then (u, v, w, z) satisfies the linear homogeneous system with constant coefficient

$$u'(\tau) = -(\kappa_1 + \gamma_1)u(\tau)$$
$$v'(\tau) = \gamma_1 u(\tau) - \kappa_2 v(\tau)$$
$$w'(\tau) = \kappa_1 u(\tau) - \alpha_1 w(\tau) - \gamma_2 w(\tau)$$
$$z'(\tau) = \gamma_2 w(\tau) + \kappa_2 v(\tau) - \alpha_2 z(\tau)$$

with initial conditions $u(0) = 1, v(0) = 0, w(0) = 0, z(0) = 0$. This system is easily solved recursively, and then the system (2.13) is an age of infection epidemic model with

$$A(\tau) = \varepsilon_E u(\tau) + \varepsilon_E \varepsilon_Q v(\tau) + w(\tau) + \varepsilon_J z(\tau), B(\tau) = u(\tau) + v(\tau) + w(\tau) + z(\tau).$$

In particular, it now follows from the argument carried out just above that $S_\infty > 0$ for the model (2.13). The proof is less complicated technically than the proof obtained for the specific model (2.13). The generalization to age of infection models both unifies the theory and makes some calculations less complicated.

References

1. R.M. Anderson, H.C. Jackson, R.M. May, A.M. Smith: Population dynamics of fox rabies in Europe. Nature, **289**, 765–771 (1981)
2. R.M. Anderson, R.M. May: Infectious Diseases of Humans. Oxford Science Publications, Oxford (1991)
3. F. Brauer, C. Castillo-Chavez: Mathematical Models in Population Biology and Epidemiology. Springer, New York (2001)
4. S. Busenberg, K.L. Cooke: Vertically Transmitted Diseases: Models and Dynamics. Biomathematics, Vol. 23. Springer, Berlin Heidelberg New York (1993)
5. C. Castillo-Chavez with S. Blower, P. van den Driessche, D. Kirschner, A.A. Yakubu (eds.): Mathematical Approaches for Emerging and Reemerging Infectious Diseases: An Introduction. Springer, Berlin Heidelberg New York (2001)
6. C. Castillo-Chavez, with S.Blower, P. van den Driessche, D. Kirschner, A.A. Yakubu (eds.): Mathematical Approaches for Emerging and Reemerging Infectious Diseases: Models, Methods and Theory. Springer, Berlin Heidelberg New York (2001)
7. C. Castillo-Chavez, K.L. Cooke, W. Huang, S.A. Levin: The role of long incubation periods in the dynamics of HIV/AIDS. Part 1: Single populations models. J. Math. Biol., **27**, 373–98 (1989)
8. C. Castillo-Chavez, H.R. Thieme: Asymptotically autonomous epidemic models. In: O. Arino, D. Axelrod, M. Kimmel, M. Langlais (eds.) Mathematical Population Dynamics: Analysis of Heterogeneity, Vol. 1: Theory of Epidemics. Wuerz, Winnipeg, pp. 33–50 (1993)
9. D.J. Daley, J. Gani: Epidemic Modelling: An Introduction. Cambridge Studies in Mathematical Biology, Vol. 16. Cambridge University Press, Cambridge (1999)
10. K. Dietz: Overall patterns in the transmission cycle of infectious disease agents. In: R.M. Anderson, R.M. May (eds.) Population Biology of Infectious Diseases. Life Sciences Research Report, Vol. 25. Springer, Berlin Heidelberg New York, pp. 87–102 (1982)
11. K. Dietz: The first epidemic model: a historical note on P.D. En'ko. Aust. J. Stat., **30**, 56–65 (1988)
12. S. Ellner, R. Gallant, J. Theiler: Detecting nonlinearity and chaos in epidemic data. In: D. Mollison (ed.) Epidemic Models: Their Structure and Relation to Data. Cambridge University Press, Cambridge, pp. 229–247 (1995)
13. A. Gumel, S. Ruan, T. Day, J. Watmough, P. van den Driessche, F. Brauer, D. Gabrielson, C. Bowman, M.E. Alexander, S. Ardal, J. Wu, B.M. Sahai: Modeling strategies for controlling SARS outbreaks based on Toronto, Hong Kong, Singapore and Beijing experience. Proc. R. Soc. Lond. B Biol. Sci., **271**, 2223–2232 (2004)
14. J.A.P. Heesterbeek, J.A.J. Metz: The saturating contact rate in marriage and epidemic models. J. Math. Biol., **31**: 529–539 (1993)
15. H.W. Hethcote: Qualitative analysis for communicable disease models. Math. Biosci., **28**, 335–356 (1976)
16. H.W. Hethcote: An immunization model for a hetereogeneous population. Theor. Popul. Biol., **14**, 338–349 (1978)
17. H.W. Hethcote: The mathematics of infectious diseases. SIAM Rev., **42**, 599–653 (2000)
18. H.W. Hethcote, S.A. Levin: Periodicity in epidemic models. In: S.A. Levin, T.G. Hallam, L.J. Gross (eds.) Applied Mathematical Ecology. Biomathematics, Vol. 18. Springer, Berlin Heidelberg New York, pp. 193–211 (1989)

19. H.W. Hethcote, H.W. Stech, P. van den Driessche: Periodicity and stability in epidemic models: a survey. In: S. Busenberg, K.L. Cooke (eds.) Differential Equations and Applications in Ecology, Epidemics and Population Problems. Academic, New York, pp. 65–82 (1981)

20. E. Hopf: Abzweigung einer periodischen Lösungen von einer stationaren Lösung eines Differentialsystems. Berlin Math-Phys. Sachsiche Akademie der Wissenschaften, Leipzig, **94**, 1–22 (1942)

21. W.O. Kermack, A.G. McKendrick: A contribution to the mathematical theory of epidemics. Proc. R. Soc. Lond. B Biol. Sci., **115**, 700–721 (1927)

22. W.O. Kermack, A.G. McKendrick: Contributions to the mathematical theory of epidemics, part. II. Proc. R. Soc. Lond. B Biol. Sci., **138**, 55–83 (1932)

23. W.O. Kermack, A.G. McKendrick: Contributions to the mathematical theory of epidemics, part. III. Proc. R. Soc. Lond. B Biol. Sci., **141**, 94–112 (1932)

24. L. Markus: Asymptotically autonomous differential systems. In: S. Lefschetz (ed.) Contributions to the Theory of Nonlinear Oscillations III. Annals of Mathematics Studies, Vol. 36. Princeton University Press, Princeton, NJ, pp. 17–29 (1956)

25. J. Mena-Lorca, H.W. Hethcote: Dynamic models of infectious diseases as regulators of population size. J. Math. Biol., **30**, 693–716 (1992)

26. W.H. McNeill: Plagues and Peoples. Doubleday, New York (1976)

27. W.H. McNeill: The Global Condition. Princeton University Press, Princeton, NJ (1992)

28. L.A. Meyers, B. Pourbohloul, M.E.J. Newman, D.M. Skowronski, R.C. Brunham: Network theory and SARS: predicting outbreak diversity. J. Theor. Biol., **232**, 71–81 (2005)

29. D. Mollison (ed.): Epidemic Models: Their Structure and Relation to Data. Cambridge University Press, Cambridge (1995)

30. M.E.J. Newman: The structure and function of complex networks. SIAM Rev., **45**, 167–256 (2003)

31. G.F. Raggett: Modeling the Eyam plague. IMA J., **18**, 221–226 (1982)

32. H.E. Soper: Interpretation of periodicity in disease prevalence. J. R. Stat. Soc. B, **92**, 34–73 (1929)

33. S.H. Strogatz: Exploring complex networks. Nature, **410**, 268–276 (2001)

34. H.R. Thieme: Asymptotically autonomous differential equations in the plane. Rocky Mt. J. Math., **24**, 351–380 (1994)

35. H.R. Thieme: Mathematics in Population Biology. Princeton University Press, Princeton, NJ (2003)

36. H.R. Thieme, C. Castillo-Chavez: On the role of variable infectivity in the dynamics of the human immunodeficiency virus. In: C. Castillo-Chavez (ed.) Mathematical and Statistical Approaches to AIDS Epidemiology. Lecture Notes in Biomathematics, Vol. 83. Springer, Berlin Heidelberg New York, pp. 200–217 (1989)

37. H.R. Thieme, C. Castillo-Chavez: How may infection-age dependent infectivity affect the dynamics of HIV/AIDS? SIAM J. Appl. Math., **53**, 1447–1479 (1989)

38. P. van den Driessche, J. Watmough: Reproduction numbers and sub-threshold endemic equilibria for compartmental models of disease transmission. Math. Biosci., **180**, 29–48 (2002)

39. G.F. Webb: Theory of Nonlinear Age-Dependent Population Dynamics. Marcel Dekker, New York (1985)

Chapter 3
An Introduction to Stochastic Epidemic Models

Linda J.S. Allen

Abstract A brief introduction to the formulation of various types of stochastic epidemic models is presented based on the well-known deterministic SIS and SIR epidemic models. Three different types of stochastic model formulations are discussed: discrete time Markov chain, continuous time Markov chain and stochastic differential equations. Properties unique to the stochastic models are presented: probability of disease extinction, probability of disease outbreak, quasistationary probability distribution, final size distribution, and expected duration of an epidemic. The chapter ends with a discussion of two stochastic formulations that cannot be directly related to the SIS and SIR epidemic models. They are discrete time Markov chain formulations applied in the study of epidemics within households (chain binomial models) and in the prediction of the initial spread of an epidemic (branching processes).

3.1 Introduction

The goals of this chapter are to provide an introduction to three different methods for formulating stochastic epidemic models that relate directly to their deterministic counterparts, to illustrate some of the techniques for analyzing them, and to show the similarities between the three methods. Three types of stochastic modeling processes are described: (1) a discrete time Markov chain (DTMC) model, (2) a continuous time Markov chain (CTMC) model, and (3) a stochastic differential equation (SDE) model. These stochastic processes differ in the underlying assumptions regarding the time and the state variables. In a DTMC model, the time and the state variables are

Department of Mathematics and Statistics, Texas Tech University, Lubbock, TX 79409-1042, USA
linda.j.allen@ttu.edu

discrete. In a CTMC model, time is continuous, but the state variable is discrete. Finally, the SDE model is based on a diffusion process, where both the time and the state variables are continuous.

Stochastic models based on the well-known SIS and SIR epidemic models are formulated. For reference purposes, the dynamics of the SIS and SIR deterministic epidemic models are reviewed in the next section. Then the assumptions that lead to the three different stochastic models are described in Sects. 3.3, 3.4, and 3.5. The deterministic and stochastic model dynamics are illustrated through several numerical examples. Some of the MatLab programs used to compute numerical solutions are provided in the last section of this chapter.

One of the most important differences between the deterministic and stochastic epidemic models is their asymptotic dynamics. Eventually stochastic solutions (sample paths) converge to the disease-free state even though the corresponding deterministic solution converges to an endemic equilibrium. Other properties that are unique to the stochastic epidemic models include the probability of an outbreak, the quasistationary probability distribution, the final size distribution of an epidemic and the expected duration of an epidemic. These properties are discussed in Sect. 3.6. In Sect. 3.7, the SIS epidemic model with constant population size is extended to one with a variable population size and the corresponding SDE model is derived.

The chapter ends with a discussion of two well-known DTMC epidemic processes that are not directly related to any deterministic epidemic model. These two processes are chain binomial epidemic processes and branching epidemic processes.

3.2 Review of Deterministic SIS and SIR Epidemic Models

In SIS and SIR epidemic models, individuals in the population are classified according to disease status, either susceptible, infectious, or immune. The immune classification is also referred to as removed because individuals are no longer spreading the disease when they are removed or isolated from the infection process. These three classifications are denoted by the variables S, I, and R, respectively.

In an SIS epidemic model, a susceptible individual, after a successful contact with an infectious individual, becomes infected and infectious, but does not develop immunity to the disease. Hence, after recovery, infected individuals return to the susceptible class. The SIS epidemic model has been applied to sexually transmitted diseases. We make some additional simplifying assumptions. There is no vertical transmission of the disease (all individuals are born susceptible) and there are no disease-related deaths. A compartmental diagram in Fig. 3.1 illustrates the dynamics of the SIS epidemic model. Solid arrows denote infection or recovery. Dotted arrows denote births or deaths.

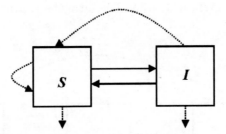

Fig. 3.1 SIS compartmental diagram

Differential equations describing the dynamics of an SIS epidemic model based on the preceding assumptions have the following form:

$$\frac{dS}{dt} = -\frac{\beta}{N}SI + (b+\gamma)I$$

$$\frac{dI}{dt} = \frac{\beta}{N}SI - (b+\gamma)I,$$

(3.1)

where $\beta > 0$ is the contact rate, $\gamma > 0$ is the recovery rate, $b \geq 0$ is the birth rate, and $N = S(t) + I(t)$ is the total population size. The initial conditions satisfy $S(0) > 0$, $I(0) > 0$, and $S(0) + I(0) = N$. We assume that the birth rate equals the death rate, so that the total population size is constant, $dN/dt = 0$. The dynamics of model (3.1) are well-known [25]. They are determined by the *basic reproduction number*. The basic reproduction number is the number of secondary infections caused by one infected individual in an entirely susceptible population [10, 26]. For model (3.1), the basic reproduction number is defined as follows:

$$\mathcal{R}_0 = \frac{\beta}{b+\gamma}.$$

(3.2)

The fraction $1/(b+\gamma)$ is the length of the infectious period, adjusted for deaths. The asymptotic dynamics of model (3.1) are summarized in the following theorem.

Theorem 1. *Let $S(t)$ and $I(t)$ be a solution to model (3.1).*

(1) If $\mathcal{R}_0 \leq 1$, then $\lim_{t\to\infty}(S(t), I(t)) = (N, 0)$ (disease-free equilibrium).

(2) If $\mathcal{R}_0 > 1$, then $\lim_{t\to\infty}(S(t), I(t)) = \left(\dfrac{N}{\mathcal{R}_0}, N\left(1 - \dfrac{1}{\mathcal{R}_0}\right)\right)$ (endemic equilibrium).

In an SIR epidemic model, individuals become infected, but then develop immunity and enter the immune class R. The SIR epidemic model has been applied to childhood diseases such as chickenpox, measles, and mumps. A compartmental diagram in Fig. 3.2 illustrates the relationship between the three classes.

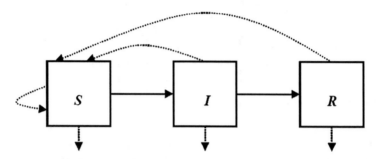

Fig. 3.2 SIR compartmental diagram

Differential equations describing the dynamics of an SIR epidemic model have the following form:

$$\frac{dS}{dt} = -\frac{\beta}{N}SI + b(I + R)$$
$$\frac{dI}{dt} = \frac{\beta}{N}SI - (b + \gamma)I \qquad (3.3)$$
$$\frac{dR}{dt} = \gamma I - bR,$$

where $\beta > 0$, $\gamma > 0$, $b \geq 0$, and the total population size satisfies $N = S(t) + I(t) + R(t)$. The initial conditions satisfy $S(0) > 0$, $I(0) > 0$, $R(0) \geq 0$, and $S(0) + I(0) + R(0) = N$. We assume that the birth rate equals the death rate so that the total population size is constant, $dN/dt = 0$.

The basic reproduction number (3.2) and the birth rate b determine the dynamics of model (3.3). The dynamics are summarized in the following theorem.

Theorem 2. *Let $S(t)$, $I(t)$, and $R(t)$ be a solution to model (3.3).*

(1) If $\mathcal{R}_0 \leq 1$, then $\lim_{t \to \infty} I(t) = 0$ (disease-free equilibrium).
(2) If $\mathcal{R}_0 > 1$, then

$$\lim_{t \to \infty} (S(t), I(t), R(t)) = \left(\frac{N}{\mathcal{R}_0}, \frac{bN}{b + \gamma}\left(1 - \frac{1}{\mathcal{R}_0}\right), \frac{\gamma N}{b + \gamma}\left(1 - \frac{1}{\mathcal{R}_0}\right) \right)$$

(endemic equilibrium).

(3) Assume $b = 0$. If $\mathcal{R}_0 \dfrac{S(0)}{N} > 1$, then there is an initial increase in the number of infected cases $I(t)$ (epidemic), but if $\mathcal{R}_0 \dfrac{S(0)}{N} \leq 1$, then $I(t)$ decreases monotonically to zero (disease-free equilibrium).

The quantity $\mathcal{R}_0 S(0)/N$ is referred to as the *initial replacement number*, the average number of secondary infections produced by an infected individual during the period of infectiousness at the outset of the epidemic [25, 26]. Since the infectious fraction changes during the course of the epidemic, the replacement number is generally defined as $\mathcal{R}_0 S(t)/N$ [25, 26]. In case (3) of Theorem 2, the disease eventually disappears from the population but if the initial replacement number is greater than one, the population experiences an outbreak.

3.3 Formulation of DTMC Epidemic Models

Let $\mathcal{S}(t)$, $\mathcal{I}(t)$, and $\mathcal{R}(t)$ denote discrete random variables for the number of susceptible, infected, and immune individuals at time t, respectively. (Calligraphic letters denote random variables.) In a DTMC epidemic model, $t \in \{0, \Delta t, 2\Delta t, \ldots\}$ and the discrete random variables satisfy

$$\mathcal{S}(t), \ \mathcal{I}(t), \ \mathcal{R}(t) \in \{0, 1, 2, \ldots, N\}.$$

The term "chain" (letter C) in DTMC means that the random variables are discrete. The term "Markov" (letter M) in DTMC is defined in the next section.

3.3.1 SIS Epidemic Model

In an SIS epidemic model, there is only one independent random variable, $\mathcal{I}(t)$, because $\mathcal{S}(t) = N - \mathcal{I}(t)$, where N is the constant total population size. The stochastic process $\{\mathcal{I}(t)\}_{t=0}^{\infty}$ has an associated probability function,

$$p_i(t) = \text{Prob}\{\mathcal{I}(t) = i\},$$

for $i = 0, 1, 2, \ldots, N$ and $t = 0, \Delta t, 2\Delta t, \ldots$, where

$$\sum_{i=0}^{N} p_i(t) = 1.$$

Let $p(t) = (p_0(t), p_1(t), \ldots, p_N(t))^T$ denote the probability vector associated with $\mathcal{I}(t)$. The stochastic process has the *Markov property* if

$$\text{Prob}\{\mathcal{I}(t + \Delta t)|\mathcal{I}(0), \mathcal{I}(\Delta t), \ldots, \mathcal{I}(t)\} = \text{Prob}\{\mathcal{I}(t + \Delta t)|\mathcal{I}(t)\}.$$

The Markov property means that the process at time $t + \Delta t$ only depends on the process at the previous time step t.

To complete the formulation for a DTMC SIS epidemic model, the relationship between the states $\mathcal{I}(t)$ and $\mathcal{I}(t + \Delta t)$ needs to be defined. This relationship is determined by the underlying assumptions in the SIS epidemic model and is defined by the transition matrix. The probability of a transition from state $\mathcal{I}(t) = i$ to state $\mathcal{I}(t + \Delta t) = j$, $i \to j$, in time Δt, is denoted as

$$p_{ji}(t + \Delta t, t) = \text{Prob}\{\mathcal{I}(t + \Delta t) = j|\mathcal{I}(t) = i\}.$$

When the transition probability $p_{ji}(t + \Delta t, t)$ does not depend on t, $p_{ji}(\Delta t)$, the process is said to be *time homogeneous*. For the stochastic SIS epidemic model, the process is time homogeneous because the deterministic model is autonomous.

To reduce the number of transitions in time Δt, we make one more assumption. The time step Δt is chosen sufficiently small such that the number of infected individuals changes by at most one during the time interval Δt, that is,

$$i \to i + 1, \quad i \to i - 1 \quad \text{or} \quad i \to i.$$

Either there is a new infection, a birth, a death, or a recovery during the time interval Δt. Of course, this latter assumption can be modified, if the time step cannot be chosen arbitrarily small. In this latter case, transition probabilities need to be defined for all possible transitions that may occur, $i \to i + 2$, $i \to i + 3$, etc. In the simplest case, with only three transitions possible, the transition probabilities are defined using the rates (multiplied by Δt) in the deterministic SIS epidemic model. This latter assumption makes the DTMC model a useful approximation to the CTMC model, described in Sect. 3.4. The transition probabilities for the DTMC epidemic model satisfy

$$p_{ji}(\Delta t) = \begin{cases} \dfrac{\beta i(N - i)}{N} \Delta t, & j = i + 1 \\ (b + \gamma)i\Delta t, & j = i - 1 \\ 1 - \left[\dfrac{\beta i(N - i)}{N} + (b + \gamma)i \right] \Delta t, & j = i \\ 0, & j \neq i + 1, i, i - 1. \end{cases}$$

The probability of a new infection, $i \to i + 1$, is $\beta i(N - i)\Delta t/N$. The probability of a death or recovery, $i \to i - 1$, is $(b + \gamma)i\Delta t$. Finally, the probability of no change in state, $i \to i$, is $1 - [\beta i(N - i)/N + (b + \gamma)i]\Delta t$. Since a birth of a susceptible must be accompanied by a death, to keep the population size constant, this probability is not needed in either the deterministic or stochastic formulations.

To simplify the notation and to relate the SIS epidemic process to a birth and death process, the transition probability for a new infection is denoted as $b(i)\Delta t$ and for a death or a recovery is denoted as $d(i)\Delta t$. Then

$$p_{ji}(\Delta t) = \begin{cases} b(i)\Delta t, & j = i+1 \\ d(i)\Delta t, & j = i-1 \\ 1 - [b(i) + d(i)]\Delta t, & j = i \\ 0, & j \neq i+1, i, i-1. \end{cases}$$

The sum of the three transitions equals one because these transitions represent all possible changes in the state i during the time interval Δt. To ensure that these transition probabilities lie in the interval $[0, 1]$, the time step Δt must be chosen sufficiently small such that

$$\max_{i \in \{1,\dots,N\}} \{[b(i) + d(i)]\Delta t\} \leq 1.$$

Applying the Markov property and the preceding transition probabilities, the probabilities $p_i(t + \Delta t)$ can be expressed in terms of the probabilities at time t. At time $t + \Delta t$,

$$p_i(t + \Delta t) = p_{i-1}(t)b(i-1)\Delta t + p_{i+1}(t)d(i+1)\Delta t + p_i(t)(1 - [b(i) + d(i)]\Delta t) \tag{3.4}$$

for $i = 1, 2, \dots, N$, where $b(i) = \beta i(N - i)/N$ and $d(i) = (b + \gamma)i$.

A transition matrix is formed when the states are ordered from 0 to N. For example, the $(1, 1)$ element in the transition matrix is the transition probability from state zero to state zero, $p_{00}(\Delta t) = 1$, and the $(N+1, N+1)$ element is the transition probability from state N to state N, $p_{NN}(\Delta t) = 1 - [b+\gamma]N\Delta t = 1 - d(N)\Delta t$. Denote the transition matrix as $P(\Delta t)$. Matrix $P(\Delta t)$ is a $(N + 1) \times (N + 1)$ tridiagonal matrix given by

$$\begin{pmatrix} 1 & d(1)\Delta t & 0 & \cdots & 0 & 0 \\ 0 & 1 - (b+d)(1)\Delta t & d(2)\Delta t & \cdots & 0 & 0 \\ 0 & b(1)\Delta t & 1 - (b+d)(2)\Delta t & \cdots & 0 & 0 \\ 0 & 0 & b(2)\Delta t & \cdots & 0 & 0 \\ \vdots & \vdots & \vdots & \ddots & \vdots & \vdots \\ 0 & 0 & 0 & \cdots & d(N-1)\Delta t & 0 \\ 0 & 0 & 0 & \cdots & 1 - (b+d)(N-1)\Delta t & d(N)\Delta t \\ 0 & 0 & 0 & \cdots & b(N-1)\Delta t & 1 - d(N)\Delta t \end{pmatrix},$$

where the notation $(b + d)(i) = [b(i) + d(i)]$ for $i = 1, 2, \dots, N$. Matrix $P(\Delta t)$ is a *stochastic matrix*, i.e., the column sums equal one.

The DTMC SIS epidemic process $\{\mathcal{I}(t)\}_{t=0}^{\infty}$ is now completely formulated. Given an initial probability vector $p(0)$, it follows that $p(\Delta t) = P(\Delta t)p(0)$. The identity (3.4) expressed in matrix and vector notation is

$$p(t + \Delta t) = P(\Delta t)p(t) = P^{n+1}(\Delta t)p(0), \tag{3.5}$$

where $t = n\Delta t$.

Difference equations for the mean and the higher order moments of the epidemic process can be obtained directly from the difference equations in (3.4). For example, the expected value for $\mathcal{I}(t)$ is $E(\mathcal{I}(t)) = \sum_{i=0}^{N} i p_i(t)$. Multiplying (3.4) by i and summing on i leads to

$$E(\mathcal{I}(t + \Delta t)) = \sum_{i=0}^{N} i p_i(t + \Delta t)$$

$$= \sum_{i=1}^{N} i p_{i-1}(t) b(i-1) \Delta t + \sum_{i=0}^{N-1} i p_{i+1}(t) d(i+1) \Delta t$$

$$+ \sum_{i=0}^{N} i p_i(t) - \sum_{i=0}^{N} i p_i(t) b(i) \Delta t - \sum_{i=0}^{N} i p_i(t) d(i) \Delta t.$$

Simplifying and substituting the expressions $\beta i(N - i)/N$ and $(b + \gamma)i$ for $b(i)$ and $d(i)$, respectively, yields

$$E(\mathcal{I}(t + \Delta t)) = E(\mathcal{I}(t)) + \sum_{i=1}^{N} p_{i-1}(t) \frac{\beta(i-1)(N - [i-1])}{N} \Delta t$$

$$- \sum_{i=0}^{N-1} p_{i+1}(t)(b + \gamma)(i+1) \Delta t$$

$$= E(\mathcal{I}(t)) + [\beta - (b + \gamma)] \Delta t E(\mathcal{I}(t)) - \frac{\beta}{N} \Delta t E(\mathcal{I}^2(t)),$$

where $E(\mathcal{I}^2(t)) = \sum_{i=0}^{N} i^2 p_i(t)$ (see, e.g., [8]). The difference equation for the mean depends on the second moment. Difference equations for the second and the higher order moments depend on even higher order moments. Therefore, these equations cannot be solved unless some additional assumptions are made regarding the higher order moments. However, because $E(\mathcal{I}^2(t)) \geq E^2(\mathcal{I}(t))$, the mean satisfies the following inequality:

$$\frac{E(\mathcal{I}(t + \Delta t)) - E(\mathcal{I}(t))}{\Delta t} \leq [\beta - (b + \gamma)] E(\mathcal{I}(t)) - \frac{\beta}{N} E^2(\mathcal{I}(t)). \qquad (3.6)$$

As $\Delta t \to 0$,

$$\frac{dE(\mathcal{I}(t))}{dt} \leq [\beta - (b + \gamma)] E(\mathcal{I}(t)) - \frac{\beta}{N} E^2(\mathcal{I}(t))$$

$$= \frac{\beta}{N} [N - E(\mathcal{I}(t))] E(\mathcal{I}(t)) - (b + \gamma) E(\mathcal{I}(t)) \qquad (3.7)$$

The right side of (3.7) is the same as the differential equation for $I(t)$ in (3.1), if, in (3.1), $I(t)$ and $S(t)$ are replaced by $E(\mathcal{I}(t))$ and $N - E(\mathcal{I}(t))$, respectively. The differential inequality implies that the mean of the random variable $\mathcal{I}(t)$ in the stochastic SIS epidemic process is less than the solution $I(t)$ to the deterministic differential equation in (3.1).

Some properties of the DTMC SIS epidemic model follow easily from Markov chain theory [6, 47]. States are classified according to their connectedness in a directed graph or digraph. The digraph of the SIS Markov chain model is illustrated in Fig. 3.3, where $i = 0, 1, \ldots, N$ are the infected states.

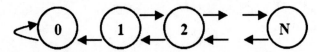

Fig. 3.3 Digraph of the stochastic SIS epidemic model

The states $\{0, 1, \ldots, N\}$ can be divided into two sets consisting of the recurrent state, $\{0\}$, and the transient states, $\{1, \ldots, N\}$. The zero state is an *absorbing state*. It is clear from the digraph that beginning from state 0 no other state can be reached; the set $\{0\}$ is closed. In addition, any state in the set $\{1, 2, \ldots, N\}$ can be reached from any other state in the set, but the set is not closed because $p_{01}(\Delta t) > 0$. For transient states it can be shown that elements of the transition matrix have the following property [6, 47]: Let $P^n = (p_{ij}^{(n)})$, where $p_{ij}^{(n)}$ is the (i, j) element of the nth power of the transition matrix, P^n, then

$$\lim_{n \to \infty} p_{ij}^{(n)} = 0$$

for any state j and any transient state i. The limit of P^n as $n \to \infty$ is a stochastic matrix; all rows are zero except the first one which has all ones. From the relationship (3.5) and Markov chain theory, it follows that

$$\lim_{t \to \infty} p(t) = (1, 0, \ldots, 0)^T,$$

where $t = n\Delta t$.

The preceding result implies, in the DTMC SIS epidemic model, that the population approaches the disease-free equilibrium (probability of absorption is one), regardless of the magnitude of the basic reproduction number. Compare this stochastic result with the asymptotic dynamics of the deterministic SIS epidemic model (Theorem 1). Because this stochastic result is asymptotic, the rate of convergence to the disease-free equilibrium can be very slow. The mean time until the disease-free equilibrium is reached (absorption) depends on the initial conditions and the parameter values, but can be extremely long (as shown in the numerical example in the next section). The expected duration of an epidemic (mean time until absorption) and the probability distribution conditioned on nonabsorption are discussed in Sect. 3.6.

3.3.2 Numerical Example

A *sample path* or *stochastic realization* of the stochastic process $\{\mathcal{I}(t)\}_{t=0}^{\infty}$ for $t \in \{0, \Delta t, 2\Delta t, \ldots\}$ is an assignment of a possible value to $\mathcal{I}(t)$ based on the probability vector $p(t)$. A sample path is a function of time, so that it can be plotted against the solution of the deterministic model. For illustrative purposes, we choose a population size of $N = 100$, $\Delta t = 0.01$, $\beta = 1$, $b = 0.25$, $\gamma = 0.25$ and an initial infected population size of $I(0) = 2$. In terms of the stochastic model,

$$\text{Prob}\{\mathcal{I}(0) = 2\} = 1.$$

In this example, the basic reproduction number is $\mathcal{R}_0 = 2$. The deterministic solution approaches an endemic equilibrium given by $\bar{I} = 50$.

Three sample paths of the stochastic model are compared to the deterministic solution in Fig. 3.4. One of the sample paths is absorbed before 200 time steps (the population following this path becomes disease-free) but two sample paths are not absorbed during 2,000 time steps. These latter sample paths follow more closely the dynamics of the deterministic solution. The horizontal axis is the number of time steps Δt. For $\Delta t = 0.01$ and 2,000 time steps, the solutions in Fig. 3.4 are graphed over the time interval $[0, 20]$. Each sample path is not continuous because at each time step, $t = \Delta t, 2\Delta t, \ldots$, the sample path either stays constant (no change in state with probability $1 - [b(i) + d(i)]\Delta t$), jumps down one integer value (with probability $d(i)\Delta t$), or jumps up one integer value (with probability $b(i)\Delta t$). For convenience, these jumps are connected with vertical line segments. Each sample path is continuous from the right but not from the left.

The entire probability distribution, $p(t)$, $t = 0, \Delta t, \ldots$, associated with this particular stochastic process can be obtained by applying (3.5). A MatLab program is provided in the last section that generates the probability distribution as a function of time (Fig. 3.5). Note that the probability distribution is bimodal, part of the distribution is at zero and the remainder of the distribution follows a path similar to the deterministic solution. Eventually, the probability distribution at zero approaches one. This bimodal distribution is important; the part of the distribution that does not approach zero (at time step 2,000) is known as the quasistationary probability distribution (see Sect. 3.6.2).

3.3.3 SIR Epidemic Model

Let $\mathcal{S}(t)$, $\mathcal{I}(t)$, and $\mathcal{R}(t)$ denote discrete random variables for the number of susceptible, infected, and immune individuals at time t, respectively. The DTMC SIR epidemic model is a bivariate process because there are two independent random variables, $\mathcal{S}(t)$ and $\mathcal{I}(t)$. The random variable $\mathcal{R}(t) =$

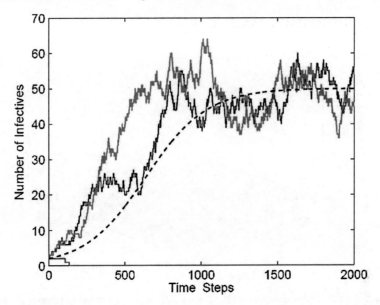

Fig. 3.4 Three sample paths of the DTMC SIS epidemic model are graphed with the deterministic solution (*dashed curve*). The parameter values are $\Delta t = 0.01$, $N = 100$, $\beta = 1$, $b = 0.25$, $\gamma = 0.25$, and $I(0) = 2$

$N - \mathcal{S}(t) - \mathcal{I}(t)$. The bivariate process $\{(\mathcal{S}(t), \mathcal{I}(t))\}_{t=0}^{\infty}$ has a joint probability function given by

$$p_{(s,i)}(t) = \text{Prob}\{\mathcal{S}(t) = s, \mathcal{I}(t) = i\}.$$

This bivariate process has the Markov property and is time-homogeneous.

Transition probabilities can be defined based on the assumptions in the SIR deterministic formulation. First, assume that Δt can be chosen sufficiently small such that at most one change in state occurs during the time interval Δt. In particular, there can be either a new infection, a birth, a death, or a recovery. The transition probabilities are denoted as follows:

$$p_{(s+k,i+j),(s,i)}(\Delta t) = \text{Prob}\{(\Delta \mathcal{S}, \Delta \mathcal{I}) = (k,j)|(\mathcal{S}(t), \mathcal{I}(t)) = (s,i)\},$$

where $\Delta \mathcal{S} = \mathcal{S}(t + \Delta t) - \mathcal{S}(t)$. Hence,

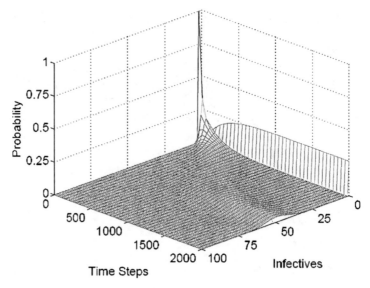

Fig. 3.5 Probability distribution of the DTMC SIS epidemic model. Parameter values are the same as in Fig. 3.4

$$P_{(s+k,i+j),(s,i)}(\Delta t) = \begin{cases} \beta i s/N \Delta t, & (k,j)=(-1,1) \\ \gamma i \Delta t, & (k,j)=(0,-1) \\ bi \Delta t, & (k,j)=(1,-1) \\ b(N-s-i)\Delta t, & (k,j)=(1,0) \\ 1-\beta i s/N \Delta t \\ \quad -[\gamma i + b(N-s)]\Delta t, & (k,j)=(0,0) \\ 0, & \text{otherwise.} \end{cases} \quad (3.8)$$

The time step Δt must be chosen sufficiently small such that each of the transition probabilities lie in the interval $[0,1]$. Because the states are now ordered pairs, the transition matrix is more complex than for the SIS epidemic model and its form depends on how the states (s,i) are ordered. However, applying the Markov property, the difference equation satisfied by the probability $p_{(s,i)}(t+\Delta t)$ can be expressed in terms of the transition probabilities,

$$p_{(s,i)}(t+\Delta t) = p_{(s+1,i-1)}(t)\frac{\beta}{N}(i-1)(s+1)\Delta t + p_{(s,i+1)}(t)\gamma(i+1)\Delta t$$
$$+p_{(s-1,i+1)}(t)b(i+1)\Delta t + p_{(s-1,i)}(t)b(N-s+1-i)\Delta t$$
$$+p_{(s,i)}(t)\left(1-\left[\frac{\beta}{N}is+\gamma i + b(N-s)\right]\Delta t\right). \quad (3.9)$$

The digraph associated with the SIR Markov chain lies on a two-dimensional lattice. It is easy to show that the state $(N,0)$ is absorbing $(p_{(N,0),(N,0)}$

$(\Delta t) = 1$) and that all other states are transient. Thus, asymptotically, all sample paths eventually are absorbed into the disease-free state $(N, 0)$. Compare this result to the deterministic SIR epidemic model (Theorem 2).

Difference equations for the mean and higher order moments can be derived from (3.9) as was done for the SIS epidemic model, e.g., $E(\mathcal{S}(t)) = \sum_{s=0}^{N} s p_{(s,i)}(t)$ and $E(\mathcal{I}(t)) = \sum_{i=0}^{N} i p_{(s,i)}(t)$. However, these difference equations cannot be solved directly because they depend on higher order moments.

3.3.4 Numerical Example

Three sample paths of the DTMC SIR model are compared to the solution of the deterministic model in Fig. 3.6. In this example, $\Delta t = 0.01$, $N = 100$, $\beta = 1$, $b = 0$, $\gamma = 0.5$, and $(S(0), I(0)) = (98, 2)$. In the stochastic model,

$$\text{Prob}\{(\mathcal{S}(0), \mathcal{I}(0)) = (98, 2)\} = 1.$$

The basic reproduction number and the initial replacement number are both greater than one; $\mathcal{R}_0 = 2$ and $\mathcal{R}_0 S(0)/N = 1.96$. According to Theorem 2 part (3), there is an epidemic (an increase in the number of cases). The epidemic is easily seen in the behavior of the deterministic solution. Each of the three sample paths also illustrate an epidemic curve.

3.4 Formulation of CTMC Epidemic Models

The CTMC epidemic processes are defined on a continuous time scale, $t \in [0, \infty)$, but the states $\mathcal{S}(t)$, $\mathcal{I}(t)$, and $\mathcal{R}(t)$ are discrete random variables, i.e.,

$$\mathcal{S}(t), \ \mathcal{I}(t), \ \mathcal{R}(t) \in \{0, 1, 2, \ldots, N\}.$$

3.4.1 SIS Epidemic Model

In the CTMC SIS epidemic model, the stochastic process depends on the collection of discrete random variables $\{\mathcal{I}(t)\}$, $t \in [0, \infty)$ and their associated probability functions $p(t) = (p_0(t), \ldots, p_N(t))^T$, where

$$p_i(t) = \text{Prob}\{\mathcal{I}(t) = i\}.$$

The stochastic process has the *Markov property*, that is,

$$\text{Prob}\{\mathcal{I}(t_{n+1}) | \mathcal{I}(t_0), \mathcal{I}(t_1), \ldots, \mathcal{I}(t_n)\} = \text{Prob}\{\mathcal{I}(t_{n+1}) | \mathcal{I}(t_n)\}$$

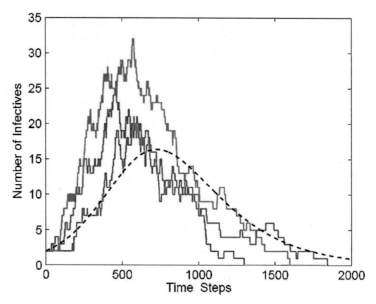

Fig. 3.6 Three sample paths of the DTMC SIR epidemic model are graphed with the deterministic solution (*dashed curve*). The parameter values are $\Delta t = 0.01$, $N = 100$, $\beta = 1$, $b = 0$, $\gamma = 0.5$, $S(0) = 98$, and $I(0) = 2$

for any sequence of real numbers satisfying $0 \le t_0 < t_1 < \cdots < t_n < t_{n+1}$. The transition probability at time t_{n+1} only depends on the most recent time t_n.

The transition probabilities are defined for a small time interval Δt. But in a CTMC model, the transition probabilities are referred to as *infinitesimal transition probabilities* because they are valid for sufficiently small Δt. Therefore, the term $o(\Delta t)$ is included in the definition $[\lim_{t \to \infty}(o(\Delta t)/\Delta t) = 0]$. The infinitesimal transition probabilities are defined as follows:

$$
p_{ji}(\Delta t) = \begin{cases}
\dfrac{\beta}{N} i(N - i)\Delta t + o(\Delta t), & j = i + 1 \\
(b + \gamma)i\Delta t + o(\Delta t), & j = i - 1 \\
1 - \left[\dfrac{\beta}{N} i(N - i) + (b + \gamma)i\right]\Delta t + o(\Delta t), & j = i \\
o(\Delta t), & \text{otherwise.}
\end{cases}
$$

Because Δt is sufficiently small, there are only three possible changes in states:

$$i \to i + 1, \quad i \to i - 1, \quad \text{or} \quad i \to i.$$

Using the same notation as for the DTMC model, let $b(i)$ denote a birth (new infection) and $d(i)$ denote a death or recovery. Then

$$p_{ji}(\Delta t) = \begin{cases} b(i)\Delta t + o(\Delta t), & j = i+1 \\ d(i)\Delta t + o(\Delta t), & j = i-1 \\ 1 - [b(i) + d(i)]\Delta t + o(\Delta t), & j = i \\ o(\Delta t), & \text{otherwise.} \end{cases}$$

Applying the Markov property and the infinitesimal transitional probabilities, a continuous time analogue of the transition matrix can be defined.
Instead of a system of difference equations, a system of differential equations
is obtained. Assume $\text{Prob}\{\mathcal{I}(0) = i_0\} = 1$. Then $p_{i,i_0}(\Delta t) = p_i(\Delta t)$ and

$$p_i(t + \Delta t) = p_{i-1}(t)b(i - 1)\Delta t + p_{i+1}(t)d(i + 1)\Delta t$$
$$+ p_i(t)(1 - [b(i) + d(i)]\Delta t) + o(\Delta t).$$

These equations are the same as the DTMC equations (3.4), except $o(\Delta t)$ is
added to the right side. Subtracting $p_i(t)$, dividing by Δt, and letting $\Delta t \to 0$,
leads to

$$\frac{dp_i}{dt} = p_{i-1}b(i - 1) + p_{i+1}d(i + 1) - p_i[b(i) + d(i)] \qquad (3.10)$$

for $i = 1, 2, \ldots, N$ and $dp_0/dt = p_1 d(1)$. These latter equations are known as
the *forward Kolmogorov differential equations* [47]. In matrix notation, they
can be expressed as

$$\frac{dp}{dt} = Qp, \qquad (3.11)$$

where $p(t) = (p_0(t), \ldots, p_N(t))^T$ and matrix Q is defined as follows:

$$Q = \begin{pmatrix} 0 & d(1) & 0 & \cdots & 0 \\ 0 & -[b(1) + d(1)] & d(2) & \cdots & 0 \\ 0 & b(1) & -[b(2) + d(2)] & \cdots & 0 \\ 0 & 0 & b(2) & \cdots & 0 \\ \vdots & \vdots & \vdots & \vdots & \vdots \\ 0 & 0 & 0 & \cdots & d(N) \\ 0 & 0 & 0 & \cdots & -d(N) \end{pmatrix},$$

$b(i) = \beta i(N - i)/N$ and $d(i) = (b + \gamma)i$. Matrix Q is referred to as the *infinitesimal generator matrix* or simply the *generator matrix* [6, 47], More generally,
the differential equations $dP/dt = QP$ are known as the forward Kolmogorov
differential equations, where $P \equiv (p_{ji}(t))$ is the matrix of infinitesimal transition probabilities. It is interesting to note that the transition matrix $P(\Delta t)$
of the DTMC model and the generator matrix Q are related as follows:

$$Q = \lim_{\Delta t \to 0} \frac{P(\Delta t) - I}{\Delta t}.$$

The generator matrix Q has a zero eigenvalue with corresponding eigenvector $(1, 0, \ldots, 0)^T$. The remaining eigenvalues are negative or have negative
real part. This can be seen by applying Gershgorin's circle theorem and the

fact that the submatrix $\tilde{Q}$ of Q, where the first row and the first column are deleted, is nonsingular [43]. Therefore, $\lim_{t\to\infty} p(t) = (1,0,0,\ldots,0)^T$. Eventual absorption occurs in the CTMC SIS epidemic model. Compare this stochastic result with Theorem 1.

Differential equations for the mean and higher order moments of $\mathcal{I}(t)$ can be derived from the differential equations (3.11). As was shown for the DTMC epidemic model, the differential equations (3.10) can be multiplied by i, then summed over i. However, we present an alternate method for obtaining the differential equations for the mean and higher order moments using generating functions. Either the probability generating function (pgf) or the moment generating function (mgf) can be used to derive the equations. The pgf for $\mathcal{I}(t)$ is defined as

$$\mathcal{P}(\theta,t) = E(\theta^{\mathcal{I}(t)}) = \sum_{i=0}^{N} p_i(t)\theta^i$$

and the mgf as

$$M(\theta,t) = E(e^{\theta\mathcal{I}(t)}) = \sum_{i=0}^{N} p_i(t)e^{i\theta}.$$

We use the mgf to derive the equations because the method of derivation is simpler than with the pgf. In addition, the moments of the distribution corresponding to $\mathcal{I}(t)$ can be easily calculated from the mgf,

$$\left.\frac{\partial^k M}{\partial \theta^k}\right|_{\theta=0} = E(\mathcal{I}^k(t))$$

for $k = 1,\ldots,n$.

First, we derive a differential equation satisfied by the mgf. Multiplying the equations in (3.10) by $e^{i\theta}$ and summing on i, leads to

$$\frac{\partial M}{\partial t} = \sum_{i=0}^{N} \frac{dp_i}{dt} e^{i\theta}$$

$$= e^{\theta} \sum_{i=1}^{N} p_{i-1} e^{(i-1)\theta} b(i-1) + e^{-\theta} \sum_{i=0}^{N-1} p_{i+1} e^{(i+1)\theta} d(i+1)$$

$$- \sum_{i=0}^{N} p_i e^{i\theta} [b(i) + d(i)].$$

Simplifying and substituting $\beta i(N-i)/N$ and $(b+\gamma)i$ for $b(i)$ and $d(i)$, respectively, yields

$$\frac{\partial M}{\partial t} = \beta(e^\theta - 1) \sum_{i=1}^{N} i p_i e^{i\theta} + (b+\gamma)(e^{-\theta} - 1) \sum_{i=1}^{N} i p_i e^{i\theta}$$

$$- \frac{\beta}{N}(e^\theta - 1) \sum_{i=1}^{N} i^2 p_i e^{i\theta}.$$

The summations on the right side of the preceding equation can be replaced with $\partial M/\partial\theta$ or $\partial^2 M/\partial\theta^2$. Then the following second order partial differential equation is obtained for the mgf:

$$\frac{\partial M}{\partial t} = [\beta(e^\theta - 1) + (b+\gamma)(e^{-\theta} - 1)]\frac{\partial M}{\partial\theta} - \frac{\beta}{N}(e^\theta - 1)\frac{\partial^2 M}{\partial\theta^2}. \qquad (3.12)$$

Bailey [13] derives a general form for the partial differential equation satisfied by the mgf and the pgf based on the infinitesimal transition probabilities for the process.

The partial differential equation for the mgf, (3.12), is used to obtain an ordinary differential equation satisfied by the mean of $\mathcal{I}(t)$. Differentiating (3.12) with respect to θ and evaluating at $\theta = 0$ yields an ordinary differential equation satisfied by the mean $E(\mathcal{I}(t))$,

$$\frac{dE(\mathcal{I}(t))}{dt} = [\beta - (b+\gamma)]E(\mathcal{I}(t)) - \frac{\beta}{N}E(\mathcal{I}^2(t)).$$

Because the differential equation for the mean depends on the second moment, it cannot be solved directly, but as was shown for the DTMC SIS epidemic model in (3.7), the mean of the stochastic SIS epidemic model is less than the deterministic solution. The differential equations for the second moment and for the variance depend on higher order moments. These higher order moments are often approximated by lower order moments by making some assumptions regarding their distributions (e.g., normality or lognormality), referred to as moment closure techniques (see, e.g., [27,34]). Then these differential equations can be solved to give approximations for the moments.

3.4.2 Numerical Example

To numerically compute a sample path of a CTMC model, we need to use the fact that the interevent time has an exponential distribution. This follows from the Markov property. The exponential distribution has what has been called the "memoryless property."

Assume $\mathcal{I}(t) = i$. Let T_i denote the interevent time, a continuous random variable for the time to the next event given the process is in state i. Let $H_i(t)$ denote the probability the process remains in state i for a period of time t. Then $H_i(t) = \text{Prob}\{T_i > t\}$. It follows that

$$H_i(t + \Delta t) = H_i(t)p_{ii}(\Delta t) = H_i(t)(1 - [b(i) + d(i)]\Delta t) + o(\Delta t).$$

Subtracting $H_i(t)$ and dividing by Δt, the following differential equation is obtained:

$$\frac{dH_i}{dt} = -[b(i) + d(i)]H_i.$$

Since $H_i(0) = 1$, the solution to the differential equation is $H_i(t) = \exp(-[b(i) + d(i)]t)$. Therefore, the interevent time T_i is an exponential random variable with parameter $b(i) + d(i)$. The cumulative distribution of T_i is

$$F_i(t) = \text{Prob}\{T_i \le t\} = 1 - \exp(-[b(i) + d(i)]t)$$

[6, 47].

The uniform random variable on $[0, 1]$ can be used to numerically compute the interevent time. Let U be a uniform random variable on $[0, 1]$. Then

$$\text{Prob}\{F_i^{-1}(U) \le t\} = \text{Prob}\{F_i(F_i^{-1}(U)) \le F_i(t)\}$$
$$= \text{Prob}\{U \le F_i(t)\} = F_i(t)$$

The interevent time T_i, given $\mathcal{I}(t) = i$, satisfies

$$T_i = F_i^{-1}(U) = -\frac{\ln(1 - U)}{b(i) + d(i)} = -\frac{\ln(U)}{b(i) + d(i)},$$

using the properties of uniform distributions.

In Fig. 3.7, three sample paths for the CTMC SIS epidemic model are compared to the deterministic solution. Parameter values are $b = 0.25$, $\gamma = 0.25$, $\beta = 1$, $N = 100$, and $I(0) = 2$. For the stochastic model,

$$\text{Prob}\{\mathcal{I}(0) = 2\} = 1.$$

The basic reproduction number is $\mathcal{R}_0 = 2$. One sample path in Fig. 3.7 is absorbed rapidly (the population following this path becomes disease-free). The sample paths for the CTMC model are not continuous for the same reasons given for the DTMC model. With each change, the process either jumps up one integer value (with probability $b(i)/[b(i) + d(i)]$) or jumps down one integer value (with probability $d(i)/[b(i) + d(i)]$). Sample paths are continuous from the right but not from the left. Compare the sample paths in Fig. 3.7 with the three sample paths in the DTMC SIS epidemic model in Fig. 3.4.

3.4.3 SIR Epidemic Model

A derivation similar to the SIS epidemic model can be applied to the SIR epidemic model. The difference, of course, is that the SIR epidemic process

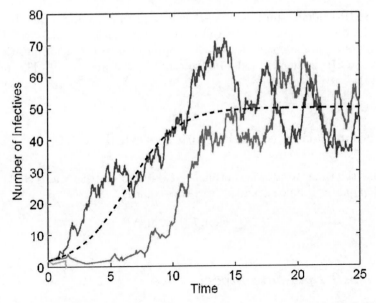

Fig. 3.7 Three samples paths of the CTMC SIS epidemic model are graphed with the deterministic solution (*dashed curve*). The parameter values are $b = 0.25$, $\gamma = 0.25$, $\beta = 1$, $N = 100$, and $I(0) = 2$. Compare with Fig. 3.4

is bivariate, $\{(\mathcal{S}(t), \mathcal{I}(t))\}$, where $\mathcal{R}(t) = N - \mathcal{S}(t) - \mathcal{I}(t)$. Assumptions similar to those for the DTMC SIR epidemic model (3.8) apply to the CTMC SIR epidemic model, except that $o(\Delta t)$ is added to each of the infinitesimal transition probabilities.

For the bivariate process, a joint probability function is associated with each pair of random variables $(\mathcal{S}(t), \mathcal{I}(t))$, $p_{(s,i)}(t) = \text{Prob}\{(\mathcal{S}(t), \mathcal{I}(t)) = (s, i)\}$. A system of forward Kolmogorov differential equations can be derived,

$$\frac{dp_{(s,i)}}{dt} = p_{(s+1,i-1)} \frac{\beta}{N}(i-1)(s+1) + p_{(s,i+1)}\gamma(i+1)$$
$$+ p_{(s-1,i+1)}b(i+1) + p_{(s-1,i)}b(N-s+1-i)$$
$$- p_{(s,i)} \left[\frac{\beta}{N}is + \gamma i + b(N-s) \right].$$

These differential equations are the limiting equations (as $\Delta t \to 0$) of the difference equations in (3.9). Differential equations for the mean and higher order moments can be derived. However, as was true for the other epidemic processes, they do not form a closed system, i.e., each successive moment depends on higher order moments. Moment closure techniques can be applied to approximate the solutions to these moment equations [27, 34].

The SIR epidemic process is Markovian and time homogeneous. In addition, the disease-free state is an absorbing state. In Sect. 3.6.3, we discuss the final size of the epidemic, which is applicable to the deterministic and stochastic SIR epidemic model in the case $\mathcal{R}_0 > 1$ and $b = 0$ (Theorem 2, part (3)).

3.5 Formulation of SDE Epidemic Models

Assume the time variable is continuous, $t \in [0, \infty)$ and the states $\mathcal{S}(t)$, $\mathcal{I}(t)$, and $\mathcal{R}(t)$ are continuous random variables, that is,

$$\mathcal{S}(t), \mathcal{I}(t), \mathcal{R}(t) \in [0, N].$$

3.5.1 SIS Epidemic Model

The stochastic SIS epidemic model depends on the number of infectives, $\{\mathcal{I}(t)\}$, $t \in [0, \infty)$, where $\mathcal{I}(t)$ has an associated probability density function (pdf), $p(x, t)$,

$$\text{Prob}\{a \leq \mathcal{I}(t) \leq b\} = \int_a^b p(x, t)dx.$$

The stochastic SIS epidemic model has the *Markov property*, i.e.,

$$\text{Prob}\{\mathcal{I}(t_n) \leq y | \mathcal{I}(t_0), \mathcal{I}(t_1), \ldots, \mathcal{I}(t_{n-1})\} = \text{Prob}\{\mathcal{I}(t_n) \leq y | \mathcal{I}(t_{n-1})\}$$

for any sequence of real numbers $0 \leq t_0 < t_1 < \cdots < t_{n-1} < t_n$. Denote the transition pdf for the stochastic process as

$$p(y, t + \Delta t; x, t),$$

where at time t, $\mathcal{I}(t) = x$, and at time $t + \Delta t$, $\mathcal{I}(t + \Delta t) = y$. The process is *time homogeneous* if the transition pdf does not depend on t but does depend on the length of time, Δt. The stochastic process is referred to as a *diffusion process* if it is a Markov process in which the infinitesimal mean and variance exist. The stochastic SIS epidemic model is a time homogeneous, diffusion process. The infinitesimal mean and variance are defined next.

For the stochastic SIS epidemic model, it can be shown that the pdf satisfies a forward Kolmogorov differential equation. This equation is a second order partial differential equation [6,21], a continuous analogue of the forward Kolmogorov differential equations for the CTMC model in (3.10). Assume $\text{Prob}\{\mathcal{I}(0) = i_0\} = 1$ and let $p(i, t; i_0, 0) = p(i, t) = p_i(t)$. Then the system of differential equations in (3.10) can be expressed as a finite difference scheme in the variable i with $\Delta i = 1$,

$$\frac{dp_i}{dt} = p_{i-1}b(i-1) + p_{i+1}d(i+1) - p_i[b(i) + d(i)]$$

$$= -\frac{\{p_{i+1}[b(i+1) - d(i+1)] - p_{i-1}[b(i-1) - d(i-1)]\}}{2\Delta i}$$

$$+ \frac{1}{2}\frac{\{p_{i+1}[b(i+1) + d(i+1)] - 2p_i[b(i) + d(i)] + p_{i-1}[b(i-1) + d(i-1)]\}}{(\Delta i)^2}.$$

Let $i = x$, $\Delta i = \Delta x$ and $p_i(t) = p(x, t)$. Then the limiting form of the preceding equation (as $\Delta x \to 0$) is the forward Kolmogorov differential equation for $p(x, t)$:

$$\frac{\partial p(x, t)}{\partial t} = \frac{\partial}{\partial x}\{[b(x) - d(x)]p(x, t)\} + \frac{1}{2}\frac{\partial^2}{\partial x^2}\{[b(x) + d(x)]\, p(x, t)\}.$$

Substituting $b(x) = \beta x(N - x)/N$ and $d(x) = (b + \gamma)x$ yields

$$\frac{\partial p(x, t)}{\partial t} = \frac{\partial}{\partial x}\left\{\left[\frac{\beta}{N}x(N - x) - (b + \gamma)x\right]p(x, t)\right\}$$

$$+ \frac{1}{2}\frac{\partial^2}{\partial x^2}\left\{\left[\frac{\beta}{N}x(N - x) + (b + \gamma)x\right]p(x, t)\right\}.$$

The coefficient of $p(x, t)$ in the first term on the right side of the preceding equation, $[\beta x(N - x)/N - (b + \gamma)x]$, is the infinitesimal mean and the coefficient of $p(x, t)$ in the second term, $[\beta x(N-x)/N+(b+\gamma)x]$, is the infinitesimal variance. More generally, the forward Kolmogorov differential equations can be expressed in terms of the transition probabilities, $p(y, s; x, t)$. To solve the differential equation requires boundary conditions for $x = 0, N$ and initial conditions for $t = 0$. An explicit solution is not possible because of the nonlinearities. We derive a SDE that is much simpler to solve numerically and whose solution is a sample path of the stochastic process.

A SDE for the SIS epidemic model can be derived from the CTMC SIS epidemic model [5]. The assumptions in the CTMC SIS epidemic model are restated in terms of $\Delta \mathcal{I} = \mathcal{I}(t + \Delta t) - \mathcal{I}(t)$. Assume

$$\text{Prob}\{\Delta\mathcal{I} = j | \mathcal{I}(t) = i\} = \begin{cases} b(i)\Delta t + o(\Delta t), & j = i + 1 \\ d(i)\Delta t + o(\Delta t), & j = i - 1 \\ 1 - [b(i) + d(i)]\Delta t + o(\Delta t), & j = i \\ o(\Delta t), & j \neq i + 1, i - 1, i \end{cases}$$

In addition, assume that $\Delta\mathcal{I}$ has an approximate normal distribution for small Δt. The expectation and the variance of $\Delta\mathcal{I}$ are computed.

$$E(\Delta\mathcal{I}) = b(\mathcal{I})\Delta t - d(\mathcal{I})\Delta t + o(\Delta t)$$

$$= [b(\mathcal{I}) - d(\mathcal{I})]\Delta t + o(\Delta t) = \mu(\mathcal{I})\Delta t + o(\Delta t).$$

$$Var(\Delta\mathcal{I}) = E(\Delta\mathcal{I})^2 - [E(\Delta\mathcal{I})]^2$$
$$= b(\mathcal{I})\Delta t + d(\mathcal{I})\Delta t + o(\Delta t)$$
$$= [b(\mathcal{I}) + d(\mathcal{I})]\Delta t + o(\Delta t) = \sigma^2(\mathcal{I})\Delta t + o(\Delta t),$$

where the notation means $b(\mathcal{I}) = \beta i(N-i)/N$ and $d(\mathcal{I}) = (b+\gamma)i$ given that $\mathcal{I}(t) = i$. Because the random variable $\Delta\mathcal{I}$ is approximately normally distributed, $\Delta\mathcal{I}(t) \sim \mathbf{N}(\mu(\mathcal{I})\Delta t, \sigma^2(\mathcal{I})\Delta t)$,

$$\mathcal{I}(t + \Delta t) = \mathcal{I}(t) + \Delta\mathcal{I}(t)$$
$$\approx \mathcal{I}(t) + \mu(\mathcal{I})\Delta t + \sigma(\mathcal{I})\sqrt{\Delta t}\,\eta,$$

where $\eta \sim \mathbf{N}(0, 1)$.

The difference equation $\mathcal{I}(t + \Delta t) = \mathcal{I}(t) + \mu(\mathcal{I})\Delta t + \sigma(\mathcal{I})\sqrt{\Delta t}\,\eta$ is Euler's method applied to the following Itô SDE:

$$\frac{d\mathcal{I}}{dt} = \mu(\mathcal{I}) + \sigma(\mathcal{I})\frac{dW}{dt},$$

where W is the Wiener process, $W(t + \Delta t) - W(t) \sim \mathbf{N}(0, \Delta t)$ [21, 31, 32]. Euler's method converges to the Itô SDE provided the coefficients, $\mu(\mathcal{I})$ and $\sigma(\mathcal{I})$, satisfy certain smoothness and growth conditions [31, 32]. The coefficients for the stochastic SIS epidemic model are $\mu(\mathcal{I}) = b(\mathcal{I}) - d(\mathcal{I})$ and $\sigma(\mathcal{I}) = \sqrt{b(\mathcal{I}) + d(\mathcal{I})}$, where

$$b(\mathcal{I}) = \frac{\beta}{N}\mathcal{I}(N - \mathcal{I}) \quad \text{and} \quad d(\mathcal{I}) = (b + \gamma)\mathcal{I}.$$

Substituting these values into the Itô SDE gives the SDE SIS epidemic model,

$$\frac{d\mathcal{I}}{dt} = \frac{\beta}{N}\mathcal{I}(N - \mathcal{I}) - (b + \gamma)\mathcal{I} + \sqrt{\frac{\beta}{N}\mathcal{I}(N - \mathcal{I}) + (b + \gamma)\mathcal{I}}\,\frac{dW}{dt}. \qquad (3.13)$$

From the Itô SDE, it can be seen that when $\mathcal{I}(t) = 0$, $d\mathcal{I}/dt = 0$. The disease-free equilibrium is an absorbing state for the Itô SDE.

We digress briefly to discuss the Wiener process $\{W(t)\}$, $t \in [0, \infty)$. The Wiener process depends continuously on t, $W(t) \in (-\infty, \infty)$. It is a diffusion process, but has some additional nice properties. The Wiener process has stationary, independent increments, that is, the increments ΔW depend only on Δt. They are independent of t and the value of $W(t)$:

$$\Delta W = W(t + \Delta t) - W(t) \sim \mathbf{N}(0, \Delta t).$$

Two sample paths of a Wiener process are graphed in Fig. 3.8.

The notation $dW(t)/dt$ is only for convenience because sample paths of $W(t)$ are continuous but nowhere differentiable [12, 21]. The Itô SDE (3.13) should be expressed as a stochastic integral equation but the SDE notation is standard.

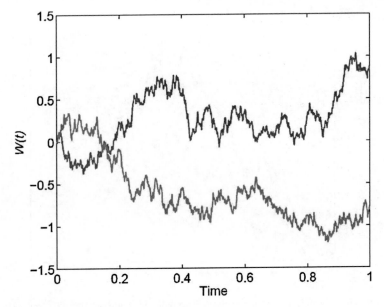

Fig. 3.8 Two sample paths of a Wiener process

3.5.2 Numerical Example

Three sample paths of the SDE SIS epidemic model are graphed in Fig. 3.9. The parameter values are $\beta = 1$, $b = \gamma = 0.25$, and $N = 100$. The initial condition is $I(0) = 2$. For the stochastic model the pdf for the initial condition is $p(x, 0) = 2\delta(x - 2)$, where $\delta(x)$ is the Dirac delta function. The basic reproduction number is $\mathcal{R}_0 = 2$, so that the deterministic solution approaches the endemic equilibrium $\bar{I} = 50$. The MatLab program which generated these sample paths is given in the last section. Compare the sample paths of the Itô SDE in Fig. 3.9 with those for the DTMC and the CTMC models in Figs. 3.4 and 3.7. The sample paths for the Itô SDE are continuous, whereas the sample paths of the DTMC and the CTMC models are discontinuous.

3.5.3 SIR Epidemic Model

A derivation similar to the Itô SDE for the SIS epidemic model can be applied to the bivariate process $\{(\mathcal{S}(t), \mathcal{I}(t))\}$ [5, 6]. Similar assumptions are made regarding the change in the random variables, $\Delta \mathcal{S}$ and $\Delta \mathcal{I}$, as in the transition probabilities for the DTMC and CTMC models. In addition, we assume that the change in these random variables is approximately normally distributed.

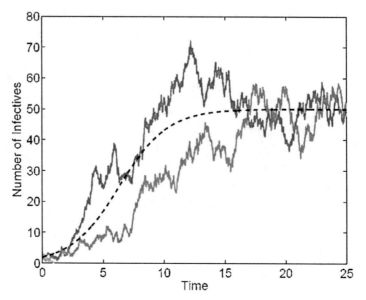

Fig. 3.9 Three sample paths of the SDE SIS epidemic model are graphed with the deterministic solution (*dashed curve*). The parameter values are $b = 0.25$, $\gamma = 0.25$, $\beta = 1$, $N = 100$, $I(0) = 2$. Compare with Figs. 3.4 and 3.7

To simplify the derivation, we assume there are no births, $b = 0$, in the SIR epidemic model.

Let $\Delta X(t) = (\Delta \mathcal{S}, \Delta \mathcal{I})^T$. Then the expectation of $\Delta X(t)$ to order Δt is

$$E(\Delta X(t)) = \begin{pmatrix} -\dfrac{\beta}{N}\mathcal{S}\mathcal{I} \\ \dfrac{\beta}{N}\mathcal{S}\mathcal{I} - \gamma\mathcal{I} \end{pmatrix} \Delta t.$$

The covariance matrix of $\Delta X(t)$ is $V(\Delta X(t)) = E(\Delta X(t)[\Delta X(t)]^T) - E(\Delta X(t))E(\Delta X(t))^T \approx E(\Delta X(t)[\Delta X(t)]^T)$ because the elements in the second term are $o([\Delta t]^2)$. Then the covariance matrix of $\Delta X(t)$ to order Δt is

$$V(\Delta X(t)) = \begin{pmatrix} \dfrac{\beta}{N}\mathcal{S}\mathcal{I} & -\dfrac{\beta}{N}\mathcal{S}\mathcal{I} \\ -\dfrac{\beta}{N}\mathcal{S}\mathcal{I} & \dfrac{\beta}{N}\mathcal{S}\mathcal{I} + \gamma\mathcal{I} \end{pmatrix} \Delta t$$

[5,6]. The random vector $X(t + \Delta t)$ can be approximated as follows:

$$X(t + \Delta t) = X(t) + \Delta X(t) \approx X(t) + E(\Delta X(t)) + \sqrt{V(\Delta X(t))}. \quad (3.14)$$

Because the covariance matrix is symmetric and positive definite, it has a unique square root $B\sqrt{\Delta t} = \sqrt{V}$ [43]. The system of equations (3.14) are an Euler approximation to a system of Itô SDEs. For sufficiently smooth coefficients, the solution $X(t)$ of (3.14) converges to the solution of the following system of Itô SDEs:

$$\frac{d\mathcal{S}}{dt} = -\frac{\beta}{N}\mathcal{SI} + B_{11}\frac{dW_1}{dt} + B_{12}\frac{dW_2}{dt}$$

$$\frac{d\mathcal{I}}{dt} = \frac{\beta}{N}\mathcal{SI} - \gamma\mathcal{I} + B_{21}\frac{dW_1}{dt} + B_{22}\frac{dW_2}{dt}$$

where W_1 and W_2 are two independent Wiener processes and $B = (B_{ij})$ [31, 32].

3.5.4 Numerical Example

Three sample paths of the SDE SIR epidemic model are graphed with the deterministic solution in Fig. 3.10. The parameter values are $\Delta t = 0.01$, $\beta = 1$, $b = 0$, $\gamma = 0.5$, and $N = 100$ with initial condition $I(0) = 2$. The basic reproduction number and initial replacement number are $\mathcal{R}_0 = 2$ and $\mathcal{R}_0 S(0)/N = 1.96$, respectively. Compare the sample paths in Fig. 3.10 with the sample paths for the DTMC SIR epidemic model in Fig. 3.6.

3.6 Properties of Stochastic SIS and SIR Epidemic Models

In the next subsections, we concentrate on some of the properties of these well-known stochastic epidemic models that distinguish them from their deterministic counterparts. Four important properties of stochastic epidemic models include the following: probability of an outbreak, quasistationary probability distribution, final size distribution of an epidemic and expected duration of an epidemic. Each of these properties depend on the stochastic nature of the process.

3.6.1 Probability of an Outbreak

An outbreak occurs when the number of cases escalates. A simple random walk model (DTMC) or a linear birth and death process (CTMC) on the set $\{0, 1, 2, \ldots\}$ can be used to estimate the probability of an outbreak. For

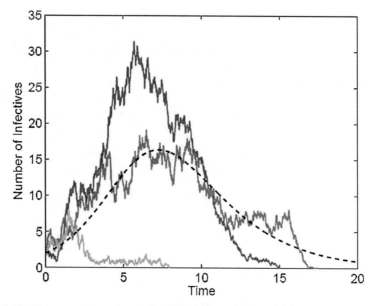

Fig. 3.10 Three sample paths of the SDE SIR epidemic model are graphed with the deterministic solution (*dashed curve*). The parameter values are $\Delta t = 0.01$, $\beta = 1$, $b = 0$, $\gamma = 0.5$, $N = 100$, and $I(0) = 2$. Compare with Fig. 3.6

example, let $X(t)$ be the random variable for the position at time t on the set $\{0, 1, 2, \ldots\}$ in a random walk model. State 0 is absorbing and the remaining states are transient. If $X(t) = x$, then in the next time interval, there is either a move to the right $x \to x + 1$ with probability p or a move to the left, $x \to x - 1$ with probability q, with the exception of state 0, where there is no movement ($p + q = 1$). In the random walk model, either the process approaches state 0 or approaches infinity. The probability of absorption into state 0 depends on p, q, and the initial position. Let $X(t) = x_0 > 0$, then it can be shown that

$$\lim_{t \to \infty} \text{Prob}\{X(t) = 0\} = \begin{cases} 1, & \text{if } p \leq q \\ \left(\dfrac{q}{p}\right)^{x_0}, & \text{if } p > q \end{cases} \qquad (3.15)$$

(e.g., [6, 13, 45]).

The identity (3.15) is also valid for a linear birth and death process in a DTMC or CTMC model, where b and d are replaced by λi and μi, where i is the position. In the linear birth and death process, the infinitesimal transition probabilities satisfy

$$p_{i+j,i}(\Delta t) = \begin{cases} \lambda i \Delta t + o(\Delta t), & j = 1 \\ \mu i \Delta t + o(\Delta t), & j = -1 \\ 1 - (\lambda + \mu)i\Delta t + o(\Delta t), & j = 0. \end{cases}$$

The identity (3.15) holds with λ replacing p and μ replacing q. The probability of absorption is one if $\lambda \leq \mu$. But if $\lambda > \mu$ the probability of absorption decreases to $(\mu/\lambda)^{x_0}$. In this latter case, the probability of population persistence is $1 - (\mu/\lambda)^{x_0}$. This identity can be used to approximate the probability of an outbreak in the DTMC and CTMC SIS and SIR epidemic models, where population persistence can be interpreted as an outbreak. The approximation improves the larger the population size N and the smaller the initial number of infected individuals.

Suppose the initial number of infected individuals i_0 is small and the population size N is large. Then the "birth" and "death" functions in an SIS epidemic model are given by

$$\text{Birth} = b(i) = \frac{\beta}{N}i(N - i) \approx \beta i$$

and

$$\text{Death} = d(i) = (b + \gamma)i.$$

Applying the identity (3.15) and the preceding approximations for the birth and death functions leads to the approximation $\mu/\lambda = (b + \gamma)/\beta = 1/\mathcal{R}_0$, that is,

$$\text{Prob}\{\mathcal{I}(t) = 0\} \approx \begin{cases} 1, & \text{if } \mathcal{R}_0 \leq 1 \\ \left(\dfrac{1}{\mathcal{R}_0}\right)^{i_0}, & \text{if } \mathcal{R}_0 > 1. \end{cases}$$

Therefore, the probability of an outbreak is

$$\text{Probability of an Outbreak} \approx \begin{cases} 0, & \text{if } \mathcal{R}_0 \leq 1 \\ 1 - \left(\dfrac{1}{\mathcal{R}_0}\right)^{i_0}, & \text{if } \mathcal{R}_0 > 1. \end{cases} \tag{3.16}$$

The estimates in (3.16) apply to the stochastic SIS and SIR epidemic models only for a range of times, $t \in [T_1, T_2]$. In the stochastic epidemic models, eventually $\lim_{t \to \infty} \text{Prob}\{\mathcal{I}(t) = 0\} = 1$ because zero is an absorbing state. The range of times for which the estimate (3.16) holds can be quite long when N is large and i_0 is small (see Fig. 3.5). In Fig. 3.5, $N = 100$, $\mathcal{R}_0 = 2$, and $i_0 = 2$, so that applying (3.16) leads to the estimate for the probability of no outbreak as $(1/2)^2 = 1/4$. The value $1/4$ is very close to the mass of the distribution concentrated at zero, $\text{Prob}\{\mathcal{I}(t) = 0\}$. In Fig. 3.11, $\text{Prob}\{\mathcal{I}(t) = 0\}$ for the DTMC SIS epidemic model is graphed for different values of $\mathcal{R}_0$. There is close agreement between the numerical values and the estimate $(1/\mathcal{R}_0)^{i_0}$ when $i_0 = 1, 2, 3$ [$(1/\mathcal{R}_0)^{i_0} = 0.5, 0.25, 0.125$].

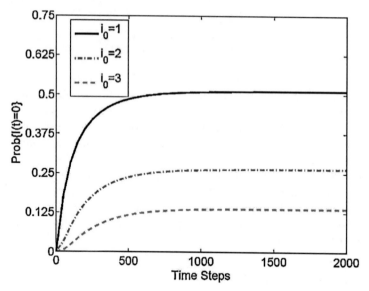

Fig. 3.11 Graphs of Prob$\{\mathcal{I}(t) = 0\}$ for $\mathcal{R}_0 = 2$, $N = 100$, and Prob$\{\mathcal{I}(0) = i_0\} = 1$, $i_0 = 1, 2, 3$

3.6.2 Quasistationary Probability Distribution

Because the zero state in the stochastic SIS epidemic models is absorbing, the unique stationary distribution approached asymptotically by the stochastic process is the disease-free equilibrium. However, as seen in the previous section and in Fig. 3.5, prior to absorption, the process approaches what appears to be a stationary distribution that is different from the disease-free equilibrium. This distribution is known as the quasistationary probability distribution (first investigated in the 1960s [18]). The quasistationary probability distribution can be obtained from the distribution conditioned on nonextinction (i.e., conditional on the disease-free equilibrium not being reached).

Let the distribution conditioned on nonextinction for the CTMC SIS epidemic model be denoted as $q(t) = (q_1(t), \ldots, q_N(t))^T$. Then $q_i(t)$ is the probability $\mathcal{I}(t) = i$ given that $\mathcal{I}(s) > 0$ for $t > s$ (the disease-free equilibrium has not been reached by time t), i.e.,

$$q_i(t) = \text{Prob}\{\mathcal{I}(t) = i | \mathcal{I}(s) > 0, \ t > s\},$$

$i = 1, 2, \ldots, N$. Because the zero state is absorbing, the probability Prob$\{\mathcal{I}(s) > 0, \ t > s\} = 1 - p_0(t)$. Therefore,

$$q_i(t) = \frac{p_i(t)}{1 - p_0(t)}, \quad i = 1, 2, \ldots, N. \tag{3.17}$$

The forward Kolmogorov differential equations for p_i given in (3.10) can be used to derive a system of differential equations for the q_i.

Differentiating the expression for q_i in (3.17) with respect to t and applying the identity for dp_i/dt in (3.10) leads to

$$\frac{dq_i}{dt} = \frac{dp_i/dt}{1 - p_0} + (b + \gamma)q_1 \frac{p_i}{1 - p_0}$$

for $i = 1, 2, \ldots, N$. In matrix notation, the system of differential equations for $q = (q_1, \ldots, q_N)^T$ are similar to the forward Kolmogorov differential equations,

$$\frac{dq}{dt} = \tilde{Q}q + (b + \gamma)q_1 q,$$

where matrix $\tilde{Q}$ is the same as matrix Q in (3.11) with the first row and column deleted. Matrix $\tilde{Q}$ is

$$\begin{pmatrix} -[b(1) + d(1)] & d(2) & \cdots & 0 \\ b(1) & -[b(2) + d(2)] & \cdots & 0 \\ 0 & b(2) & \cdots & 0 \\ \vdots & \vdots & \vdots & \vdots \\ 0 & 0 & \cdots & d(N) \\ 0 & 0 & \cdots & -d(N) \end{pmatrix},$$

where $b(i) = \beta i(N - i)/N$ and $d(i) = (b + \gamma)i$.

Now, the quasistationary probability distribution can be defined. The *quasistationary probability distribution* is the stationary distribution (time-independent solution) $q^* = (q_1^*, \ldots, q_N^*)^T$ satisfying

$$\tilde{Q}q^* = -(b + \gamma)q_1^* q^*. \tag{3.18}$$

Although q^* cannot be solved directly from the system of equations (3.18), it can be solved indirectly via an iterative scheme (see, e.g., [38, 39]).

The quasistationary distribution is related to the eigenvalues of the original matrix Q, where $dp/dt = Qp$. The solution to the forward Kolmogorov differential equations (3.11) satisfy

$$p(t) = v_0 + v_1 e^{r_1 t} + \cdots + v_N e^{r_N t},$$

where $v_0 = (1, 0, 0, \ldots, 0)^T$ [28, 38, 39]. Since matrix Q is the same as $\tilde{Q}$, with the first row and column deleted, the vector $v_1 = (-1, q_1^*, q_2^*, \ldots, q_N^*)^T$ is an eigenvector of Q corresponding to the eigenvalue $r_1 = -(b + \gamma)q_1^*$, that is,

$$Qv_1 = r_1 v_1$$

so that

$$p(t) = (1,0,0,\ldots,0)^T + (-1,q_1^*,q_2^*,\ldots,q_N^*)^T e^{r_1 t} + \cdots + v_N e^{r_N t}.$$

Nasell discusses two approximations to the quasistationary probability distribution [38–40]. One approximation assumes $d(1) = 0$. For this approximation, the system of differential equations for q simplify to

$$\frac{dq}{dt} = Q_1 q, \tag{3.19}$$

where

$$Q_1 = \begin{pmatrix} -b(1) & d(2) & \cdots & 0 \\ b(1) & -[b(2) + d(2)] & \cdots & 0 \\ 0 & b(2) & \cdots & 0 \\ \vdots & \vdots & \vdots & \vdots \\ 0 & 0 & \cdots & d(N) \\ 0 & 0 & \cdots & -d(N) \end{pmatrix}.$$

System (3.19) has a unique stable stationary distribution, $p^1 = (p_1^1,\ldots,p_N^1)^T$, where $Q_1 p^1 = \mathbf{0}$. Because matrix Q_1 is tridiagonal, p^1 has an explicit solution given by

$$p_i^1 = p_1^1 \frac{(N-1)!}{i(N-i)!} \left(\frac{R_0}{N}\right)^{i-1}, \quad i = 2,\ldots,N,$$

$$p_1^1 = \left[\sum_{k=1}^{N} \frac{(N-1)!}{k(N-k)!} \left(\frac{R_0}{N}\right)^{k-1}\right]^{-1}.$$

[8, 38–40] A simple recursion formula can be easily applied to find this approximation:

$$p_{i+1}^1 = \frac{b(i)}{d(i+1)} p_i^1$$

with the property that $\sum_{i=1}^{N} p_i^1 = 1$. The exact quasistationary distribution and the first approximation (for the DTMC and the CTMC epidemic models) are graphed for different values of R_0 in Fig. 3.12. Note that the agreement between the exact quasistationary distribution and the approximation improves as R_0 increases. In addition, note that the mean values are close to the stable endemic equilibrium of the deterministic SIS epidemic model.

The second approximation to the quasistationary probability distribution replaces $d(i)$ by $d(i-1)$. Then the differential equations for q simplify to

$$\frac{dq}{dt} = Q_2 q,$$

where

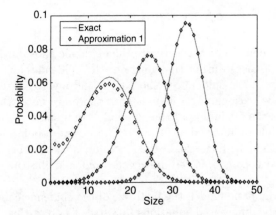

Fig. 3.12 Exact quasistationary distribution and the first approximation to the quasistationary distribution for $\mathcal{R}_0 = 1.5, 2$, and 3 when $N = 50$

$$
Q_2 = \begin{pmatrix}
-b(1) & d(1) & \cdots & 0 \\
b(1) & -[b(2) + d(1)] & \cdots & 0 \\
0 & b(2) & \cdots & 0 \\
\vdots & \vdots & \vdots & \vdots \\
0 & 0 & \cdots & d(N-1) \\
0 & 0 & \cdots & -d(N)
\end{pmatrix} .
$$

The stable stationary solution is the unique solution p^2 to $Q_2 p^2 = \mathbf{0}$. An explicit solution for p^2 is given by

$$
p_i^2 = p_1^2 \frac{(N-1)!}{(N-i)!} \left(\frac{\mathcal{R}_0}{N}\right)^{i-1} , \quad i = 2, \ldots, N,
$$

$$
p_1^2 = \left[\sum_{k=1}^{N} \frac{(N-1)!}{(N-k)!} \left(\frac{\mathcal{R}_0}{N}\right)^{k-1}\right]^{-1}
$$

(see [8, 38–40]).

3.6.3 Final Size of an Epidemic

In the SIR epidemic model, eventually the epidemic ends. Of interest is the total number of cases during the course of the epidemic, i.e., the final size of the epidemic. If the epidemic is short term and involves a relatively small population, it is reasonable to assume that there are no births nor deaths. In addition, at the beginning of the epidemic, we assume all individuals are either susceptible or infected, $R(0) = 0$. The initial population size is $N = S(0) + I(0)$. Then the final size of the epidemic is the number of susceptible individuals that become infected during the epidemic plus the initial number infected.

In the deterministic model, the final size of the epidemic can be computed directly from the differential equations (3.3) (see Chap. 2, this volume). Integrating the differential equation $dI/dS = -1 + N\gamma/\beta S$, leads to

$$I(t) + S(t) = I(0) + S(0) + \frac{N\gamma}{\beta} \ln \frac{S(t)}{S(0)}.$$

Letting $t \to \infty$,

$$S(\infty) = I(0) + S(0) + \frac{N\gamma}{\beta} \ln \frac{S(\infty)}{S(0)}.$$

The final size of the epidemic is

$$R(\infty) = N - S(\infty).$$

The final sizes in the deterministic SIR epidemic model are summarized in Table 3.1 when $I(0) = 1$ and $\gamma = 1$ for various values of $\mathcal{R}_0$ and N.

Table 3.1 Final size of an epidemic when $\gamma = 1$ and $I(0) = 1$ for the deterministic SIR epidemic model

$\mathcal{R}_0$	N		
	20	100	1,000
0.5	1.87	1.97	2.00
1	5.74	13.52	44.07
2	16.26	80.02	797.15
5	19.87	99.31	993.03
10	20.00	100.00	999.95

In the stochastic SIR epidemic model there is a distribution associated with the final size of the epidemic. Let (s, i) denote the ordered pairs of values for the susceptible and infected individuals in the CTMC model. The epidemic ends when $\mathcal{I}(t)$ reaches zero. When the epidemic ends, the random variable for the number of susceptible individuals ranges from 0 to $N - \mathcal{I}(0) = N - i_0$. In particular, the set $\{(s, 0)\}_{s=0}^{N-i_0}$ is absorbing,

$$\lim_{t \to \infty} \sum_{s=0}^{N-i_0} P_{(s,0)}(t) = 1.$$

Daley and Gani [17] discuss two different methods to compute the probability distribution associated with the final size. The simpler method, originally developed by Foster [20], depends on the embedded Markov chain, that is, the DTMC model associated with the CTMC model. To apply this method, the transition matrix for the embedded Markov chain needs to be computed. This requires computing the probability of a transition between the states (s, i), where the states lie in the set $\{(s, i) : s = 0, 1, \ldots, N; i = 0, 1, \ldots, N - s\}$. In the embedded Markov chain for the final size, the times between transitions are not important, only the probabilities.

For example, suppose $N = 3$, then the states in the transition matrix are

$$(s, i) \in \{(3,0), (2,0), (1,0), (0,0), (2,1), (1,1), (0,1), (1,2), (0,2), (0,3)\}, \tag{3.20}$$

i.e., there are 10 ordered pairs of states. There are only two types of transitions, either an infected individual recovers, $(s, i) \to (s, i-1)$ or a susceptible individual becomes infected, $(s, i) \to (s - 1, i + 1)$. In the first type of transition, an infected individual recovers with probability

$$p_s = \frac{\gamma i}{\gamma i + (\beta/N)is} = \frac{\gamma}{\gamma + (\beta/N)s}, \quad s = 0, 1, 2.$$

In the second type of transition, a susceptible individual becomes infected with probability $1 - p_s$. If the 10 states are ordered as in (3.20), then the transition matrix for the embedded Markov chain is a 10×10 matrix with the following form:

$$T = \begin{pmatrix} 1 & 0 & 0 & 0 & 0 & 0 & 0 & 0 & 0 & 0 \\ 0 & 1 & 0 & 0 & p_2 & 0 & 0 & 0 & 0 & 0 \\ 0 & 0 & 1 & 0 & 0 & p_1 & 0 & 0 & 0 & 0 \\ 0 & 0 & 0 & 1 & 0 & 0 & p_0 & 0 & 0 & 0 \\ 0 & 0 & 0 & 0 & 0 & 0 & 0 & 0 & 0 & 0 \\ 0 & 0 & 0 & 0 & 0 & 0 & 0 & p_1 & 0 & 0 \\ 0 & 0 & 0 & 0 & 0 & 0 & 0 & 0 & p_0 & 0 \\ 0 & 0 & 0 & 0 & 1-p_2 & 0 & 0 & 0 & 0 & 0 \\ 0 & 0 & 0 & 0 & 0 & 1-p_1 & 0 & 0 & 0 & p_0 \\ 0 & 0 & 0 & 0 & 0 & 0 & 0 & 1-p_1 & 0 & 0 \end{pmatrix}$$

The upper left 4×4 corner of matrix T is the identity matrix because these are the four absorbing states. The first four rows are the transitions into these four absorbing states. Matrix T is a stochastic matrix, whose column sums equal one (note that $p_0 = 1$). Given the initial distribution for the states $p(0)$, then the distribution for the final size can be found from the first four entries of $\lim_{t \to \infty} T^t p(0)$ (the remaining entries are zero). However, it is not necessary to compute the limit as $t \to \infty$, since the limit converges by time $t = 2N - 1$. For this example, it is straightforward to compute the final size distribution. The final size is either 1, 2, or 3 with corresponding probabilities p_2, $p_1^2(1 - p_2)$ and $(1 - p_1^2)(1 - p_2)$, respectively. In Fig. 3.13, there are graphs of three final size distributions for different values of $\mathcal{R}_0$ when $\gamma = 1$, $\text{Prob}\{\mathcal{I}(0) = 1\} = 1$, and $N = 20$.

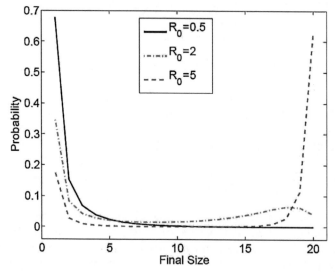

Fig. 3.13 Distribution for the final size of an epidemic for three different values of $\mathcal{R}_0$ when $\gamma = 1$, $N = 20$, and $\text{Prob}\{\mathcal{I}(0) = 1\} = 1$

When $\mathcal{R}_0$ is less than one or very close to one, then the final size distribution is skewed to the right, but if $\mathcal{R}_0$ is much greater than one, then the distribution is skewed to the left. The average final sizes for the stochastic SIR when $N = 20$ and $N = 100$ are listed in Table 3.2. Compare the values in Table 3.2 to those in Table 3.1. For values of $\mathcal{R}_0$ less than one or much greater than one, the average final sizes for the stochastic SIR epidemic model are closer to the values of the final sizes for the deterministic model.

Table 3.2 Average final size of an epidemic when $\gamma = 1$, $b = 0$, and $\text{Prob}\{\mathcal{I}(0) = 1\} = 1$ for the stochastic SIR epidemic model

$\mathcal{R}_0$	N	
	20	100
0.5	1.76	1.93
1	3.34	6.10
2	8.12	38.34
5	15.66	79.28
10	17.98	89.98

3.6.4 Expected Duration of an Epidemic

The duration of an epidemic corresponds to the time until absorption, i.e., the time T until $\mathcal{I}(T) = 0$. For the stochastic SIS epidemic model, the probability of absorption is one, regardless of the value of $\mathcal{R}_0$. However, depending on the initial number infected, i, the population size N, and the value of $\mathcal{R}_0$, the time until absorption can be very short or very long. Here, we derive a system of equations that can be solved to find the expected time until absorption for a stochastic SIS epidemic model.

Let T_i denote the random variable for the time until absorption and let

$$\tau_i = E(T_i)$$

denote the expected time until absorption beginning from an initial infected population size of i, $i = 0, 1, \ldots, N$. Let the higher order moments for the time until absorption be denoted as

$$\tau_i^r = E(T_i^r),$$

$i = 0, 1, \ldots, N$. Note that $\tau_0 = 0 = \tau_0^r$. Then, considered as a birth and death process, the mean time until absorption in the DTMC SIS epidemic model satisfies the following difference equation:

$$\tau_i = b(i)\Delta t(\tau_{i+1} + \Delta t) + d(i)\Delta t(\tau_{i-1} + \Delta t)$$
$$+ (1 - [b(i) + d(i)]\Delta t)(\tau_i + \Delta t), \quad i = 1, \ldots, N \qquad (3.21)$$

The CTMC SIS epidemic model satisfies the same relationship as (3.21), except that a term $o(\Delta t)$ is added to the right side of each equation. Simplifying the equations in (3.21) leads to a system of difference equations for the expected duration of an epidemic (for both the CTMC and the DTMC models),

$$d(i)\tau_{i-1} - [b(i) + d(i)]\tau_i + b(i)\tau_{i+1} = -1,$$

where $b(i) = i(N - i)(\beta i/N)$ and $d(i) = (b + \gamma)i$ [7,33]. Similar difference equations apply to the higher order moment τ_i^r in the CTMC SIS epidemic model,

$$d(i)\tau_{i-1}^r - [b(i) + d(i)]\tau_i^r + b(i)\tau_{i+1}^r = -r\tau_i^{r-1}$$

[7, 22, 41, 42].

The mean and higher order moments can be expressed in matrix form. Let $\tau = (\tau_1, \tau_2, \ldots, \tau_N)^T$, $\tau^r = (\tau_1^r, \tau_2^r, \ldots, \tau_N^r)^T$ and $\tau^1 = \tau$. Then

$$D\tau = -1 \quad \text{and} \quad D\tau^r = -r\tau^{r-1}.$$

where $1 = (1, \ldots, 1)^T$ and

$$D = \begin{pmatrix} -[b(1) + d(1)] & b(1) & 0 & \cdots & 0 & 0 \\ d(2) & -[b(2) + d(2)] & b(2) & \cdots & 0 & 0 \\ \vdots & \vdots & \vdots & \vdots & \vdots & \vdots \\ 0 & 0 & 0 & \cdots & d(N) & -d(N) \end{pmatrix}.$$

Matrix D is nonsingular because it is irreducibly diagonally dominant [43]. Hence, the solutions τ and τ^r are unique.

A solution for the expected time until absorption, based on a system of SDEs, can also be derived [7]. The relationship satisfied by τ follows from the backward Kolmogorov differential equations. Let $\tau(y)$ denote the expected time until absorption beginning from an infected population size of $y \in (0, N)$. Then it can be shown that $\tau(y)$ is the solution to the following boundary value problem:

$$[b(y) - d(y)]\frac{d\tau(y)}{dy} + \frac{[b(y) + d(y)]}{2}\frac{d^2\tau(y)}{dy^2} = -1, \tag{3.22}$$

where

$$\tau(0) = 0 \quad \text{and} \quad \frac{d\tau(y)}{dy}\bigg|_{y=N} = 0,$$

$b(y) = (N - y)(\beta y/N)$ and $d(y) = (b + \gamma)y$ in the SDE SIS epidemic model [7].

It is interesting to note that if the derivatives in the boundary value problem for $\tau(y)$ in (3.22) are approximated by finite difference formulas, then the difference equations for τ_i, given in (3.21), for the CTMC and DTMC epidemic models are obtained [7]. For $y \in [i, i + 1]$, let

$$\frac{d\tau(y)}{dy} \approx \frac{\tau_{i+1} - \tau_{i-1}}{2},$$

where $\tau_i = \tau(i)$ and $\tau_{i+1} = \tau(i + 1)$. In addition, let

$$\frac{d^2\tau(y)}{dy^2} \approx \tau_{i+1} - 2\tau_i + \tau_{i-1}.$$

With these approximations, the boundary value problem for $\tau(y)$ in (3.22) is approximated by the difference equations for τ_i in (3.21).

The expected duration of an SIS epidemic can be calculated from the solution to (3.21) or (3.22). Allen and Allen [7] compare the mean and the variance for the time until population extinction for the three different types of stochastic formulations considered here. However, their stochastic models were applied to a population with logistic growth (similar to the SIS epidemic model).

As an example, consider the expected duration for an SIS epidemic, based on the DTMC or CTMC model. Because the DTMC and CTMC models satisfy the same set of equations for the expected duration, these results apply to both models. With a population size of $N = 25$ and either $\mathcal{R}_0 = 2$ or $\mathcal{R}_0 = 1.5$. The solution $\tau = -D^{-1}\mathbf{1}$ is graphed in Fig. 3.14. If the population size is increased to $N = 50$ or $N = 100$ with the same value for $\mathcal{R}_0 = 1.5$, the expected duration for large i increases to $\tau_i \approx 160$ and $\tau_i \approx 3,500$, respectively. At population sizes of $N = 50$ and $N = 100$ but a basic reproduction of $\mathcal{R}_0 = 2$, the expected duration for large i is much larger, $\tau_i \approx 25,000$ and $\tau_i \approx 2.6 \times 10^8$, respectively. Of course, the expected duration depends on the particular time units of the model. For example, if the time units are days, then $\tau_i \approx 160 \approx 5.3$ months and $\tau_i \approx 25,000 \approx 68.5$ years. This latter estimate is much longer than a reasonable epidemiological time frame, implying that the disease does not die out but persists. Hence, for these examples, when $N \geq 100$ and $\mathcal{R}_0 \geq 2$, if the outbreak begins with a sufficient number of infected individuals, then the results for the stochastic SIS epidemic are in close agreement with the predictions of the deterministic SIS epidemic model; the disease becomes endemic.

3.7 Epidemic Models with Variable Population Size

Suppose the population size N is not constant but varies according to some population growth law. To formulate an epidemic model, an assumption must be made concerning the population birth and death rates which depend on the population size N. Here, we assume, for simplicity, that the birth rate and death rates have a logistic form,

$$\lambda(N) = bN \quad \text{and} \quad \mu(N) = b\frac{N^2}{K},$$

respectively. Then the total population size satisfies the logistic differential equation

$$\frac{dN}{dt} = \lambda(N) - \mu(N) = bN\left(1 - \frac{N}{K}\right),$$

where $K > 0$ is the carrying capacity. There are many functional forms that can be chosen for the birth and death rates [7]. Their choice should depend on the dynamics of the particular population being modeled. For example, in

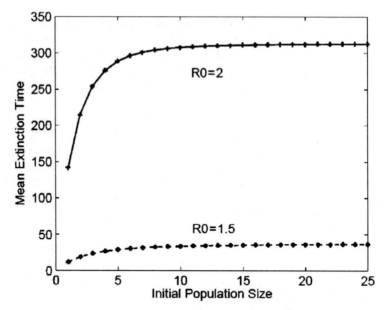

Fig. 3.14 Expected duration of an SIS epidemic with a population size of $N = 25$; $\mathcal{R}_0 = 1.5$ ($b = 1/3$, $\gamma = 1/3$ and $\beta = 1$) and $\mathcal{R}_0 = 2$ ($b = 1/4$, $\gamma = 1/4$ and $\beta = 1$)

animal diseases (e.g., rabies in canine populations [37, 46] and hantavirus in rodent populations [2, 3, 9, 44]), logistic growth is assumed, then the choice of $\lambda(N)$ and $\mu(N)$ depends on whether the births and deaths are density-dependent. For human diseases, a logistic growth assumption may not be very realistic.

A deterministic SIS epidemic model is formulated for a population satisfying the logistic differential equation. Again, for simplicity, we assume there are no disease-related deaths and no vertical transmission of the disease; all newborns are born susceptible. Then the deterministic SIS epidemic model has the form:

$$\frac{dS}{dt} = \frac{S}{N}(\lambda(N) - \mu(N)) - \frac{\beta}{N}SI + (b + \gamma)I$$

$$\frac{dI}{dt} = -\frac{I}{N}\mu(N) + \frac{\beta}{N}SI - \gamma I,$$

(3.23)

where $S(0) > 0$ and $I(0) > 0$. It is straightforward to show that the solution to this system of differential equations depends on the basic reproduction number $\mathcal{R}_0 = \beta/(b + \gamma)$.

Theorem 3. *Let $S(t)$ and $I(t)$ be a solution to model (3.23).*

(1) If $\mathcal{R}_0 \leq 1$, then $\lim_{t \to \infty} (S(t), I(t)) = (K, 0)$.

(2) If $\mathcal{R}_0 > 1$, then $\lim\limits_{t\to\infty}(S(t), I(t)) = (K/\mathcal{R}_0, K(1 - 1/\mathcal{R}_0))$.

Stochastic epidemic models for each of the three types (CTMC, DTMC, and SDE models) can be formulated. Because $S(t) + I(t) = N(t)$, the process is bivariate. We derive an SDE model and compare the graph of a sample path for the stochastic model to the solution of the deterministic model.

Let $\mathcal{S}(t)$ and $\mathcal{I}(t)$ be continuous random variables for the number of susceptible and infected individuals at time t,

$$S(t), \; \mathcal{I}(t) \in [0, \infty).$$

Then, applying the same methods as for the SDE SIS and SIR epidemic models [5, 6],

$$\frac{d\mathcal{S}}{dt} = \frac{\mathcal{S}}{\mathcal{N}}(\lambda(\mathcal{N}) - \mu(\mathcal{N})) - \frac{\beta}{\mathcal{N}}\mathcal{SI} + (b + \gamma)\mathcal{I} + B_{11}\frac{dW_1}{dt} + B_{12}\frac{dW_2}{dt}$$

$$\frac{d\mathcal{I}}{dt} = -\frac{\mathcal{I}}{\mathcal{N}}\mu(\mathcal{N}) + \frac{\beta}{\mathcal{N}}\mathcal{SI} - \gamma\mathcal{I} + B_{21}\frac{dW_1}{dt} + B_{22}\frac{dW_2}{dt},$$

where $B = (B_{ij})$ is the square root of the following covariance matrix:

$$\begin{pmatrix} \frac{\mathcal{S}}{\mathcal{N}}(\lambda(\mathcal{N}) + \mu(\mathcal{N})) + \frac{\beta}{\mathcal{N}}\mathcal{SI} + (b + \gamma)\mathcal{I} & -\frac{\beta}{\mathcal{N}}\mathcal{SI} - \gamma\mathcal{I} \\ -\frac{\beta}{\mathcal{N}}\mathcal{SI} - \gamma\mathcal{I} & \frac{\mathcal{I}}{\mathcal{N}}\mu(\mathcal{N}) + \frac{\beta}{\mathcal{N}}\mathcal{SI} + \gamma\mathcal{I} \end{pmatrix}.$$

The variables W_1 and W_2 are two independent Wiener processes. The absorbing state for the bivariate process is total population extinction, $\mathcal{N} = 0$.

3.7.1 Numerical Example

As might be anticipated, the variability in the population size results in an increase in the variability in the number of infected individuals. As an example, let $\beta = 1$, $\gamma = 0.25 = b$, and $K = 100$. Then the basic reproduction number is $\mathcal{R}_0 = 2$. The SDE SIS epidemic model with constant population size, $N = 100$, is compared to the SDE SIS epidemic model with variable population size, $\mathcal{N}(t)$, in Fig. 3.15. One sample path of the SDE epidemic model is graphed against the deterministic solution.

More realistic stochastic epidemic models can be derived based on their deterministic formulations. Excellent references for a variety of recent deterministic epidemic models include the books by Anderson and May [10], Brauer and Castillo-Chavez [15], Diekmann and Heesterbeek [19], and Thieme [48] and the review articles by Hethcote [26] and Brauer and van den Driessche [16].

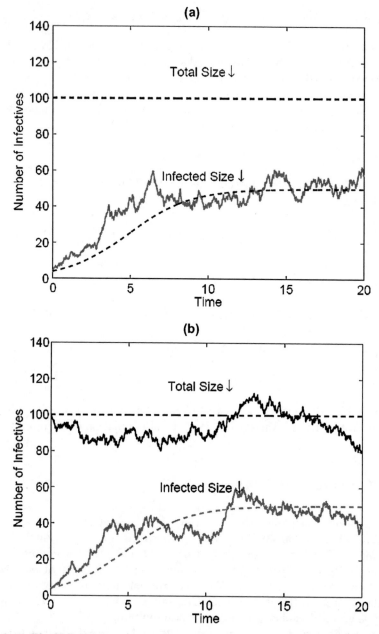

Fig. 3.15 The SDE SIS epidemic model (a) with constant population size, $N = 100$ and (b) with variable population size, $\mathcal{N}(t)$. The parameter values are $\beta = 1$, $\gamma = 0.25 = b$, $K = 100$, and $\mathcal{R}_0 = 2$

In this chapter, the simplest types of epidemic models were chosen as an introduction to the methods of derivation for various types of stochastic models (DTMC, CTMC, and SDE models). In many cases these three stochastic formulations produce similar results, if the time step Δt is small [7]. There are advantages numerically in applying the discrete time approximations (DTMC model and the Euler approximation to the SDE model) in that the discrete simulations generally have a shorter computational time than the CTMC model. Mode and Sleeman [36] discuss some computational methods in stochastic processes in epidemiology. The most important consideration in modeling, however, is to choose a model that best represents the demographics and epidemiology of the population being modeled.

We conclude this chapter with a discussion of some well-known stochastic epidemic models that are not based on any deterministic epidemic model.

3.8 Other Types of DTMC Epidemic Models

Two other types of DTMC epidemic models are discussed briefly that are not directly related to any deterministic epidemic model. These models are chain binomial epidemic models and epidemic branching processes.

3.8.1 Chain Binomial Epidemic Models

Two well-known DTMC models are the Greenwood and the Reed–Frost models. These models were developed to help understand the spread of disease within a small population such as a household. They are referred to as chain binomial epidemic models because a binomial distribution is used to determine the number of new infectious individuals. The Greenwood model developed in 1931, was named after its developer [23]. The Reed–Frost model, developed in 1928, was named for two medical researchers, who developed the model for teaching purposes at Johns Hopkins University. It wasn't until 1952 that the Reed–Frost model was published [1,17].

Let $\mathcal{S}_t$ and $\mathcal{I}_t$ be discrete random variables for the number of susceptible and infected individuals in the household at time t. Initially, the models assume that there are $\mathcal{I}_0 = i_0 \geq 1$ infected individuals and $\mathcal{S}_0 = s_0$ susceptible individuals. The progression of the disease is followed by keeping track of the number of susceptible individuals over time. At time t, infected individuals are in contact with all the susceptible members of the household to whom they may spread the disease. However, it is not until time $t+1$ that susceptible individuals who have contracted the disease are infectious. The period of time from t to $t+1$ is the latent period and the infectious period is contracted to a point. Only at time t can the infectious individuals $\mathcal{I}_t$ infect susceptible members S_t. After that time, they are no longer infectious. It follows that the newly infectious individuals at time $t + 1$ satisfy

$$S_{t+1} + I_{t+1} = S_t.$$

These models are bivariate Markov chain models that depend on the two random variables, S_t and I_t, $\{(S_t, I_t)\}$.

The models of Greenwood and Reed–Frost differ in the assumption regarding the probability of infection. Suppose there are a total of $I_t = i$ infected individuals at time t. Let p_i be the probability that a susceptible individual does not become infected at time t. The Greenwood model assumes that $p_i = p$ is a constant and the Reed–Frost model assumes that $p_i = p^i$. For each model, the transition probability from state (s_t, i_t) to (s_{t+1}, i_{t+1}) is assumed to have a binomial distribution. Sample paths are denoted as $\{s_0, s_1, \ldots, s_{t-1}, s_t\}$. The epidemic stops at time t when $s_{t-1} = s_t$ because there are no more infectious individuals to spread the disease, $i_t = s_{t-1} - s_t = 0$.

3.8.1.1 Greenwood Model

In the Greenwood model, the random variable S_{t+1} is a binomial random variable that depends on S_t and p, $S_{t+1} \sim b(S_t, p)$. The probability of a transition from (s_t, i_t) to (s_{t+1}, i_{t+1}) depends only on s_t, s_{t+1}, and p. It is defined as follows:

$$p_{s_{t+1}, s_t} = \binom{s_t}{s_{t+1}} p^{s_{t+1}} (1 - p)^{s_t - s_{t+1}}.$$

The conditional mean and variance of S_{t+1} and I_{t+1} are given by

$$E(S_{t+1}|S_t) = pS_t, \quad E(I_{t+1}|S_t) = (1 - p)S_t$$

and

$$\text{Var}(S_{t+1}|S_t) = p(1 - p)S_t = \text{Var}(I_{t+1}|S_t).$$

Four sample paths of the Greenwood model when $s_0 = 6$ and $i_0 = 1$ are illustrated in Fig. 3.16. Applying the preceding transition probabilities, it is clear that the sample path $\{6, 6\}$ occurs with probability $p_{6,6} = p^6$ and the sample path $\{6, 5, 5\}$ occurs with probability $p_{6,5}p_{5,5} = 6p^{10}(1-p)$. The probability distributions associated with the size and the duration of epidemics in the chain binomial models can be easily defined, once the probability distributions associated with each sample path are determined. The discrete random variable $W = S_0 - S_t$ is the size of the epidemic and the discrete random variable T is the length of the path, e.g., if $\{s_0, s_1, \ldots, s_{t-1}, s_t\}$, then $T = t$.

Table 3.3 summarizes the probabilities associated with the Greenwood and Reed–Frost epidemic models when $s_0 = 3$ and $i_0 = 1$ (see [17]).

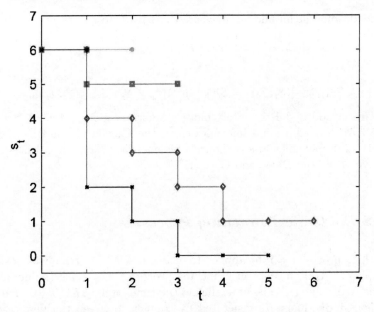

Fig. 3.16 Four sample paths for the Greenwood chain binomial model when $s_0 = 6$ and $i_0 = 1$: $\{6, 6\}$, $\{6, 5, 5\}$, $\{6, 4, 3, 2, 1, 1\}$, and $\{6, 2, 1, 0, 0\}$

Table 3.3 Sample paths, size T, and duration W for the Greenwood and Reed–Frost models when $s_0 = 3$ and $i_0 = 1$

Sample paths $\{s_0, \ldots, s_{t-1}, s_t\}$	Duration T	Size W	Greenwood model	Reed–Frost model
3 3	1	0	p^3	p^3
3 2 2	2	1	$3(1-p)p^4$	$3(1-p)p^4$
3 2 1 1	3	2	$6(1-p)^2p^4$	$6(1-p)^2p^4$
3 1 1	2	2	$3(1-p)^2p^2$	$3(1-p)^2p^3$
3 2 1 0 0	4	3	$6(1-p)^3p^3$	$6(1-p)^3p^3$
3 2 0 0	3	3	$3(1-p)^3p^2$	$3(1-p)^3p^2$
3 1 0 0	3	3	$3(1-p)^3p$	$3(1-p)^3p(1+p)$
3 0 0	2	3	$(1-p)^3$	$(1-p)^3$
Total			1	1

3.8.1.2 Reed–Frost Model

In the Reed–Frost model, the random variable $\mathcal{S}_{t+1}$ is binomially distributed and satisfies $\mathcal{S}_{t+1} \sim b(\mathcal{S}_t, p^{\mathcal{I}_t})$. The probability of a transition from (s_t, i_t) to (s_{t+1}, i_{t+1}) is defined as follows:

$$p_{(s,i)_{t+1},(s,i)_t} = \binom{s_t}{s_{t+1}} (p^{i_t})^{s_{t+1}} (1 - p^{i_t})^{s_t - s_{t+1}}.$$

The conditional mean and variance associated with $\mathcal{S}_{t+1}$ are

$$E(\mathcal{S}_{t+1}|(\mathcal{S}_t, \mathcal{I}_t)) = \mathcal{S}_t p^{\mathcal{I}_t}, \quad E(\mathcal{I}_{t+1}|(\mathcal{S}_t, \mathcal{I}_t)) = \mathcal{S}_t(1 - p^{\mathcal{I}_t})$$

and

$$\mathrm{Var}(\mathcal{S}_{t+1}|(\mathcal{S}_t, \mathcal{I}_t)) = \mathcal{S}_t(1 - p^{\mathcal{I}_t})p^{\mathcal{I}_t} = \mathrm{Var}(\mathcal{I}_{t+1}|(\mathcal{S}_t, \mathcal{I}_t)).$$

The Greenwood and Reed–Frost models differ when $\mathcal{I}_t > 1$ for $t > 0$ (see Table 3.3). For additional information on the Greenwood and Reed–Frost models, and epidemics among households consult Ackerman et al. [4], Ball and Lyne [14], and Daley and Gani [17].

3.8.2 Epidemic Branching Processes

Branching processes can be applied to epidemics. We illustrate with a simple example of a Galton–Watson branching processes. Let $\mathcal{I}_t$ be the number of new cases at time t. We assume during the time interval t to $t+1$ that new infectious individuals are generated by contacts between the new cases at time t and the susceptible population. Suppose each infected individual infects on the average $\mathcal{R}_0$ susceptible individuals. In a Galton–Watson process, the simplifying assumption is that each infected individual is independent from all other infected individuals.

Let $\{p_k\}_{k=0}^{\infty}$ be the probabilities associated with the number of new infections per infected individual. Then the probability generating function (pgf) for the number of new infections is

$$f(t) = \sum_{k=0}^{\infty} p_k t^k$$

with mean $f'(1) = \mathcal{R}_0$.

An important result from the theory of branching processes states that the probability of extinction (probability the epidemic eventually ends), $\lim_{t\to\infty} \mathrm{Prob}\{\mathcal{I}_t = 0\}$, depends on the pgf $f(t)$. If $0 \le p_0 + p_1 < 1$ and $\mathcal{R}_0 > 1$, then there exists a unique fixed point $q \in [0, 1)$ such that $f(q) = q$. The assumption $0 \le p_0 + p_1 < 1$ guarantees that there is a positive probability of infecting more than one individual. It is the value of q and the initial number of infected individuals in the population that determine the probability of extinction. The next theorem summarizes the main result concerning the probability of extinction. For a proof of this result and extensions, please consult the references [6, 24, 29, 30, 35, 45].

Theorem 4. *Suppose the pgf $f(t)$ satisfies $0 \le f(0) + f'(0) < 1$ and $\mathrm{Prob}\{\mathcal{I}_0 = i_0\} = 1$, where $i_0 > 0$.*

(1) If $\mathcal{R}_0 \le 1$, then $\lim_{t\to\infty} \mathrm{Prob}\{\mathcal{I}_t = 0\} = 1$.

(2) If $\mathcal{R}_0 > 1$, then $\lim\limits_{t \to \infty} \text{Prob}\{\mathcal{I}_t = 0\} = q^{i_0}$, where q is the unique fixed point in $[0, 1)$ such that $f(q) = q$.

As a consequence of this theorem, the probability the epidemic persists in the population (the disease becomes endemic) is $1 - q^{i_0}$, provided $\mathcal{R}_0 > 1$.

Antia et al. [11] assume that the number of cases $\mathcal{I}_t$ follows a Poisson distribution with mean $\mathcal{R}_0$. The pgf of a Poisson probability distribution satisfies

$$f(t) = \sum_{k=0}^{\infty} \exp(-\mathcal{R}_0) \frac{\mathcal{R}_0^k}{k!} t^k = \exp(-\mathcal{R}_0(1 - t)).$$

Applying Theorem 4, we can estimate the probability the disease becomes endemic. If $\mathcal{R}_0 > 1$, the fixed point of f satisfies

$$q = \exp(-\mathcal{R}_0(1 - q)).$$

For example, if $\mathcal{R}_0 = 1.5$ and $\text{Prob}\{\mathcal{I}_0 = 1\} = 1$, then $1 - q = 0.583$, but if $\text{Prob}\{\mathcal{I}_0 = 2\} = 1$, then $1 - q^2 = 0.826$. If $\mathcal{R}_0 = 2$ and $\text{Prob}\{\mathcal{I}_0 = 2\} = 1$, then $1 - q^2 = 0.959$.

3.9 MatLab Programs

The following three MatLab programs were used to generate sample paths and the probability distribution associated with the stochastic SIS epidemic model. MatLab Program # 1 computes the probability distribution for the DTMC SIS epidemic model. MatLab Programs # 2 and # 3 compute sample paths associated with CTMC and SDE SIS epidemic models, respectively.

```
% MatLab Program # 1
% Discrete Time Markov Chain
% SIS Epidemic Model
% Transition Matrix and Graph of Probability Distribution
clear all
set(gca,'FontSize',18);
set(0,'DefaultAxesFontSize',18);
time=2000;
dtt=0.01; % Time step
beta=1*dtt;
b=0.25*dtt;
gama=0.25*dtt;
N=100; % Total population size
en=50; % plot every enth time interval
T=zeros(N+1,N+1); % T is the transition matrix, defined below
v=linspace(0,N,N+1);
```

```
p=zeros(time+1,N+1);
p(1,3)=1; % Two individuals initially infected.
bt=beta*v.*(N-v)/N;
dt=(b+gama)*v;
for i=2:N % Define the transition matrix
    T(i,i)=1-bt(i)-dt(i); % diagonal entries
    T(i,i+1)=dt(i+1); % superdiagonal entries
    T(i+1,i)=bt(i); % subdiagonal entries
end
T(1,1)=1;
T(1,2)=dt(2);
T(N+1,N+1)=1-dt(N+1);
for t=1:time
    y=T*p(t,:)';
    p(t+1,:)=y';
end
pm(1,:)=p(1,:);
for t=1:time/en;
    pm(t+1,:)=p(en*t,:);
end
ti=linspace(0,time,time/en+1);
st=linspace(0,N,N+1);
mesh(st,ti,pm);
xlabel('Number of Infectives');
ylabel('Time Steps');
zlabel('Probability');
view(140,30);
axis([0,N,0,time,0,1]);


% Matlab Program # 2
% Continuous Time Markov Chain
% SIS Epidemic Model
% Three Sample Paths and the Deterministic Solution
clear
set(0,'DefaultAxesFontSize', 18);
set(gca,'fontsize',18);
beta=1;
b=0.25;
gam=0.25;
N=100;
init=2;
time=25;
sim=3;
for k=1:sim
    clear t s i
```

```
      t(1)=0;
      i(1)=init;
      s(1)=N-init;
      j=1;
      while i(j)>0 & t(j)<time
         u1=rand; % uniform random number
         u2=rand; % uniform random number
         a=(beta/N)*i(j)*s(j)+(b+gam)*i(j);
         probi=(beta*s(j)/N)/(beta*s(j)/N+b+gam);
         t(j+1)=t(j)-log(u1)/a;
         if u2 <= probi
            i(j+1)=i(j)+1;
            s(j+1)=s(j)-1;
         else
            i(j+1)=i(j)-1;
            s(j+1)=s(j)+1;
         end
         j=j+1;
      end
      plot(t,i,'r-','LineWidth',2)
      hold on
end


% Matlab Program # 3
% Stochastic Differential Equation
% SIS Epidemic Model
% Three Sample Paths and the Deterministic Solution
clear
beta=1;
b=0.25;
gam=0.25;
N=100;
init=2;
dt=0.01;
time=25;
sim=3;
for k=1:sim
    clear i, t
    j=1;
    i(j)=init;
    t(j)=dt;
    while i(j)>0 & t(j)<25
       mu=beta*i(j)*(N-i(j))/N-(b+gam)*i(j);
       sigma=sqrt(beta*i(j)*(N-i(j))/N+(b+gam)*i(j));
       rn=randn; % standard normal random number
```

```
        i(j+1)=i(j)+mu*dt+sigma*sqrt(dt)*rn;
        t(j+1)=t(j)+dt;
        j=j+1;
    end
    plot(t,i,'r-','Linewidth',2);
    hold on
end
% Euler's method applied to the deterministic SIS model.
y(1)=init;
for k=1:time/dt
    y(k+1)=y(k)+dt*(beta*(N-y(k))*y(k)/N-(b+gam)*y(k));
end
plot([0:dt:time],y,'k--','LineWidth',2);
axis([0,time,0,80]);
xlabel('Time');
ylabel('Number of Infectives');
hold off
```

References

1. Abbey, H.: An examination of the Reed–Frost theory of epidemics. Hum. Biol., **24**, 201–233 (1952)
2. Abramson, G., Kenkre, V. M.: Spatiotemporal patterns in hantavirus infection. Phys. Rev. E, **66**, 011912-1–5 (2002)
3. Abramson, G., Kenkre, V. M., Yates, T. L., Parmenter, R. R.: Traveling waves of infection in the hantavirus epidemics. Bull. Math. Biol., **65**, 519–534 (2003)
4. Ackerman, E., Elveback, L. R., Fox, J. P.: Simulation of Infectious Disease Epidemics. Charles C. Thomas, Springfield, IL (1984)
5. Allen, E. J.: Stochastic differential equations and persistence time for two interacting populations. Dyn. Contin. Discrete Impulsive Syst., **5**, 271–281 (1999)
6. Allen, L. J. S.: An Introduction to Stochastic Processes with Applications to Biology. Prentice Hall, Upper Saddle River, NJ (2003)
7. Allen, L. J. S., Allen, E. J.: A comparison of three different stochastic population models with regard to persistence time. Theor. Popul. Biol., **64**, 439–449 (2003)
8. Allen, L. J. S., Burgin, A. M.: Comparison of deterministic and stochastic SIS and SIR models in discrete time. Math. Biosci., **163**, 1–33 (2000)
9. Allen, L. J. S., Langlais, M., Phillips, C.: The dynamics of two viral infections in a singlehost population with applications to hantavirus. Math. Biosci., **186**, 191–217 (2003)
10. Anderson, R. M., May, R. M.: Infectious Diseases of Humans. Oxford University Press, Oxford (1992)
11. Antia, R., Regoes, R. R., Koella, J. C., Bergstrom, C. T.: The role of evolution in the emergence of infectious diseases. Nature, **426**, 658–661 (2003)
12. Arnold, L.: Stochastic Differential Equations: Theory and Applications. Wiley, New York (1974)
13. Bailey, N. T. J.: The Elements of Stochastic Processes with Applications to the Natural Sciences. Wiley, New York (1990)

14. Ball, F. G., Lyne, O. D.: Epidemics among a population of households. In: Castillo-Chavez, C., Blower, S., van den Driessche, P., Kirschner, D., Yakubu, A. -A. (eds.) Mathematical Approaches for Emerging and Reemerging Infectious Diseases: An Introduction. Springer, Berlin Heidelberg New York, pp. 115–142 (2002)
15. Brauer, F., Castillo-Chávez, C.: Mathematical Models in Population Biology and Epidemiology. Springer, Berlin Heidelberg New York (2001)
16. Brauer, F., van den Driessche, P.: Some directions for mathematical epidemiology. In: Ruan, S., Wolkowicz, G. S. K., Wu, J. (eds.) Dynamical Systems and Their Applications to Biology. Fields Institute Communications 36, AMS, Providence, RI, pp. 95–112 (2003)
17. Daley, D. J., Gani, J.: Epidemic Modelling: An Introduction. Cambridge Studies in Mathematical Biology, Vol. 15. Cambridge University Press, Cambridge (1999)
18. Darroch, J. N., Seneta, E.: On quasi-stationary distributions in absorbing continuous-time finite Markov chains. J. Appl. Probab., **4**, 192–196 (1967)
19. Diekmann, O., Heesterbeek, J. A. P.: Mathematical Epidemiology of Infectious Diseases: Model Building, Analysis and Interpretation. Wiley, New York (2000)
20. Foster, F. G.: A note on Bailey's and Whittle's treatment of a general stochastic epidemic. Biometrika, **42**, 123–125 (1955)
21. Gard, T. C.: Introduction to Stochastic Differential Equations. Marcel Dekker, New York (1988)
22. Goel, N. S., Richter-Dyn, N.: Stochastic Models in Biology. Academic, New York (1974)
23. Greenwood, M.: On the statistical measure of infectiousness. J. Hyg. Cambridge **31**, 336–351 (1931)
24. Harris, T. E.: The Theory of Branching Processes. Springer, Berlin Heidelberg New York (1963)
25. Hethcote, H. W.: Qualitative analyses of communicable disease models. Math. Biosci. **28**, 335–356 (1976)
26. Hethcote, H. W.: The mathematics of infectious diseases. SIAM Rev., **42**, 599–653 (2000)
27. Isham, V.: Assessing the variability of stochastic epidemics. Math. Biosci., **107**, 209–224 (1991)
28. Jacquez, J. A., Simon, C. P.: The stochastic SI epidemic model with recruitment and deaths I. Comparison with the closed SIS model. Math. Biosci., **117**, 77–125 (1993)
29. Jagers, P.: Branching Processes with Biological Applications. Wiley, London (1975)
30. Kimmel, M., Axelrod, D. E.: Branching Processes in Biology. Springer, Berlin Heidelberg New York (2002)
31. Kloeden, P. E., Platen, E.: Numerical Solution of Stochastic Differential Equations. Springer, Berlin Heidelberg New York (1992)
32. Kloeden, P. E., Platen, E., Schurz, H.: Numerical Solution of SDE Through Computer Experiments. Springer, Berlin Heidelberg New York (1997)
33. Leigh, E. G.: The average lifetime of a population in a varying environment. J. Theor. Biol., **90**, 213–219 (1981)
34. Lloyd, A. L.: Estimating variability in models for recurrent epidemics: assessing the use of moment closure techniques. Theor. Popul. Biol., **65**, 49–65 (2004)
35. Mode, C. J.: Multitype Branching Processes. Elsevier, New York (1971)
36. Mode, C. J., Sleeman, C. K.: Stochastic Processes in Epidemiology. HIV/AIDS, Other Infectious Diseases and Computers. World Scientific, Singapore (2000)
37. Murray, J. D., Stanley, E. A., Brown, D. L.: On the spatial spread of rabies among foxes. Proc. R. Soc. Lond. B, **229**, 111–150 (1986)
38. Nasell, I.: The quasi-stationary distribution of the closed endemic SIS model. Adv. Appl. Probab., **28**, 895–932 (1996)
39. Nasell, I.: On the quasi-stationary distribution of the stochastic logistic epidemic. Math. Biosci., **156**, 21–40 (1999)

40. Nasell, I.: Endemicity, persistence, and quasi-stationarity. In: Castillo-Chavez, C. with Blower, S., van den Driessche, P., Kirschner, D., Yakubu, A. -A. (eds.) Mathematical Approaches for Emerging and Reemerging Infectious Diseases an Introduction. Springer, Berlin Heidelberg New York, pp. 199–227 (2002)
41. Nisbet, R. M., Gurney, W. S. C.: Modelling Fluctuating Populations. Wiley, Chichester (1982)
42. Norden, R. H.: On the distribution of the time to extinction in the stochastic logistic population model. Adv. Appl. Probab., **14**, 687–708 (1982)
43. Ortega, J. M.: Matrix Theory a Second Course. Plenum, New York (1987)
44. Sauvage, F., Langlais, M., Yoccoz, N. G., Pontier, D.: Modelling hantavirus in fluctuating populations of bank voles: the role of indirect transmission on virus persistence. J. Anim. Ecol., **72**, 1–13 (2003)
45. Schinazi, R. B.: Classical and Spatial Stochastic Processes. Birkhäuser, Boston (1999)
46. Suppo, Ch., Naulin, J. M., Langlais, M., Artois, M.: A modelling approach to vaccination and contraception programmes for rabies control in fox populations. Proc. R. Soc. Lond. B, **267**, 1575–1582 (2000)
47. Taylor, H. M., Karlin, S.: An Introduction to Stochastic Modeling, 3rd edn. Academic, San Diego (1998)
48. Thieme, H. R.: Mathematics in Population Biology. Princeton University Press, Princeton (2003)

Part II
Advanced Modeling and Heterogeneities

Chapter 4
An Introduction to Networks in Epidemic Modeling

Fred Brauer

Abstract We use a stochastic branching process to describe the beginning of a disease outbreak. Unlike compartmental models, if the basic reproduction number is greater than one there may be a minor outbreak or a major epidemic with a probability depending on the nature of the contact network. We use a network approach to determine the distribution of outbreak and epidemic sizes.

4.1 Introduction

The Kermack–McKendrick compartmental epidemic model assumes that the sizes of the compartments are large enough that the mixing of members is homogeneous, or at least that there is homogeneous mixing in each subgroup if the population is stratified by activity levels. However, at the beginning of a disease outbreak, there is a very small number of infective individuals and the transmission of infection is a stochastic event depending on the pattern of contacts between members of the population; a description should take this pattern into account.

It has often been observed in epidemics that there is a small number of "superspreaders" who transmit infection to many other members of the population, while most infectives do not transmit infections at all or transmit infections to very few others [17]. This suggests that homogeneous mixing at the beginning of an epidemic may not be a good assumption. The SARS epidemic of 2002–2003 spread much more slowly than would have been expected on the basis of the data on disease spread at the start of the epidemic. Early

Department of Mathematics, University of British Columbia, 1984, Mathematics Road, Vancouver BC, Canada V6T 1Z2 brauer@math.ubc.ca

in the SARS epidemic of 2002–2003 it was estimated that $\mathcal{R}_0$ had a value between 2.2 and 3.6. At the beginning of an epidemic, the exponential rate of growth of the number of infectives is approximately $(\mathcal{R}_0 - 1)/\alpha$, where $1/\alpha$ is the generation time of the epidemic, estimated to be approximately 10 days for SARS . This would have predicted at least 30,000 cases of SARS in China during the first four months of the epidemic. In fact, there were fewer than 800 cases reported in this time. An explanation for this discrepancy is that the estimates were based on transmission data in hospitals and crowded apartment complexes. It was observed that there was intense activity in some locations and very little in others. This suggests that the actual reproduction number (averaged over the whole population) was much lower, perhaps in the range 1.2–1.6, and that heterogeneous mixing was a very important aspect of the epidemic.

4.2 The Probability of a Disease Outbreak

Our approach will be to give a stochastic branching process description of the beginning of a disease outbreak to be applied so long as the number of infectives remains small, distinguishing a (minor) disease outbreak confined to this stage from a (major) epidemic which occurs if the number of infectives begins to grow at an exponential rate . Once an epidemic has started we may switch to a deterministic compartmental model, arguing that in a major epidemic contacts would tend to be more homogeneously distributed. However, if we continue to follow the network model we would obtain a somewhat different estimate of the final size of the epidemic. Simulations suggest that the assumption of homogeneous mixing in a compartmental model may lead to a higher estimate of the final size of the epidemic than the prediction of the network model.

We describe the network of contacts between individuals by a graph with members of the population represented by vertices and with contacts between individuals represented by edges. The study of graphs originated with the abstract theory of Erdös and Rényi of the 1950s and 1960s [3–5], and has become important more recently in many areas, including social contacts and computer networks, as well as the spread of communicable diseases. We will think of networks as bi-directional, with disease transmission possible in either direction along an edge.

An edge is a contact between vertices that can transmit infection. The number of edges of a graph at a vertex is called the *degree* of the vertex. The degree distribution of a graph is $\{p_k\}$, where p_k is the fraction of vertices having degree k. The degree distribution is fundamental in the description of the spread of disease. Initially, we assume that all contacts between an infective and a susceptible transmit infection, but we will relax this assumption in Sect. 4.3.

We think of a small number of infectives in a population of susceptibles large enough that in the initial stage we may neglect the decrease in the size of the susceptible population. Our development begins along the lines of that of [7] and then develops along the lines of [6, 14, 16]. We assume that the infectives make contacts independently of one another and let p_k denote the probability that the number of contacts by a randomly chosen individual is exactly k, with $\sum_{k=0}^{\infty} p_k = 1$. In other words, $\{p_k\}$ is the degree distribution of the vertices of the graph corresponding to the population network.

For convenience, we define the *generating function*

$$G_0(z) = \sum_{k=0}^{\infty} p_k z^k.$$

Since $\sum_{k=0}^{\infty} p_k = 1$, this power series converges for $0 \leq z \leq 1$, and may be differentiated term by term. Thus

$$p_k = \frac{G_0^{(k)}(0)}{k!}, \quad k = 0, 1, 2, \cdots.$$

It is easy to verify that the generating function has the properties

$$G_0(0) = p_0, \quad G_0(1) = 1, \quad G_0'(z) > 0, \quad G_0''(z) > 0.$$

The mean degree, which we denote by $< k >$, is

$$< k >= \sum_{k=1}^{\infty} k p_k = G_0'(1).$$

More generally, we define the moments

$$< k^j >= \sum_{k=1}^{\infty} k^j p_k, \quad j = 1, 2, \cdots \infty.$$

When a disease is introduced into a network, we think of it as starting at a vertex (patient zero) who transmits infection to every individual to whom this individual is connected, that is, along every edge of the graph from the vertex corresponding to this individual. We assume that this initial vertex has been infected by a contact outside the population (component of the network) being studied. For transmission of disease after this initial contact we need to use the *excess degree* of a vertex. If we follow an edge to a vertex, the excess degree of this vertex is one less than the degree. We use the excess degree because infection can not be transmitted back along the edge whence it came. The probability of reaching a vertex of degree k, or excess degree $(k - 1)$, by following a random edge is proportional to k, and thus the probability that a vertex at the end of a random edge has excess degree $(k - 1)$ is a constant multiple of $k p_k$ with the constant chosen to make the sum over k

of the probabilities equal to 1. Then the probability that a vertex has excess degree $(k - 1)$ is

$$q_{k-1} = \frac{kp_k}{<k>}.$$

This leads to a generating function $G_1(z)$ for the excess degree

$$G_1(z) = \sum_{k=1}^{\infty} q_{k-1} z^{k-1} = \sum_{k=1}^{\infty} \frac{kp_k}{<k>} z^{k-1} = \frac{1}{<k>} G_0'(z),$$

and the mean excess degree, which we denote by $<k_e>$, is

$$<k_e> = \frac{1}{<k>} \sum_{k=1}^{\infty} k(k-1) p_k$$

$$= \frac{1}{<k>} \sum_{k=1}^{\infty} k^2 p_k - \frac{1}{<k>} \sum_{k=1}^{\infty} kp_k$$

$$= \frac{<k^2>}{<k>} - 1 = G_1'(1).$$

We let $\mathcal{R}_0 = G_1'(1)$, the mean excess degree. This is the mean number of secondary cases infected by patient zero and is the basic reproduction number as usually defined; the threshold for an epidemic is determined by $\mathcal{R}_0$.

Our next goal is to calculate the probability that the infection will die out and will not develop into a major epidemic. We begin by assuming that patient zero is a vertex of degree k of the network. Suppose patient zero transmits infection to a vertex of degree j. We let z_n denote the probability that this infection dies out within the next n generations. For the infection to die out in n generations each of these j secondary infections must die out in $(n - 1)$ generations. The probability of this is z_{n-1} for each secondary infection, and the probability that all secondary infections will die out in $(n - 1)$ generations is z_{n-1}^j. Now z_n is the sum over j of these probabilities, weighted by the probability q_j of j secondary infections. Thus

$$z_n = \sum_{j=0}^{\infty} q_j z_{n-1}^j = G_1(z_{n-1}).$$

Since $G_1(z)$ is an increasing function, the sequence z_n is an increasing sequence and has a limit z_∞, which is the probability that this infection will die out eventually. Then z_∞ is the limit as $n \to \infty$ of the solution of the difference equation

$$z_n = G_1(z_{n-1}), \quad z_0 = 0.$$

Thus z_∞ must be an equilibrium of this difference equation, that is, a solution of $z = G_1(z)$. Let w be the smallest positive solution of $z = G_1(z)$. Then, because $G_1(z)$ is an increasing function of z, $z \leq G_1(z) \leq G_1(w) = w$ for

$0 \leq z \leq w$. Since $z_0 = 0 < w$ and $z_{n-1} \leq w$ implies

$$z_n = G_1(z_{n-1}) \leq G_1(w) = w,$$

it follows by induction that

$$z_n \leq w, \; n = 0, 1, \cdots \infty.$$

From this we deduce that
$$z_\infty = w.$$

The equation $G_1(z) = z$ has a root $z = 1$ since $G_1(1) = 1$. Because the function $G_1(z) - z$ has a positive second derivative, its derivative $G_1'(z) - 1$ is increasing and can have at most one zero. This implies that the equation $G_1(z) = z$ has at most two roots in $0 \leq z \leq 1$. If $\mathcal{R}_0 < 1$ the function $G_1(z) - z$ has a negative first derivative

$$G_1'(z) - 1 \leq G_1'(1) - 1 = \mathcal{R}_1 - 1 < 0$$

and the equation $G_1(z) = z$ has only one root, namely $z = 1$. On the other hand, if $\mathcal{R}_0 > 1$ the function $G_1(z) - z$ is positive for $z = 0$ and negative near $z = 1$ since it is zero at $z = 1$ and its derivative is positive for $z < 1$ and z near 1. Thus in this case the equation $G_1(z) = z$ has a second root $z_\infty < 1$.

The probability that the disease outbreak will die out eventually is the sum over k of the probabilities that the initial infection in a vertex of degree k will die out, weighted by the degree distribution $\{p_k\}$ of the original infection, and this is

$$\sum_{k=0}^{\infty} p_k z_\infty^k = G_0(z_\infty).$$

To summarize this analysis, we see that if $\mathcal{R}_0 < 1$ the probability that the infection will die out is 1. On the other hand, if $\mathcal{R}_0 > 1$ there is a solution $z_\infty < 1$ of

$$G_1(z) = z$$

and there is a probability $1 - G_0(z_\infty) > 0$ that the infection will persist, and will lead to an epidemic. However, there is a positive probability $G_0(z_\infty)$ that the infection will increase initially but will produce only a minor outbreak and will die out before triggering a major epidemic. This distinction between a minor outbreak and a major epidemic, and the result that if $\mathcal{R}_0 > 1$ there may be only a minor outbreak and not a major epidemic are aspects of stochastic models not reflected in deterministic models.

4.3 Transmissibility

Contacts do not necessarily transmit infection. For each contact between individuals of whom one has been infected and the other is susceptible there is a probability that infection will actually be transmitted. This probability depends on such factors as the closeness of the contact, the infectivity of the member who has been infected, and the susceptibility of the susceptible member. We assume that there is a mean probability T, called the *transmissibility*, of transmission of infection. The transmissibility depends on the rate of contacts, the probability that a contact will transmit infection, the duration time of the infection, and the susceptibility. The development in Sect. 4.2 assumed that all contacts transmit infection, that is, that $T = 1$.

In this section, we will continue to assume that there is a network describing the contacts between members of the population whose degree distribution is given by the generating function $G_0(z)$, but we will assume in addition that there is a mean transmissibility T.

When disease begins in a network, it spreads to some of the vertices of the network. Edges that are infected during a disease outbreak are called *occupied*, and the size of the disease outbreak is the cluster of vertices connected to the initial vertex by a continuous chain of occupied edges.

The probability that exactly m infections are transmitted by an infective vertex of degree k is

$$\binom{k}{m} T^m (1 - T)^{k-m}.$$

We define $\Gamma_0(z, T)$ be the generating function for the distribution of the number of occupied edges attached to a randomly chosen vertex, which is the same as the distribution of the infections transmitted by a randomly chosen individual for any (fixed) transmissibility T. Then

$$\Gamma_0(z, T) = \sum_{m=0}^{\infty} \left[\sum_{k=m}^{\infty} p_k \binom{k}{m} T^m (1 - T)^{(k-m)} \right] z^m$$

$$= \sum_{k=0}^{\infty} p_k \left[\sum_{m=0}^{k} \binom{k}{m} (zT)^m (1 - T)^{(k-m)} \right] \qquad (4.1)$$

$$= \sum_{k=0}^{\infty} p_k [zT + (1 - T)]^k = G_0(1 + (z - 1)T).$$

In this calculation we have used the binomial theorem to see that

$$\sum_{m=0}^{k} \binom{k}{m} (zT)^m (1 - T)^{(k-m)} = [zT + (1 - T)]^k.$$

Note that

$$\Gamma_0(0,T) = G_0(1-T), \quad \Gamma_0(1,T) = G_0(1) = 1, \quad \Gamma_0'(z,T) = TG_0'(1+(z-1)T).$$

For secondary infections we need the generating function $\Gamma_1(z,T)$ for the distribution of occupied edges leaving a vertex reached by following a randomly chosen edge. This is obtained from the excess degree distribution in the same way,

$$\Gamma_1(z,T) = G_1(1+(z-1)T)$$

and

$$\Gamma_1(0,T) = G_1(1-T), \quad \Gamma_1(1,T) = G_1(1) = 1, \quad \Gamma_1'(z,T) = TG_1'(1+(z-1)T).$$

The basic reproduction number is now

$$\mathcal{R}_0 = \Gamma_1'(1,T) = TG_1'(1).$$

The calculation of the probability that the infection will die out and will not develop into a major epidemic follows the same lines as the argument in Sect. 4.2 for $T = 1$. The result is that if $\mathcal{R}_0 = TG_1'(1) < 1$ the probability that the infection will die out is 1. If $\mathcal{R}_0 > 1$ there is a solution $z_\infty(T) < 1$ of

$$\Gamma_1(z,T) = z,$$

and a probability $1-\Gamma_0(z_\infty(T),T) > 0$ that the infection will persist, and will lead to an epidemic. However, there is a positive probability $\Gamma_1(z_\infty(T),T)$ that the infection will increase initially but will produce only a minor outbreak and will die out before triggering a major epidemic.

Another interpretation of the basic reproduction number is that there is a *critical transmissibility* T_c defined by

$$T_c G_1'(1) = 1.$$

In other words, the critical transmissibility is the transmissibility that makes the basic reproduction number equal to 1. If the mean transmissibility can be decreased below the critical transmissibility, then an epidemic can be prevented.

The measures used to try to control an epidemic may include contact interventions, that is, measures affecting the network such as avoidance of public gatherings and rearrangement of the patterns of interaction between caregivers and patients in a hospital, and transmission interventions such as careful hand washing or face masks to decrease the probability that a contact will lead to disease transmission.

4.4 The Distribution of Disease Outbreak and Epidemic Sizes

We define $H_0(z,T)$ to be the generating function for the distribution of outbreak sizes corresponding to a randomly chosen vertex. In a corresponding way, we define $H_1(z,T)$ to be the generating function for the sizes of the clusters of connected vertices reached by following a randomly chosen edge.

For the generating function $H_1(z,T)$, it is easy to verify that $[H_1(z,T)]^2$ represents the distribution function for the sum of the infected cluster sizes for two vertices, and similarly for higher powers. If we begin on a randomly chosen edge, the probability that the vertex at the end of this edge has degree k is q_k, and each of the k vertices connected to it has a distribution of infected cluster sizes given by $H_1(z,T)$. Then

$$
\begin{aligned}
H_1(z,T) &= z \sum_{m=0}^{\infty} \left[\sum_{k=m}^{\infty} q_k \binom{k}{m} T^m (1-T)^{(k-m)} \right] z(H_1(z,T))^m \\
&= z \sum_{k=0}^{\infty} q_k \left[\sum_{m=0}^{k} \binom{k}{m} (TH_1(z,T))^m (1-T)^{(k-m)} \right] \\
&= z \sum_{k=0}^{\infty} q_k [TH_1(z,T) + (1-T)]^k = zG_1(1 + (H_1(z,T) - 1)T).
\end{aligned}
$$

Thus

$$
H_1(z,T) = z\Gamma_1(H_1(z,T),T). \tag{4.2}
$$

Similarly, the size of the cluster reachable from a randomly chosen vertex is distributed according to

$$
H_0(z,T) = z\Gamma_0(H_1(z,T),T). \tag{4.3}
$$

The mean size of the disease outbreak is $H_0'(1,T)$. We calculate this by implicit differentiation of (4.2) after using implicit differentiation of (4.3) to calculate $H_1'(z,t)$.

Implicit differentiation of (4.2) gives

$$
\begin{aligned}
H_1'(z,T) &= \Gamma_1(H_1(z,T),T) + z\Gamma_1'(H_1(z,T),T)H_1'(z,T) \\
&= \frac{\Gamma_1(H_1(z,T),T)}{1 - z\Gamma_1'(H_1(z,T),T)} \\
H_1'(1,T) &= \frac{\Gamma_1(H_1(1,T),T)}{1 - \Gamma_1'(H_1(1,T),T)}.
\end{aligned} \tag{4.4}
$$

Then implicit differentiation of (4.3) using (4.4) gives

$$H_0'(z,T) = \Gamma_0(H_1(z,T),T) + z\Gamma_0'(H_1(z,T),T)H_1'(z,T) \tag{4.5}$$
$$= \Gamma_0(H_1(z,T),T) + z\Gamma_0'(H_1(z,T),T)\frac{\Gamma_1(H_1(z,T),T)}{1 - z\Gamma_1'(H_1(z,T),T)}.$$

Because
$$H_1(1,T) = 1, \quad \Gamma_1(H_1(1,T),T) = \Gamma_1(1,T) = 1,$$

this reduces to

$$H_0'(1,T) = 1 + \frac{\Gamma_0'(1,T)}{1 - \Gamma_1'(1,T)} = 1 + \frac{TG_0'(1)}{1 - TG_1'(1)} = 1 + \frac{TG_0'(1)}{1 - \mathcal{R}_0}.$$

This expression for the mean outbreak size is valid if $\mathcal{R}_0 = TG_1'(1) < 1$.

There is a phase transition at $\mathcal{R}_0 = 1$. A "giant" component of the graph appears, and there is a major epidemic. If $\mathcal{R}_0 \geq 1$, we exclude the "giant" component of the graph from the definition of $H_1(z,T)$ and then $H_1(1,T) < 1$. Because of (4.2) we must have

$$H_1(1,T) = \Gamma_1(H_1(1,T))$$

and therefore $H_1(1,T)$ must be the second root $z_\infty(T)$ of

$$\Gamma_1(z,T) = z$$

as found in Sect. 4.3. In this case, $\Gamma_0(z_\infty(T))$ is the probability that there will be only a small disease outbreak and $1 - \Gamma_0(z_\infty(T))$ is the probability that there will be an epidemic.

If $\mathcal{R}_0 < 1$, $H_1(1,T) = 1$, $z_\infty(T) = 1$, and the probability of an epidemic is 0. If there is an epidemic, we define $S(T)$ to be the fraction of the graph affected by the infection, the epidemic size. Above the epidemic threshold,

$$H_0(1,T) = 1 - S(T),$$

and

$$S(T) = 1 - H_0(1,T) = 1 - \Gamma_0(H_1(1,T),T) = 1 - \Gamma_0(z_\infty(T),T),$$

where $z_\infty(T) = \Gamma_1(z_\infty(T),T) = H_1(1,T)$. Thus the size of the epidemic, if an epidemic occurs, is equal to the probability of an epidemic.

Compartmental models assume homogeneous mixing, corresponding to a Poisson network. As we shall see in the next section, for a Poisson network,

$$\Gamma_0(z,T) = \Gamma_1(z,T) = e^{\mathcal{R}_0(z-1)}.$$

Then the equation $\Gamma_1(z,T) = z$ is

$$e^{\mathcal{R}_0(z-1)} = z,$$

and the size of the epidemic is $1 - z_\infty(T)$. This is equivalent to the final size relation for a deterministic compartmental model [7, Sect. 1.3].

More sophisticated network analysis makes it possible to predict such quantities as the probability that an individual will set off an epidemic, the risk for an individual of becoming infected, the probability that a cluster of infections will set off a small disease outbreak when the transmissibility is less than the critical transmissibility, and how the probability of an epidemic depends on the degree of patient zero, the initial disease case [12, 14].

4.5 Some Examples of Contact Networks

The above analysis assumes that there is a known generating function $G_0(z)$ or, equivalently, a degree distribution $\{p_k\}$. In studying a disease outbreak, we need to know the degree distribution of the network. If we know the degree distribution we can calculate the basic reproduction number and also the probability of an epidemic. What kinds of networks are observed in practice in social interactions? There are some standard examples.

If contacts between members of the population are random, corresponding to the assumption of mass action in the transmission of disease, then the probabilities p_k are given by the *Poisson distribution*

$$p_k = \frac{e^{-c} c^k}{k!}$$

for some constant c. To show this, we think of a probability of contact $c\Delta t$ in a time interval Δt, and we let

$$n = \frac{1}{\Delta t}.$$

Then the probability of k contacts in a time interval Δt is

$$\binom{n}{k} (\frac{c}{n})^k (1 - \frac{c}{n})^{n-k},$$

where

$$\binom{n}{k} = \frac{n!}{k!(n-k)!}$$

is the *binomial coefficient*. We rewrite this probability as

$$\frac{n(n-1)(n-2)\cdots(n-k+1)}{n^k} \frac{c^k}{k!} \frac{(1 - \frac{c}{n})^n}{(1 - \frac{c}{n})^k}.$$

We let $\Delta t \to 0$, or $n \to \infty$. Since

$$\frac{n(n-1)(n-2)\cdots(n-k+1)}{n^k} \to 1, \quad (1-\frac{c}{n})^k \to 1,$$

and

$$(1-\frac{c}{n})^n \to e^{-c},$$

the limiting probability that there are k contacts is

$$p_k = \frac{e^{-c}c^k}{k!}.$$

Then the generating function is

$$G_0(z) = e^{-c}\sum_{k=0}^{\infty}\frac{c^k}{k!}z^k = e^{-c}e^{cz} = e^{c(z-1)},$$

and

$$G_0'(z) = ce^{c(z-1)}, \quad G_0'(1) = c.$$

The generating function for the Poisson distribution is $e^{c(z-1)}$. Then $G_1(z) = G_0(z)$, and $\mathcal{R}_0 = TG_1'(1) = cT$, so that

$$\Gamma_1(z,T) = G_1(1+(z-1)T) = e^{\mathcal{R}_0(z-1)}.$$

The commonly observed situation that most infectives do not pass on infection but there are a few "superspreading events" [17] corresponds to a probability distribution that is quite different from a Poisson distribution, and could give a quite different probability that an epidemic will occur. For example, taking $T = 1$ for simplicity, if $\mathcal{R}_0 = 2.5$ the assumption of a Poisson distribution gives $z_\infty = 0.107$ and $G_0(z_\infty) = 0.107$, so that the probability of an epidemic is 0.893. The assumption that nine out of ten infectives do not transmit infection while the tenth transmits 25 infections gives

$$G_0(z) = (z^{25} + 9)/10, \quad G_1(z) = z^{24}, \quad z_\infty = 0, \quad G_0(z_\infty) = 0.9,$$

from which we see that the probability of an epidemic is 0.1. Another example, possibly more realistic, is to assume that a fraction $(1-p)$ of the population follows a Poisson distribution with constant r while the remaining fraction p consists of superspreaders each of whom makes L contacts. This would give a generating function

$$G_0(z) = (1-p)e^{r(z-1)} + pz^L$$
$$G_1(z) = \frac{r(1-p)e^{r(z-1)} + pLz^{L-1}}{r(1-p)+pL},$$

and

$$\mathcal{R}_0 = \frac{r^2(1-p)+pL(L-1)}{r(1-p)+pL}.$$

For example, if $r = 2.2, L = 10, p = 0.01$, numerical simulation gives

$$\mathcal{R}_0 = 2.5, \quad z_\infty = 0.146,$$

so that the probability of an epidemic is 0.849.

These examples demonstrate that the probability of a major epidemic depends strongly on the nature of the contact network. Simulations suggest that for a given value of the basic reproduction number the Poisson distribution is the one with the maximum probability of a major epidemic.

It has been observed that in many situations there is a small number of long range connections in the graph, allowing rapid spread of infection. There is a high degree of clustering (some vertices with many edges) and there are short path lengths. Such a situation may arise if a disease is spread to a distant location by an air traveller. This type of network is called a *small world* network. Long range connections in a network can increase the likelihood of an epidemic dramatically.

A third kind of network frequently observed is a *scale free* network. In a random network, the quantity p_k approaches zero very rapidly (exponentially) as $k \to \infty$. A scale free network has a "fatter tail", with p_k approaching zero as $k \to \infty$ but more slowly than in a random network. In an epidemic setting it corresponds to a situation in which there is an active core group but there are also "superspreaders" making many contacts. In a scale free network, p_k is proportional to $k^{-\alpha}$ with α a constant. In practice, α is usually between 2 and 3. Often an exponential cutoff is introduced in applications of scale free networks in order to make $G_0'(1) < \infty$ for every choice of α, so that

$$p_k = Ck^{-\alpha}e^{-k/\theta}.$$

The constant C, chosen so that $\sum_{k=0}^{\infty} p_k = 1$, can be expressed in terms of logarithmic integrals.

These examples indicate that the probability of an epidemic depends strongly on the contact network at the beginning of a disease outbreak. The study of complex networks is a field which is developing very rapidly. Some basic references are [15, 18], and other references to particular kinds of networks include [1, 2, 13, 19]. Examination of the contact network in a disease outbreak situation may lead to an estimate of the probability distribution for the number of contacts [11, 12], and thus to a prediction of the course of the disease outbreak.

A recent development in the study of networks in epidemic modeling is the construction of very detailed networks by observation of particular locations. The data that goes into such a network includes household sizes, age distributions, travel to schools, workplaces, and other public locations. The networks constructed are very complex but may offer a great deal of realism. However, it is very difficult to estimate how sensitive the predictions obtained from a model using such a complex network will be to small changes in the network. Nevertheless, simulations based on complicated networks are the primary models currently being used for developing strategies to cope with a potential influenza pandemic. This approach has been followed in [8–10].

An alternative to simulations based on a very detailed network would be to analyze the behaviour of a model based on a simpler network, such as a random network or a scale-free network with parameters chosen to match the reproduction number corresponding to the detailed network. A truncated scale free network would have superspreaders and thus may be closer than a random network to what is often observed in actual epidemics.

4.6 Conclusions

We have described the beginning of a disease outbreak in terms of the degree distribution of a branching process, and have related this to a contact network. There is a developing theory of network epidemic models which is not confined to the early stages [12,14]. This involves more complicated considerations, such as the way in which a contact network may change over the course of an epidemic. We have restricted our attention to the beginning of an epidemic in order not to have to examine these complications. There are many aspects of network models for epidemics that have not yet been studied.

While we have suggested using a deterministic compartmental model once an epidemic is underway, it may be reasonable to go beyond the simplest Kermack–McKendrick epidemic model. Heterogeneity of contact rates, age structure, and other aspects of an actual epidemic can be modeled. Ideally, for the initial stages of an epidemic we would like to use a network somewhere between the over-simplification of a random network and the extreme complication of an individual-based model.

References

1. R. Albert & A.-L. Barabási: Statistical mechanics of complex networks, Rev. Mod. Phys. **74**, 47–97 (2002).
2. A.-L. Barabási & R. Albert: Emergence of scaling in random networks, Science **286**, 509–512 (1999).
3. P. Erdös & A. Rényi: On random graphs, Publ. Math. **6**, 290–297 (1959).
4. P. Erdös & A. Rényi: On the evolution of random graphs, Pub. Math. Inst. Hung. Acad. Sci. **5**, 17–61 (1960).
5. P. Erdös & A. Rényi: On the strengths of connectedness of a random graph, Acta Math. Sci. Hung. **12**, 261–267 (1961).
6. D.S. Callaway, M.E.J. Newman, S.H. Strogatz & D.J. Watts: Network robustness and fragility: Percolation on random graphs, Phys. Rev. Lett. **85**, 5468–5471 (2000).
7. O. Diekmann & J.A.P. Heesterbeek: Mathematical Epidemiology of Infectious Diseases, Wiley, Chichester (2000).

8. N.M. Ferguson, D.A.T. Cummings, S. Cauchemez, C. Fraser, S. Riley, A. Meeyai, S. Iamsirithaworn & D.S. Burke: Strategies for containing an emerging influenza pandemic in Southeast Asia, Nature **437**, 209–214 (2005).

9. I.M. Longini, M.E. Halloran, A. Nizam & Y. Yang: Containing pandemic influenza with antiviral agents, Am. J. Epidem. **159**, 623–633 (2004).

10. I.M. Longini, A. Nizam, S. Xu, K. Ungchusak, W. Hanshaoworakul, D.A.T. Cummings & M.E. Halloran: Containing pandemic influenza at the source, Science **309**, 1083–1087 (2005).

11. M.J. Keeling & K.T.D. Eames: Networks and epidemic models, J. R. Soc. Interface **2**, 295–307 (2006).

12. L.A. Meyers, B. Pourbohloul, M.E.J. Newman, D.M. Skowronski & R.C. Brunham: Network theory and SARS: Predicting outbreak diversity. J. Theor. Biol. **232**, 71–81 (2005).

13. A.L. Lloyd & R.M. May: Epidemiology: How viruses spread among computers and people, Science **292**, 1316–1317 (2001).

14. M.E.J. Newman: The spread of epidemic disease on networks, Phys. Rev. E **66**, 016128 (2002).

15. M.E.J. Newman: The structure and function of complex networks. SIAM Rev. **45**, 167–256 (2003).

16. M.E.J. Newman, S.H. Strogatz & D.J. Watts: Random graphs with arbitrary degree distributions and their applications, Phys. Rev. E **64**, 026118 (2001).

17. S. Riley, C. Fraser, C.A. Donnelly, A.C. Ghani, L.J. Abu-Raddad, A.J. Hedley, G.M. Leung, L.-M. Ho, T.-H. Lam, T.Q. Thach, P. Chau, K.-P. Chan, S.-V. Lo, P.-Y. Leung, T. Tsang, W. Ho, K.-H. Lee, E.M.C. Lau, N.M. Ferguson & R.M. Anderson: Transmission dynamics of the etiological agent of SARS in Hong Kong: Impact of public health interventions, Science **300**, 1961–1966 (2003).

18. S.H. Strogatz: Exploring complex networks, Nature **410**, 268–276 (2001).

19. D.J. Watts & S.H. Strogatz: Collective dynamics of 'small world' networks, Nature **393**, 440–442 (1998).

Chapter 5
Deterministic Compartmental Models: Extensions of Basic Models

P. van den Driessche

Abstract The basic compartmental models for disease transmission are extended to include three separate biological features. The first such feature is vertical transmission of disease, for which two ordinary differential equation models (SIR and SEIR) are formulated and analyzed. In particular, vertical transmission is shown to increase the basic reproduction number. Immigration of infective individuals is considered as a second feature, and the resulting model has a unique endemic equilibrium (with no disease-free state). An illustration is provided that includes screening and isolating infectives to reduce the spread of disease. A constant period of temporary immunity is introduced in an SIRS model as the third feature. This results in an integro-differential equation for the fraction of infectives. Analysis shows that, for some parameter values, Hopf bifurcation can give rise to periodic solutions.

5.1 Introduction

Basic deterministic compartmental models are introduced and discussed in chapters by Allen [1] and Brauer [2]; the latter also describes models with demographic effects and models with infectivity depending on the age of infection. For some diseases and situations, it is desirable to include other biological features, and to investigate whether these can qualitatively change the model results.

In this chapter three such features are considered separately, namely vertical transmission, immigration of infectives, and temporary immunity upon recovery (which is introduced briefly in Sect. 4.5 of [2]). Models from the

Department of Mathematics and Statistics, University of Victoria, P.O. BOX 3060 STN CSC, Victoria, BC, Canada V8W 3R4
pvdd@math.uvic.ca

literature are summarized: readers are encouraged to consult the original papers for more details and references. These are but a few examples of how the basic models can be extended to better represent certain diseases. More examples are given in subsequent chapters.

5.2 Vertical Transmission

5.2.1 Kermack–McKendrick SIR Model

This section is based on a model in the authoritative book on vertically transmitted diseases by Busenberg and Cooke [4, Chap. 2 especially Sect. 2.8]. In the models of the chapter by Brauer [2], disease is transmitted horizontally between infective and susceptible individuals. By contrast, vertical disease transmission is the direct transfer of a disease from an infective parent to an unborn or newly born offspring. The latter can occur, for example, from breast-feeding. Chagas' disease, hepatitis B and HIV/AIDS are examples of diseases that can be transmitted vertically [4,8]. The SIR model considered here is based on the special simple case of that proposed by Kermack and McKendrick in 1932, but includes input of infectives due to vertical transmission from infective parents.

The model is formulated with the assumptions as in the chapter by Brauer [2] with the following extensions. A fraction q of offspring of infective individuals are assumed infected at birth; thus a fraction $p = 1 - q$ of such offspring are susceptible. The birth and death rate constant for the susceptible and recovered compartments is $b > 0$, whereas $\tilde{b} > 0$ is the birth and death rate constant for the infective compartment. The disease is assumed to be non-fatal, thus the total population size $K = S + I + R$ remains constant, where S, I, R denotes the number in the susceptible, infective, recovered compartment, respectively. Such a model has the form

$$
\begin{aligned}
S' &= -\beta SI + p\tilde{b}I + b(S + R) - bS \\
I' &= \beta SI + q\tilde{b}I - \tilde{b}I - \gamma I \\
R' &= \gamma I - bR.
\end{aligned}
\tag{5.1}
$$

Recall that mass action incidence is assumed, with each individual making $\beta K > 0$ contacts sufficient to transmit infection per unit time, and that γ is the recovery rate constant for infectives.

The variable R can be eliminated from system (5.1), which reduces to the 2-dimensional system

$$S' = -\beta SI + p\tilde{b}I + b(K - S - I)$$
$$I' = \beta SI - p\tilde{b}I - \gamma I \tag{5.2}$$

for which $\{(S, I) \in R_2^+ : S + I \leq K\}$ is invariant. For $p = 1$ (no vertical transmission) and $\tilde{b} = b$, system (5.2) reduces to a model given by Brauer [2, Sect. 2.1].

To begin analysis of (5.2), first consider the two equilibria. These are the disease-free equilibrium $(S, I) = (K, 0)$ and the endemic equilibrium (S_∞, I_∞) with

$$S_\infty = \frac{p\tilde{b} + \gamma}{\beta}, \quad I_\infty = \frac{b(K\beta - p\tilde{b} - \gamma)}{(\gamma + b)\beta} \tag{5.3}$$

provided that $K > (p\tilde{b} + \gamma)/\beta$. This condition gives a lower bound on the population size needed to sustain the disease. The basic reproduction number $\mathcal{R}_0$ can be easily found from the I' equation as

$$\mathcal{R}_0 = \frac{\beta K + q\tilde{b}}{\gamma + \tilde{b}} \tag{5.4}$$

which satisfies $\mathcal{R}_0 > 1$ if and only if $K > (p\tilde{b} + \gamma)/\beta$, that is the endemic equilibrium exists. The expression for $\mathcal{R}_0$ given in (5.4) comes from accounting for all new infections (due to horizontal and vertical transmission) and multiplying by the average infective period, namely $1/(\gamma + \tilde{b})$. The vertical transmission has the effect of increasing $\mathcal{R}_0$ by a factor of $q\tilde{b}/(\gamma + \tilde{b})$.

It is easy to show that if $\mathcal{R}_0 < 1$, then the disease-free equilibrium is globally asymptotically stable, and the disease dies out. If $\mathcal{R}_0 > 1$, then the disease-free equilibrium is unstable and the endemic equilibrium exists. In this case a special method using a Lyapunov function due to Beretta and Capasso, see [5, page 11] or [4, Theorem 2.8], can be used to show that the endemic equilibrium is globally asymptotically stable, and so the disease remains in the population. Thus $\mathcal{R}_0 = 1$ gives a sharp disease threshold.

Here an alternative method is presented to show that (S_∞, I_∞) attracts all solutions with initial values $(S(0), I(0))$ in $\{(S, I) : S \geq 0, I > 0\}$ if $\mathcal{R}_0 > 1$. From (5.2)

$$\frac{\partial}{\partial S}\left(\frac{S'}{I}\right) + \frac{\partial}{\partial I}\left(\frac{I'}{I}\right) = -\beta - \frac{b}{I} < 0.$$

Thus by the Bendixon–Dulac criterion (see e.g., [4, page 72]) there are no periodic orbits with $I > 0$. An application of the Poincaré- Bendixon Theorem (see e.g., [4, page 72]) completes the proof.

5.2.2 SEIR Model

Consider now a more general model that includes vertical transmission and also contains an exposed (latent) compartment. Infected individuals are exposed before becoming infective, and the length of this exposed period depends on the disease. This gives an SEIR model, in which E denotes the exposed compartment.

Li, Smith and Wang [15] formulate such a model that includes a fraction of the offsprings of infected hosts (both exposed and infective) that are infected at birth and so enter the exposed compartment, giving vertical transmission of the disease. They state that their model is appropriate for rubella and the SEI limit (with no recovery) is appropriate for Chagas' disease. Since the total population is constant in their model, they work with *fractions* in each compartment; thus S, E, I, R denote the fraction in the susceptible, exposed, infective, recovered compartment, respectively, with $S + E + I + R = 1$. The natural birth and death rate constant is denoted by b, exposed individuals become infective with rate constant ϵ, and infective individuals recover with rate constant γ. Horizontal incidence is assumed to be of the bilinear mass action form βSI. A fraction $p \in [0, 1]$ of the offspring from exposed individuals and a fraction $q \in [0, 1]$ of the offspring from infective individuals are born into the exposed compartment. Thus vertical transmission gives a term $pbE + qbI$ entering the exposed compartment and a similar reduction in the birth of susceptibles. The model is given by the following system [15]

$$
\begin{aligned}
S' &= b - \beta SI - pbE - qbI - bS \\
E' &= \beta SI + pbE + qbI - (\epsilon + b)E \\
I' &= \epsilon E - (\gamma + b)I \\
R' &= \gamma I - bR.
\end{aligned}
\tag{5.5}
$$

Note that if $p = q = 0$, then the system reduces to the classical SEIR model with mass action.

Let

$$
\Omega : \left\{ (S, E, I, R) \in R_+^4 : S + E + I + R = 1 \right\}.
$$

Any solution starting in Ω does not leave R_+^4 by crossing one of its faces. Since also $(S + E + I + R)' = 0$, the solution remains in Ω for all $t \geq 0$. Thus Ω is a positively invariant set that is biologically feasible. Using the relation $R = 1 - S - E - I$, (5.5) can be reduced to the equivalent 3-dimensional system, given by the first three equations in (5.5) on the closed invariant set

$$
\Gamma : \left\{ (S, E, I) \in R_+^3 : S + E + I \leq 1 \right\}.
$$

The 3-dimensional system has the disease-free equilibrium $(S, I, R) = (1, 0, 0)$ and an endemic equilibrium $(S_\infty, I_\infty, R_\infty)$ with

$$S_\infty = ((\epsilon + b)(\gamma + b) - pb(\gamma + b) - qb\epsilon)/\beta\epsilon$$

provided this is less than one. In this case $I_\infty = \epsilon b(1 - S^\infty)/((\epsilon + b)(\gamma + b))$ and $E_\infty = (\gamma + b)I_\infty/\epsilon$.

The authors define a basic reproduction number [15, equation(2.3)]

$$\mathcal{R}_0(p, q) = \frac{\beta\epsilon}{(\epsilon + b)(\gamma + b) - pb(\gamma + b) - qb\epsilon} \tag{5.6}$$

and show that if $\mathcal{R}_0(p, q) \leq 1$, then the disease-free equilibrium is globally stable in Γ; whereas if $\mathcal{R}_0(p, q) > 1$, then the unique endemic equilibrium is globally stable in the interior of Γ. Thus $\mathcal{R}_0(p, q)$ is a sharp threshold, determining whether the disease dies out or persists at an endemic level. The interesting and highly nontrivial proof in [15] for $\mathcal{R}_0(p, q) > 1$ employs a geometric approach introduced by Li and Muldowney [14]. The authors [15] state that a key step in their proof is the construction of a suitable Lyapunov function for the second additive compound of the Jacobian matrix of the system. Interested readers please consult [15, Sect. 4] for details of the proof.

The contribution to $\mathcal{R}_0(p, q)$ in (5.6) of vertical transmission is given in [15, Sect. 5]. However, we present here an alternate basic reproduction number, denoted by $\mathcal{R}_0$, that is derived from the next generation matrix method [6] as elaborated in [16]. From the 3-dimensional system taking E and I as infecteds, and horizontal and vertical transmission as giving new infecteds, the matrices F and V defined in [16] (which should be consulted for more details) are

$$F = \begin{bmatrix} pb & qb + \beta \\ 0 & 0 \end{bmatrix}, \quad V = \begin{bmatrix} \epsilon + b & 0 \\ -\epsilon & \gamma + b \end{bmatrix}.$$

Then $\mathcal{R}_0 = \rho(FV^{-1})$, where ρ denotes the spectral radius, is easily found (since F has rank 1) as

$$\mathcal{R}_0 = \frac{\beta\epsilon + pb(\gamma + b) + qb\epsilon}{(\epsilon + b)(\gamma + b)}. \tag{5.7}$$

Here the first term in the numerator comes from horizontal transmission and the second and third terms come from vertical transmission, which increases the value of $\mathcal{R}_0$. The term $pb/(\epsilon + b)$ accounts for vertical transmission from exposed individuals with $1/(\epsilon + b)$ the average exposed period. The term $qb\epsilon/(\epsilon + b)(\gamma + b)$ accounts for vertical transmission from infected individuals with $\epsilon/(\epsilon + b)$ giving the fraction surviving the exposed compartment, and $1/(\gamma + b)$ the average time in the infective compartment. Comparing (5.6) and (5.7) it follows that $\mathcal{R}_0(p, q) = 1$ precisely where $\mathcal{R}_0 = 1$. Thus the sharp threshold is given by either number, but the biological interpretation and the dependence on the model parameters is better given by $\mathcal{R}_0$ in (5.7).

5.3 Immigration of Infectives

In the previous section, newborn infecteds enter the infected population. Consider now a communicable disease introduced into a population by infectives immigrating from outside. Given such a situation, a model can be formulated to describe the dynamical spread of disease and to suggest possible control strategies. The SIS models considered here are related to the basic models described in the chapter by Brauer [2, Sect. 2.2] and are taken from [3]. Consider a constant flow A into the population per unit time with a fraction $p \in (0, 1]$ infective. The per capita natural death rate constant is denoted by $d > 0$. Letting S, I denote the number of susceptible, infective individuals, respectively, the total population $N = S + I$ varies with time. Taking mass action incidence and denoting the recovery rate constant and the disease death rate constant by γ and α, respectively, the model equations are

$$S' = (1 - p)A - \beta SI - dS + \gamma I$$
$$I' = pA + \beta SI - (d + \gamma + \alpha)I \qquad (5.8)$$
$$N' = A - dN - \alpha I.$$

For nonnegative initial values, the model is well posed with $N \leq A/d$. From the second equation, it follows that with immigration of infectives there is no disease-free equilibrium. Working in I, N variables, and eliminating N, at an endemic equilibrium

$$G(I) = \beta(d + \alpha)I^2 - \sigma I - pdA = 0$$

where $\sigma = \beta A - d(d + \gamma + \alpha)$. Thus there is a unique equilibrium given by

$$I_\infty = \frac{\sigma + \sqrt{\sigma^2 + 4\beta A dp(d + \alpha)}}{2\beta(d + \alpha)}, \qquad N_\infty = \frac{A - \alpha I_\infty}{d}. \qquad (5.9)$$

This model can be generalized by replacing mass action incidence by the assumption that each individual makes $\beta(N)N$ contacts sufficient to transmit infection per unit time; see [2]. It is biologically reasonable to assume that $\beta(N)N$ is a nondecreasing function of N and $\beta(N)$ is a nonincreasing function of N. These assumptions are satisfied by mass action incidence ($\beta(N) = \beta$), standard incidence ($\beta(N) = \lambda/N$) and saturating incidence ($\beta(N) = a/(1 + bN)$). The model equations are now as in system (5.4) with β replaced by $\beta(N)$. It is more convenient to write the equilibrium equation in terms of N, namely

$$[(d + \alpha)N - A]\,\beta(N) = -\frac{\alpha^2 pA}{A - dN} + \alpha(d + \gamma + \alpha).$$

The left side of this equation is zero at $N_1 = A/(d + \alpha)$ and is increasing, whereas the right side is positive at N_1, decreases and is zero at

$$N = N_0 = \frac{A(d + \gamma + \alpha(1 - p))}{d(d + \gamma + \alpha)}.$$

Thus there is a unique endemic equilibrium given by $N_\infty \in [N_1, N_0]$, and $I_\infty > I_0 = pA/(d + \gamma + \alpha)$.

To investigate the stability, linearize system (5.8) about this equilibrium to give the Jacobian matrix

$$\begin{bmatrix} -\frac{pA}{I} - I\beta(N) & (N\beta(N))'I - \beta'(N)I^2 \\ -\alpha & -d \end{bmatrix}$$

at (I_∞, N_∞). By considering the signs of each entry (noting that the $(1,2)$ entry is nonnegative), this matrix is sign stable; see, e.g., [12]. Thus for any values of the parameters, the endemic equilibrium is locally asymptotically stable. A Bendixson–Dulac calculation as in Sect. 2.1 shows that there are no period orbits with $I > 0$. The Poincaré–Bendixson theorem then completes the proof that the endemic equilibrium is globally asymptotically stable, so solutions of the SIS model with immigration of infectives converge to the endemic equilibrium (I_∞, N_∞). For mass action incidence, this equilibrium is given explicitly by (5.9).

If $\mathcal{R}_0$ is defined in the usual way with mass action incidence as

$$\mathcal{R}_0 = \frac{\beta A}{(d + \gamma + \alpha)d}$$

then for p close to 0

$$I_\infty \approx \begin{cases} \frac{Ad}{|\sigma|}p & \text{if } \mathcal{R}_0 < 1 \ (\sigma < 0) \\ \frac{Ad}{\sigma}p + \frac{\sigma}{\beta(d+\alpha)} & \text{if } \mathcal{R}_0 > 1 \ (\sigma > 0). \end{cases}$$

The limiting infective population is a smooth function of $\mathcal{R}_0$, with the threshold $\mathcal{R}_0 = 1$ not as sharp as in the classical case (except in the limit as $p \to 0^+$).

Gani et al. [7, Sect. 2] formulated models for the spread of HIV in a constant population prison, and considered a program of screening with quarantining of prisoners found to be HIV positive. Note that quarantining is used here to mean the isolation of infective individuals. A continuous analog of their SI model is formulated in [3, Sect. 5]. This simple model indicates that such a program can reduce the infective population size, but a more detailed model including more realistic assumptions and data on HIV is needed to give quantitative predictions.

Consider model (5.8) in the limit with $\gamma = 0$ and $\alpha > 0$ (since HIV is a fatal disease). The demographics now refer to incarceration (at rate A with a fraction p infective) and release of prisoners (with rate constant d). Taking one month as the time unit, $A = 25$, $d = 1/24$, $p = 0.1$ giving a prison population carrying capacity as 600. The mean infective period is assumed to be ten years (so $\alpha = 1/120$). With $\beta = 1/3000$ (1 contact every 5 months/infective), $I_\infty = 391$ and $N_\infty = 521$. Thus 75% are infective at equilibrium, which (by numerical simulation) is reached in 2.5 years.

For constant τ with $0 \le \tau \le 1$, $\tau(S+I)$ prisoners are screened in unit time and the τI found to be infective are moved to a quarantine compartment. The number in this compartment is denoted by $Q(t)$. This gives the model [3, Sect. 5]

$$
\begin{aligned}
S' &= (1-p)A - \beta SI - dS \\
I' &= pA + \beta SI - (d + \alpha + \tau)I \\
Q' &= \tau I - (d + \alpha)Q.
\end{aligned}
\tag{5.10}
$$

The total population $N = S + I + Q$ thus satisfies

$$
N' = A - dN - \alpha I - \alpha Q.
$$

Analysis of this model is similar to that of (5.8) and shows that there is a unique endemic equilibrium $(S_\infty, I_\infty, Q_\infty)$ with

$$
I_\infty = \frac{\sigma \sqrt{\sigma^2 + 4\beta Adp(d + \alpha + \tau)}}{2\beta(d + \alpha + \tau)}
$$

with $\sigma = \beta A - d(d + \alpha + \tau)$, and that this equilibrium is locally asymptotically stable. For global stability, note that the first two equations of (5.10) do not contain Q, and so by analogy with (5.8) it follows that $(S, I) \to (S_\infty, I_\infty)$ as $t \to \infty$. The third equation then shows that $Q \to Q_\infty$.

Taking parameters as for the prison population above, if $\tau = 0.1$ (so 42 prisoners screened per month), then $I_\infty = 71$; whereas if $\tau = 0.2$ (so 95 prisoners screened per month), then $I_\infty = 25$. It takes about two years to reach these equilibria. Thus, from this model, a considerable reduction of infectives occurs with screening and quarantining of infectives.

5.4 General Temporary Immunity

For diseases that confer only temporary immunity, for example strains of influenza, an SIRS model is appropriate. If the SIR Kermack–McKendrick model is assumed with the addition of a recovered period that is exponentially distributed, then an ordinary differential equation model results. For this model, the basic reproduction number gives a sharp threshold, determining whether the disease dies out or goes to an endemic value.

To allow for a more general recovered period, let $P(t)$ be the fraction of recovered individuals remaining in the recovered class t units after recovery from infection. It is reasonable to assume that $P(t)$ is nonincreasing, $P(0^+) = 1$, $\lim_{t \to \infty} P(t) = 0$ and the average period of immunity $\int_0^\infty P(v)dv = \omega$ is finite. Assuming that the average infectious period is $1/\gamma$ and neglecting demographics, gives the system for the fractions in the infective, recovered and susceptible compartment as

$$I(t) = I(0)e^{-\gamma t} + \int_0^t \beta S(x)I(x)e^{-\gamma(t-x)}dx$$

$$R(t) = R_0(t) + \int_0^t \gamma I(x)P(t-x)dx \qquad (5.11)$$

$$S(t) = 1 - I(t) - R(t)$$

where $R_0(t)$ is the number initially removed and still removed at t, with $R_0(\infty) = 0$. This model is formulated and analyzed in [11], and has richer dynamics than the corresponding ordinary differential equation SIRS model.

System (5.11) is equivalent to the integrodifferential equation

$$I'(t) = \gamma I(t) + \beta I(t)[1 - I(t) - R_0(t) - \gamma \int_0^t I(t+u)P(-u)du]. \qquad (5.12)$$

By standard theorems on retarded functional differential equations [10, 13], there exists a unique solution of (5.12) for all $t \geq 0$. Here $\mathcal{R}_0 = \beta/\gamma$, and it is shown in [11] that if $\mathcal{R}_0 \leq 1$, then all solutions tend to the disease-free equilibrium; but if $\mathcal{R}_0 > 1$, the disease-free equilibrium is unstable and a unique endemic equilibrium (S_∞, I_∞) exists that is given by

$$S_\infty = \frac{1}{\mathcal{R}_0}, \qquad I_\infty = \frac{1 - 1/\mathcal{R}_0}{1 + \omega\gamma}.$$

For further analysis with $\mathcal{R}_0 > 1$, assume a constant period of temporary immunity, thus

$$P(t) = \begin{cases} 1 \text{ for } 0 \leq t < \omega \\ \\ 0 \text{ for } t \geq \omega. \end{cases}$$

Then for $t \geq \omega$, equation (5.12) becomes

$$I'(t) = -\gamma I(t) + \beta I(t)[1 - I(t) - \gamma \int_{-\omega}^0 I(t+u)du].$$

Translating I_∞ to the origin by using $I(t) = I_\infty(1 + X(t))$ and letting $t = \omega\tau$ gives

$$X'(\tau) = \frac{-\omega\gamma(\mathcal{R}_0 - 1)}{1 + \omega\gamma}(X(\tau) + 1)[X(\tau) + \omega\gamma \int_{-1}^0 X(\tau + v)dv].$$

Linearizing about $X = 0$ and setting $X(\tau)$ proportional to $e^{z\tau}$ yields the quasi-polynomial characteristic equation

$$z + \frac{\omega\gamma(\mathcal{R}_0 - 1)}{1 + \omega\gamma}[1 + \omega\gamma \int_{-1}^{0} e^{zv}dv] = 0. \tag{5.13}$$

The assumption of a constant recovery period (through the step function $P(t)$) has resulted in a difficult stability problem, even for the linearized equation about the endemic equilibrium. However, it is possible to find purely imaginary roots of (5.13) by setting $z = i\mu$ for $\mu > 0$, which on equating real and imaginary parts becomes

$$\frac{\sin\mu}{\mu} = -\frac{1}{\omega\gamma} \quad \text{and} \quad \mu^2 = \frac{(\omega\gamma)^2(\mathcal{R}_0 - 1)(1 - \cos\mu)}{1 + \omega\gamma}.$$

This gives a family of imaginary root curves for $\mu \in ((2k - 1)\pi, 2k\pi)$, $k = 1, 2, \dots$. For $\omega\gamma < 1$, all roots have negative real parts, so the endemic equilibrium is locally asymptotically stable below the lowest imaginary root curve $k = 1$. Assume $\mathcal{R}_0 > 1$ is fixed and that $z = i\mu_c$ solves (5.13) when $\omega\gamma = c$, then there is a *Hopf bifurcation* from $X = 0$ for small $|\omega\gamma - c|$ of the form

$$X(\tau) = |A(\mu_c)(\omega\gamma - c)|^{\frac{1}{2}}\,[\cos(\mu_c\tau) + o(|\omega\gamma - c|^{\frac{1}{2}})]$$

where $\omega\gamma > c$ and $A \neq 0$. If the bifurcation point $(\mathcal{R}_0, c)$ is on the lowest imaginary root curve, then the periodic solution is locally asymptotically stable and has period between ω and 2ω. If the bifurcation point is on a higher curve ($k = 2, 3, \dots$), then the periodic solution is unstable. Details of the Hopf bifurcation theorem can be found in [9] and [13].

Thus a constant period of temporary immunity can lead, for some parameter values, to solutions of this SIRS model that oscillate about the endemic equilibrium. For more details of this and oscillatory solutions for an ordinary differential equation model that has at least three removed classes (corresponding to a gamma-distributed time delay in the recovered class), please consult [11]. It is interesting to note that an alternate SIRS model with an arbitrarily distributed time delay in the infectious compartment and an exponentially distributed delay in the removed compartment does not exhibit periodic solutions [11, Sect. 5]. For epidemic models that include delays and vertical transmission see [4, Chap. 4].

Mechanisms that can lead to oscillatory solutions either autonomously or through external forcing in epidemic models are discussed in [10]. In addition to delays in the recovered compartment, these mechanisms include nonlinear incidence, age structure and periodic incidence. Such oscillations are often seen in disease incidence data; thus models that predict this phenomenon are useful in understanding disease spread and suggesting possible control measures.

References

1. Allen, L.J.S.: An introduction to stochastic epidemic models. Chapter 3 of Mathematical Epidemiology (this volume)
2. Brauer, F.: An introduction to compartmental models in epidemiology. Chapter 2 of Mathematical Epidemiology (this volume)
3. Brauer, F. and van den Driessche, P.: Models for transmission of disease with immigration of infectives. Math. Biosci. **171**, 143–154 (2001)
4. Busenberg, S. and Cooke, K.: Vertically Transmitted Diseases. Biomath. **23**, Springer, Berlin Heidelberg New York (1993)
5. Capasso, V.: Mathematical Structures of Epidemic Systems. Lect. Notes Biomath. **97**, Springer, Berlin Heidelberg New York (1993)
6. Diekmann, O. and Heesterbeek, J.A.P.: Mathematical Epidemiology of Infectious Diseases. Model Building, Analysis and Interpretation, Wiley (2000)
7. Gani, J. Yakowitz, S. and Blount, M.: The spread and quarantine of HIV infection in a prison system. SIAM J. Appl. Math. **57**, 1510–1530 (1997)
8. Hart, C.A.: Microterrors. Firefly Books. Axis (2004)
9. Hale, J.K. and Verduyn Lunel, S.M.: Introduction to Functional Differential Equations. Springer, Berlin Heidelberg New York(1993)
10. Hethcote, H.W., and Levin, S.A.: Periodicity in epidemiological models. In: Levin, S.A., Hallam, T.G., Gross, L.J. (eds) Applied Mathematical Ecology, Biomath. Texts. **18**, Springer, Berlin Heidelberg New York 193–211 (1989)
11. Hethcote, H.W., Stech, H.W., and van den Driessche, P.: Non-linear oscillations in epidemic models. SIAM J. Appl. Math. **40**, 1–9 (1981)
12. Jeffries, C., Klee, V., and van den Driessche, P.: When is a matrix sign stable? Can. J. Math. **29**, 315–326 (1977)
13. Kuang, Y.: Delay Differential Equations with Applications in Population Dynamics. Academic (1993)
14. Li, M.Y., and Muldowney, J.S.: A geometric approach to global- stability problems. SIAM. J. Math. Anal. **27**, 1070–1083 (1996)
15. Li, M.Y., Smith, H.L., and Wang, L.: Global dynamics of an SEIR epidemic model with vertical transmission. SIAM J. Appl. Math. **62**, 58–69 (2001)
16. van den Driessche, P., and Watmough, J.: Reproduction numbers and sub-threshold endemic equilibria for compartmental models of disease transmission. Math. Bios. **180**, 29–48 (2002)

Chapter 6
Further Notes on the Basic Reproduction Number

P. van den Driessche and James Watmough

Abstract The basic reproduction number, $\mathcal{R}_0$ is a measure of the potential for disease spread in a population. Mathematically, $\mathcal{R}_0$ is a threshold for stability of a disease-free equilibrium and is related to the peak and final size of an epidemic. The purpose of these notes is to give a precise definition and algorithm for obtaining $\mathcal{R}_0$ for a general compartmental ordinary differential equation model of disease transmission. Several examples of calculating $\mathcal{R}_0$ are included, and the epidemiological interpretation of this threshold parameter is connected to the local and global stability of a disease-free equilibrium.

6.1 Introduction

The *basic reproduction number*, $\mathcal{R}_o$ is defined as the expected number of secondary infections produced by an index case in a completely susceptible population [1,8]. This number is a measure of the potential for disease spread within a population. If $\mathcal{R}_o < 1$, then a few infected individuals introduced into a completely susceptible population will, on average, fail to replace themselves, and the disease will not spread. If, on the other hand, $\mathcal{R}_o > 1$, then the number of infected individuals will increase with each generation and the disease will spread. Note that the basic reproduction number is a threshold parameter for invasion of a disease organism into a completely susceptible population; once the disease has begun to spread, conditions favouring spread will change and $\mathcal{R}_o$ may no longer be a good measure of disease transmission. However, in many disease transmission models, the peak prevalence of

Department of Mathematics and Statistics, University of Victoria, P.O. BOX 3060 STN CSC, Victoria, BC, Canada V8W 3R4 pvdd@math.uvic.ca

Department of Mathematics and Statistics, University of New Brunswick, NB, Canada E3B 5A3 watmough@unb.ca infected hosts and the final size of the epidemic are increasing functions of $\mathcal{R}_o$, making it a useful measure of spread.

Many researchers use *reproductive* in place of *reproduction* and *rate* or *ratio* in place of *number*. Convincing arguments can be made for each combination: $\mathcal{R}_o$ can be specified as either a *ratio* of *rates*, or a *number* of secondary cases per index case. In the context of differential equation models (or, more generally, evolution equation models), $\mathcal{R}_o$ arises through dimensional analysis as a dimensionless *rate* of transmission. At the time this manuscript was being prepared, a search of the Biological Abstracts indicated that each combination was equally popular!

The purpose of this chapter is threefold:

- to give a mathematical definition of $\mathcal{R}_o$ for compartmental ordinary differential equation (ODE) models,
- to show the connection between $\mathcal{R}_o$ and the local and global asymptotic stability of an ODE model, and
- to illustrate the possible bifurcations of the solution sets of the ODE models as $\mathcal{R}_o$ passes through the threshold.

This chapter is based on the papers of Castillo-Chavez et al. [6] and van den Driessche and Watmough [18] and the book of Diekmann and Heesterbeek [8]. Results on the theory of nonnegative matrices are taken from Berman and Plemmons [3]. An excellent review of basic compartmental disease transmission models is given by Hethcote [12]. A recent review of $\mathcal{R}_o$ in a broader context is given by Heffernan et al. [11].

6.2 Compartmental Disease Transmission Models

This chapter focuses on compartmental models for disease transmission. Individuals are characterized by a single, discrete state variable and are sorted into compartments based on this state. A compartment is called a *disease compartment* if the individuals therein are infected. Note that this use of the term *disease* is broader than the clinical definition and includes asymptotic stages of infection as well as symptomatic. Suppose there are n disease compartments and m nondisease compartments, and let $x \in \mathbb{R}^n$ and $y \in \mathbb{R}^m$ be the subpopulations in each of these compartments. Further, denote by $\mathcal{F}_i$ the rate secondary infections increase the ith disease compartment and by $\mathcal{V}_i$ the rate disease progression, death and recovery decrease the ith compartment. The compartmental model can then be written in the following form:

$$x_i' = \mathcal{F}_i(x, y) - \mathcal{V}_i(x, y) , \quad i = 1, \ldots, n, \tag{6.1a}$$
$$y_j' = g_j(x, y) , \quad j = 1, \ldots, m, \tag{6.1b}$$

where $'$ denotes differentiation with respect to time. Note that the decomposition of the dynamics into $\mathcal{F}$ and $\mathcal{V}$ and the designation of compartments as infected or uninfected may not be unique; different decompositions correspond to different epidemiological interpretations of the model. The definitions of $\mathcal{F}$ and $\mathcal{V}$ used here differ slightly from those in [18].

The derivation of the basic reproduction number is based on the linearization of the ODE model about a disease-free equilibrium. The following assumptions are made to ensure the existence of this equilibrium and to ensure the model is well posed.

(A1) Assume $\mathcal{F}_i(0, y) = 0$ and $\mathcal{V}_i(0, y) = 0$ for all $y \geq 0$ and $i = 1, \ldots, n$. All new infections are secondary infections arising from infected hosts; there is no immigration of individuals into the disease compartments.

(A2) Assume $\mathcal{F}_i(x, y) \geq 0$ for all nonnegative x and y and $i = 1, \ldots, n$. The function $\mathcal{F}$ represents new infections and cannot be negative.

(A3) Assume $\mathcal{V}_i(x, y) \leq 0$ whenever $x_i = 0$, $i = 1, \ldots, n$. Each component, $\mathcal{V}_i$, represents a net outflow from compartment i and must be negative (inflow only) whenever the compartment is empty.

(A4) Assume $\sum_{i=1}^{n} \mathcal{V}_i(x, y) \geq 0$ for all nonnegative x and y. This sum represents the total outflow from all infected compartments. Terms in the model leading to increases in $\sum_{i=1}^{n} x_i$ are assumed to represent secondary infections and therefore belong in $\mathcal{F}$.

(A5) Assume the disease-free system $y' = g(0, y)$ has a unique equilibrium that is asymptotically stable. That is, all solutions with initial conditions of the form $(0, y)$ approach a point $(0, y_o)$ as $t \to \infty$. We refer to this point as the disease-free equilibrium.

Assumption (A1) ensures that the disease-free set, which consists of all points of the form $(0, y)$, is invariant. That is, any solution with no infected individuals at some point in time will be free of infection for all time. This in turn ensures that the disease-free equilibrium is also an equilibrium of the full system.

Suppose a single infected person is introduced into a population originally free of disease. The initial ability of the disease to spread through the population is determined by an examination of the linearization of (6.1a) about the disease-free equilibrium $(0, y_o)$. Using Assumption (A1), it can be shown that

$$\frac{\partial \mathcal{F}_i}{\partial y_j}(0, y_o) = \frac{\partial \mathcal{V}_i}{\partial y_j}(0, y_o) = 0$$

for every pair (i, j). This implies that the linearized equations for the disease compartments, x, are decoupled from the remaining equations and can be written as

$$x' = (F - V)x, \tag{6.2}$$

where F and V are the $n \times n$ matrices with entries

$$F = \frac{\partial \mathcal{F}_i}{\partial x_j}(0, y_o) \quad \text{and} \quad V = \frac{\partial \mathcal{V}_i}{\partial x_j}(0, y_o).$$

Using Assumption (A5), linear stability of the system (6.1) is completely determined by the linear stability of $(F - V)$ in (6.2); see Sect. 6.5.

6.3 The Basic Reproduction Number

The number of secondary infections produced by a single infected individual can be expressed as the product of the expected duration of the infectious period and the rate secondary infections occur. For the general model with n disease compartments, these are computed for each compartment for a hypothetical index case. The expected time the index case spends in each compartment is given by the integral $\int_0^\infty \phi(t, x_o) \, dt$, where $\phi(t, x_o)$ is the solution to (6.2) with $F = 0$ (no secondary infections) and nonnegative initial conditions, x_o, representing an infected index case:

$$x' = -Vx, \qquad x(0) = x_o. \tag{6.3}$$

In effect, this solution shows the path of the index case through the disease compartments from the initial exposure through to death or recovery with the ith component of $\phi(t, x_o)$ interpreted as the probability that the index case (introduced at time $t = 0$) is in disease state i at time t. The solution to (6.3) is $\phi(t, x_o) = e^{-Vt}x_o$, where the exponential of a matrix is defined by the Taylor series

$$e^A = I + A + \frac{A^2}{2} + \frac{A^3}{3!} + \cdots + \frac{A^k}{k!} + \cdots$$

This series converges for all t (see, for example, [13]). Thus $\int_0^\infty \phi(t, x_o) \, dt = V^{-1}x_o$, and the (i, j) entry of the matrix V^{-1} can be interpreted as the expected time an individual initially introduced into disease compartment j spends in disease compartment i.

The (i, j) entry of the matrix F is the rate secondary infections are produced in compartment i by an index case in compartment j. Hence, the expected number of secondary infections produced by the index case is given by

$$\int_0^\infty Fe^{-Vt}x_o \, dt = FV^{-1}x_o.$$

Following Diekmann and Heesterbeek [8], the matrix $K = FV^{-1}$ is referred to as the *next generation matrix* for the system at the disease-free equilibrium. The (i, j) entry of K is the expected number of secondary infections in compartment i produced by individuals initially in compartment j, assuming, of course, that the environment seen by the individual remains homogeneous for the duration of its infection.

As we shall see in Sect. 6.5, the next generation matrix, $K = FV^{-1}$, is nonnegative and therefore has a nonnegative eigenvalue, $\mathcal{R}_o = \rho(FV^{-1})$, such that there are no other eigenvalues of K with modulus greater than $\mathcal{R}_o$ and there is a nonnegative eigenvector ω associated with $\mathcal{R}_o$ [3, Theorem 1.3.2]. This eigenvector is in some sense the distribution of infected individuals that produces the greatest number, $\mathcal{R}_o$, of secondary infections per generation. Thus, $\mathcal{R}_o$ and the associated eigenvector ω suitably define a "typical" infective and the basic reproduction number can be rigorously defined as the spectral radius of the next generation matrix, K. The *spectral radius* of a matrix K, denoted $\rho(K)$, is the maximum of the moduli of the eigenvalues of K. If K is irreducible, then $\mathcal{R}_o$ is a simple eigenvalue of K. However, if K is reducible, which is often the case for diseases with multiple strains, then K may have several positive real eigenvectors corresponding to reproduction numbers for each competing strain of the disease.

6.4 Examples

For a given model, neither the next generation matrix, K, nor the basic reproduction number, $\mathcal{R}_o$, are uniquely defined; there may be several possible decompositions of the dynamics into the components $\mathcal{F}$ and $\mathcal{V}$ and thus many possibilities for K. Usually only a single decomposition has a realistic epidemiological interpretation. These ideas are illustrated by the following examples.

6.4.1 The SEIR Model

In the SEIR model for a childhood disease such as measles, the population is divided into four compartments: susceptible (S), exposed and latently infected (E), infectious (I) and recovered with immunity (R). Let S, E, I and R denote the subpopulations in each compartment. The usual SEIR model is written as follows:

$$S' = \Pi - \mu S - \beta SI, \qquad (6.4a)$$
$$E' = \beta SI - (\mu + \kappa)E, \qquad (6.4b)$$
$$I' = \kappa E - (\mu + \alpha)I, \qquad (6.4c)$$
$$R' = \alpha I - \mu R, \qquad (6.4d)$$

together with nonnegative initial conditions.

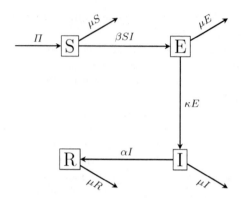

Fig. 6.1 Progression of infection from susceptible (S) individuals through the exposed (E), infected (I), and treated (R) compartments for the simple SEIR model

The progression through the compartments is illustrated in Fig. 6.1. New infections in compartment E arise by contacts between susceptible and infected individuals in compartments S and I at a rate βSI. Individuals progress from compartment E to I at a rate κ and develop immunity at a rate α. In addition, natural mortality claims individuals at a rate μ. For simplicity, the model assumes a constant recruitment, Π, of susceptible individuals. With incidence βSI and β constant this is commonly referred to as the mass action model. More generally, β may be taken as a function of the total population $N = S + E + I + R$.

The system has a unique disease-free equilibrium, with $S_o = \Pi/\mu$. Taking the infected compartments to be E and I gives

$$\mathcal{F} = \begin{pmatrix} \beta SI \\ 0 \end{pmatrix} \qquad \text{and} \qquad \mathcal{V} = \begin{pmatrix} (\mu + \kappa)E \\ -\kappa E + (\mu + \alpha)I \end{pmatrix}.$$

Hence,

$$F = \begin{pmatrix} 0 & \beta S_o \\ 0 & 0 \end{pmatrix},$$

$$V = \begin{pmatrix} (\mu + \kappa) & 0 \\ -\kappa & (\mu + \alpha) \end{pmatrix},$$

and the next generation matrix is

$$K = FV^{-1} = \begin{pmatrix} \dfrac{\kappa\beta S_o}{(\mu+\kappa)(\mu+\alpha)} & \dfrac{\beta S_o}{\mu+\alpha} \\ 0 & 0 \end{pmatrix}. \tag{6.5}$$

The $(1,2)$ entry of K is the expected number of secondary infections produced in compartment E by an individual initially in compartment I over the course of its infection. To interpret this term, recall that βS_o is the rate of infection for our single infected individual in a population of S_o susceptible individuals, and $1/(\mu+\alpha)$ is the expected duration of the infectious period. The ratio $\kappa/(\mu+\kappa)$ is the fraction of individuals that progress from E to I. Hence, the $(1,1)$ entry of K is the expected number of secondary infections produced in compartment E by an infected individual originally in compartment E. The eigenvalues of K are 0 and

$$\mathcal{R}_o = \frac{\kappa\beta S_o}{(\mu+\kappa)(\mu+\alpha)}. \tag{6.6}$$

6.4.2 A Variation on the Basic SEIR Model

In the basic SEIR model of Sect. 6.4.1, suppose that individuals in compartment E are mildly infectious and produce secondary infections at the reduced rate $\epsilon\beta SE$ with $0 < \epsilon < 1$. This gives rise to an additional nonzero entry in F, and K becomes

$$K = FV^{-1} = \begin{pmatrix} \dfrac{\epsilon\beta S_o}{\mu+\kappa} + \dfrac{\kappa\beta S_o}{(\mu+\kappa)(\mu+\alpha)} & \dfrac{\beta S_o}{\mu+\alpha} \\ 0 & 0 \end{pmatrix}. \tag{6.7}$$

The reproduction number is now

$$\mathcal{R}_o = \frac{\epsilon\beta S_o}{\mu+\kappa} + \frac{\kappa\beta S_o}{(\mu+\kappa)(\mu+\alpha)}. \tag{6.8}$$

The two terms of $\mathcal{R}_o$ are the number of secondary infections produced by an index case initially in compartment E, just as with the model of Sect. 6.4.1. The first term is the number of secondary infections during the earlier, mildly infectious stage and the second term is the number of secondary infections during the fully infectious stage.

6.4.3 A Simple Treatment Model

To illustrate the mathematical ambiguity in the choice of $\mathcal{R}_o$, consider the basic SEI model with treatment of infective individuals. Suppose infectious individuals are treated at a rate r_2, but that treatment is only partially effective: a fraction q of treated infectious individuals recover with partial immunity, and a fraction $p = 1 - q$ return to a latent stage of infection. The ambiguity in $\mathcal{R}_o$ arises from the two possible interpretations of treatment failure. Treatment of latently infected individuals, at a rate r_1, is also included in the model and always results in recovery.

The dynamics of the model are illustrated in Fig. 6.2. The model maintains the basic structure of the SEIR model of Sect. 6.4.1, but the R compartment is replaced by a compartment of treated individuals (T) and standard (rather than mass action) incidence is assumed. Since treatment confers only partial immunity, treated individuals are reinfected at a rate $\beta_2 T/N$, where $N = S + E + I + T$. The constant recruitment rate used in the previous example is generalized to a density dependent rate, but all other parameters retain their earlier interpretations. The disease transmission model consists of the following differential equations together with nonnegative initial conditions:

$$S' = b(N) - \mu S - \beta_1 SI/N, \tag{6.9a}$$

$$E' = \beta_1 SI/N + \beta_2 TI/N - (\mu + \kappa + r_1)E + pr_2 I, \tag{6.9b}$$

$$I' = \kappa E - (\mu + r_2)I, \tag{6.9c}$$

$$T' = -\mu T + r_1 E + qr_2 I - \beta_2 TI/N. \tag{6.9d}$$

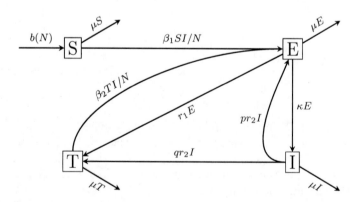

Fig. 6.2 Progression of infection from susceptible (S) individuals through the exposed (E), infected (I), and treated (T) compartments for the treatment model (6.9)

This model is a caricature of the more complex models for tuberculosis proposed by Blower et al. [4] and Castillo-Chavez and Feng [5]. Further analysis and discussion can be found in those papers.

The disease compartments are E and I, as before, and treatment failure, modelled by the term pr_2I in the second equation, is interpreted as part of progression of an infected individual through the disease compartments, rather than a new infection. With this interpretation, $\mathcal{F}$ and $\mathcal{V}$ are as follows:

$$\mathcal{F} = \begin{pmatrix} \beta_1 SI/N + \beta_2 TI/N \\ 0 \end{pmatrix}, \qquad \mathcal{V} = \begin{pmatrix} (\mu + \kappa + r_1)E - pr_2I \\ -\kappa E + (\mu + r_2)I \end{pmatrix}.$$

An equilibrium solution with $E = I = 0$ must have $T = 0$ and $S = S_o$, where S_o is any positive solution of $b(S_o) = \mu S_o$. This will be locally stable, and therefore a DFE, if $b'(S_o) < \mu$. Assuming this to be the case, evaluating the derivatives of $\mathcal{F}$ and $\mathcal{V}$ at $S = S_o$, $E = I = T = 0$ leads to the following expressions for F and V.

$$F = \begin{pmatrix} 0 & \beta_1 \\ 0 & 0 \end{pmatrix}, \qquad V = \begin{pmatrix} \mu + \kappa + r_1 & -pr_2 \\ -\kappa & \mu + r_2 \end{pmatrix}.$$

As with the SEIR model, FV^{-1} has rank one, and a straightforward calculation shows the spectral radius to be

$$\mathcal{R}_c = \frac{\beta_1 \kappa}{(\mu + \kappa + r_1)(\mu + r_2) - \kappa pr_2}. \qquad (6.10)$$

The notation $\mathcal{R}_c$ is used to denote the reproduction number with control measures in place. A heuristic derivation of $\mathcal{R}_c$ is given in [18]. Briefly, $\mathcal{R}_c$ can be written as the geometric series $(h_1 + h_1^2 h_2 + h_1^3 h_2^2 + \ldots)\beta_1/(\mu + r_2)$, where $h_1 = \kappa/(\mu + \kappa + r_1)$ is the fraction of individuals leaving compartment E who progress to compartment I, and $h_2 = pr_2/(\mu + r_2)$ is the fraction of individuals leaving compartment I who re-enter compartment E. The product $h_1^k h_2^{k-1}$ is the fraction of exposed individuals who pass through compartment I at least k times, and the sum of these products is the expected number of times an exposed individual passes through compartment I. Multiplying by $\beta_1/(\mu + r_2)$ gives $\mathcal{R}_c$, since each time an individual enters the infectious compartment I, they spend, on average, $1/(\mu + r_2)$ time units there producing, on average, $\beta_1/(\mu + r_2)$ secondary infections.

In contrast, if treatment failure is considered to be a new infection, then $\mathcal{F}$ and $\mathcal{V}$ are as follows:

$$\mathcal{F} = \begin{pmatrix} \beta_1 SI/N + \beta_2 TI/N + pr_2I \\ 0 \end{pmatrix}, \qquad \mathcal{V} = \begin{pmatrix} (\mu + \kappa + r_1)E \\ -\kappa E + (\mu + r_2)I \end{pmatrix}.$$

Differentiation at the disease-free equilibrium then leads to the following expressions for F, V and the spectral radius:

$$F = \begin{pmatrix} 0 & \beta_1 + pr_2 \\ 0 & 0 \end{pmatrix}, \qquad V = \begin{pmatrix} \mu + \kappa + r_1 & 0 \\ -\kappa & \mu + r_2 \end{pmatrix},$$

$$\rho(FV^{-1}) = \frac{\beta_1 \kappa + pr_2 \kappa}{(\mu + \kappa + r_1)(\mu + r_2)}. \tag{6.11}$$

Note that since $T = 0$ at the disease-free equilibrium, the reinfection term does not appear in either linearization and the choice of whether to place the $\beta_2 TI/N$ term in $\mathcal{F}$ or $\mathcal{V}$ is of little practical consequence.

Mathematically the two resulting thresholds are equivalent since the conditions $\mathcal{R}_c < 1$ from (6.10) and $\rho(FV^{-1}) < 1$ from (6.11) yield the same portion of parameter space. The difference between the two expressions (6.10) and (6.11) lies in their epidemiological interpretation. For example, in the second interpretation, the infection rate is $\beta_1 + pr_2$ and an exposed individual is expected to spend $\kappa/((\mu + \kappa + r_1)(\mu + r_2))$ time units in compartment I. The flaw in this reasoning is that treatment failure does not give rise to a *newly infected* individual, but only changes the infection status of an already infected individual. If the conditions are used simply as threshold parameters, then the difference between the two choices does not matter. However, any analysis of the sensitivity of $\mathcal{R}_c$ to control parameters should be done on an epidemiological meaningful threshold.

6.4.4 A Vaccination Model

Consider the following SI vaccination model proposed by Gandon et al. [10].

$$S' = (1 - p)\Pi - \mu S - (\beta I + \beta_v I_v) S,$$
$$S_v' = p\Pi - \mu S_v - (1 - r)(\beta I + \beta_v I_v) S_v,$$
$$I' = (\beta I + \beta_v I_v) S - (\mu + \nu)I,$$
$$I_v' = (1 - r)(\beta I + \beta_v I_v) S_v - (\mu + \nu_v)I_v.$$

S, I, S_v and I_v denote the subpopulations in the unvaccinated susceptible, unvaccinated infectious, vaccinated susceptible and vaccinated infectious compartments, respectively. Susceptible individuals are recruited at a rate Π and a fraction, p, of these recruits are vaccinated immediately. Individuals leave the population at a rate μ, with additional disease-induced host mortality at the rates ν and ν_v. Vaccination of infectious individuals reduces the transmission rate from β to β_v and vaccination of susceptible individuals reduces the probability of transmission by a fraction r.

The system has a unique disease-free equilibrium, given by $S_o = (1-p)N_o$ and $S_{vo} = pN_o$, where $N_o = \Pi/\mu$. The disease compartments are I and I_v, V is the diagonal matrix

$$V = \begin{pmatrix} \mu + \nu & 0 \\ 0 & \mu + \nu_v \end{pmatrix},$$

and F is a rank one matrix that can be expressed as the product of the two vectors $\omega = \left(S_o, \ (1-r)S_{vo}\right)^T$ and $\beta = \left(\beta, \ \beta_v\right)^T$ as follows:

$$F = \omega\beta^T = \begin{pmatrix} \beta S_o & \beta_v S_o \\ (1-r)\beta S_{vo} & (1-r)\beta_v S_{vo} \end{pmatrix}. \tag{6.12}$$

Since F has rank one, the next generation matrix also has rank one. The spectral radius of a rank one matrix is its trace. Hence,

$$\mathcal{R}_c = \rho\left(FV^{-1}\right) = \beta^T V^{-1}\omega = \frac{\beta S_o}{\mu + \nu} + \frac{(1-r)\beta_v S_{vo}}{\mu + \nu_v}.$$

The simplest interpretation to place on this number is that it is the sum of the number of secondary infections of unvaccinated susceptible individuals produced by an index case in I and the number of secondary infections of vaccinated susceptible individuals produced by an index case in I_v. This simple interpretation is misleading. The correct, although not immediately obvious, interpretation is that $\mathcal{R}_c$ is the number of secondary infections, both vaccinated and unvaccinated, produced by an "index case", ω, distributed in both infectious compartments, with one part in I and $(1-r)S_{vo}/S_o$ parts in I_v. Quite simply, $\mathcal{R}_c$ is the eigenvalue of K with largest modulus and ω is an associated eigenvector.

This simple vaccination model assumes the effects of the vaccine on susceptible and infectious individuals are separable, which leads to a rank one next generation matrix and a simple expression for $\mathcal{R}_c$. Replacing the four incidence parameter combinations β, β_v, $(1-r)\beta$ and $(1-r)\beta_v$, with the four parameters β_{uu}, β_{uv}, β_{vu} and β_{vv} respectively, leads to the next generation matrix

$$K = \begin{pmatrix} \dfrac{\beta_{uu} S_o}{\mu + \nu} & \dfrac{\beta_{uv} S_o}{\mu + \nu_v} \\[2mm] \dfrac{\beta_{vu} S_{ov}}{\mu + \nu} & \dfrac{\beta_{vv} S_{ov}}{\mu + \nu_v} \end{pmatrix}. \tag{6.13}$$

Denoting the four entries of K as $\mathcal{R}_{uu}$, $\mathcal{R}_{uv}$, $\mathcal{R}_{vu}$ and $\mathcal{R}_{vv}$, the spectral radius of K is

$$\mathcal{R}_c = \frac{\mathcal{R}_{uu} + \mathcal{R}_{vv}}{2} + \frac{1}{2}\sqrt{(\mathcal{R}_{uu} + \mathcal{R}_{vv})^2 - 4\mathcal{R}_{uu}\mathcal{R}_{vv} + 4\mathcal{R}_{uv}\mathcal{R}_{vu}}. \tag{6.14}$$

Although this expression defies interpretation as anything other than the spectral radius of K, the threshold condition

$$\mathcal{R}_c < 1$$

is equivalent to the pair of conditions

$$\frac{1}{2}\left(\mathcal{R}_{uu} + \mathcal{R}_{vv}\right) < 1,$$

$$\mathcal{R}_{uu} + \mathcal{R}_{vv} + \mathcal{R}_{uv}\mathcal{R}_{vu} - \mathcal{R}_{uu}\mathcal{R}_{vv} < 1.$$

Note that these conditions only hold for nonnegative matrices and differ slightly from the more general Jury conditions. Several authors [7, 14] have interpreted the left hand side of the second inequality as the reproduction number for the model. The danger in this interpretation is that the magnitude of this expression does not give any insight into the solutions of the model. As Roberts and Heesterbeek [17] point out, this distinction is important if $\mathcal{R}_c$ is used as a measure of the effectiveness of disease control measures.

6.4.5 A Vector-Host Model

Some diseases, notably Dengue fever, malaria and West Nile virus, are not transmitted directly from host to host, but through a vector. The simplest vector-host model couples a simple SIS model for the hosts with an SI model for the vectors. Susceptible hosts (S_h) become infectious hosts (I_h) at a rate $\beta_h S_h I_v$ through contact with infected vectors (I_v). Similarly, susceptible vectors (S_v) become infectious vectors (I_h) at a rate $\beta_v S_v I_h$ by contacts with infected hosts. The model is given by the following equations together with nonnegative initial conditions:

$$I'_h = \beta_h S_h I_v - (\mu_h + \gamma)I_h, \qquad (6.15\text{a})$$

$$I'_v = \beta_v S_v I_h - \mu_v I_v, \qquad (6.15\text{b})$$

$$S'_h = \Pi_h - \mu_h S_h - \beta_h S_h I_v + \gamma I_h, \qquad (6.15\text{c})$$

$$S'_v = \Pi_v - \mu_v S_v - \beta_v S_v I_h. \qquad (6.15\text{d})$$

As before, μ_h and μ_v represent removal rates and Π_h and Π_v recruitment rates. The parameter γ is the recovery rate for infected hosts. Vectors are assumed to remain infected for life. This simple model forms the core of many vector-host models. More detailed analyses and discussions of vector-host models can be found in such papers as Feng and Velasco-Hernández [9] on Dengue fever, and Wonham et al. [20] on West Nile virus.

The two disease compartments are I_h and I_v. The disease-free equilibrium has host and vector populations of $S_{ho} = \Pi_h/\mu_h$ and $S_{vo} = \Pi_v/\mu_v$ respectively. Hence

$$F = \begin{pmatrix} 0 & \beta_h S_{ho} \\ \beta_v S_{vo} & 0 \end{pmatrix}, \quad V = \begin{pmatrix} (\mu_h + \gamma) & 0 \\ 0 & \mu_v \end{pmatrix}, \quad K = \begin{pmatrix} 0 & \dfrac{\beta_h S_{ho}}{\mu_v} \\ \dfrac{\beta_v S_{vo}}{\mu_h + \gamma} & 0 \end{pmatrix}.$$

The entries of K are interpreted as the number of secondary infections produced by infected vectors and hosts during the course of their infection. Note that infected hosts produce infected vectors and *vise versa*. The positive eigenvalue of K is

$$\mathcal{R}_o = \sqrt{\frac{\beta_h \beta_v S_{ho} S_{vo}}{(\mu_h + \gamma)\mu_v}}.$$

The square root arises since it takes two generations for infected hosts to produce new infected hosts. That is

$$K^2 = \begin{pmatrix} \dfrac{\beta_h \beta_v S_{ho} S_{vo}}{(\mu_h + \gamma)\mu_v} & 0 \\ 0 & \dfrac{\beta_h \beta_v S_{ho} S_{vo}}{(\mu_h + \gamma)\mu_v} \end{pmatrix}$$

In practise, what we have given as K^2 is often taken as K and the square root is left off the reproduction number. Indeed, this was the original interpretation of Macdonald (see [1]).

6.4.6 A Model with Two Strains

The reproduction number for models with multiple strains is usually the larger of the reproduction numbers for the two strains in isolation. However, many such models also poses multiple endemic equilibria, and there is a threshold similar to the basic reproduction number connected with the ability of one strain to invade and outcompete another. As a simple example, consider the special case of the n-strain SIR model of Andreasen et al. [2] given by the following system of equations together with nonnegative initial conditions:

$$S' = \Pi - \mu S - \beta_1 S(I_2 + I_{12}) - \beta_1 S(I_1 + I_{21}), \tag{6.16a}$$
$$I_1' = \beta_1 S(I_1 + I_{21}) - (\mu + \gamma_1)I_1, \tag{6.16b}$$
$$I_2' = \beta_1 S(I_2 + I_{12}) - (\mu + \gamma_2)I_2, \tag{6.16c}$$
$$S_1' = \gamma_1 I_1 - \sigma_1 \beta_2 S_1(I_2 + I_{12}) - \mu S_1, \tag{6.16d}$$
$$S_2' = \gamma_2 I_2 - \sigma_2 \beta_1 S_2(I_1 + I_{21}) - \mu S_2, \tag{6.16e}$$
$$I_{21}' = \sigma_2 \beta_1 S_2(I_1 + I_{21}) - (\mu + \gamma_1)I_{21}, \tag{6.16f}$$
$$I_{12}' = \sigma_1 \beta_2 S_1(I_2 + I_{12}) - (\mu + \gamma_2)I_{12}. \tag{6.16g}$$

Naive individuals (S) are infected with strain one at a rate $\beta_1(I_1 + I_{21})S$ or strain two at a rate $\beta_2(I_2 + I_{12})S$. Individuals in compartment S_1 have recovered, at rate γ_1, from an infection with strain one, with full immunity to reinfection with strain one and partial immunity, modelled by the factor σ_1, to infection with strain two. Upon infection with strain two, which occurs at a rate $\sigma_1 \beta_2(I_2 + I_{12})S_1$, they enter compartment I_{12}. Thus I_{12} is the number of individuals who are currently infected with strain two and had a previous infection with strain one.

The model has four equilibria. We will only concern ourselves with two of the equilibria in this discussion. Further analysis of a more detailed model including treatment can be found in Nuño et al. [16]. Linearizing the model equations about the disease-free equilibrium, $S = S_o = \Pi/\mu$, $I_1 = S_1 = I_2 = S_2 = I_{21} = I_{12} = 0$, leads to the following expressions for F and V.

$$F = \begin{pmatrix} \beta_1 S_o & 0 & \beta_1 S_o & 0 \\ 0 & \beta_2 S_o & 0 & \beta_2 S_o \\ 0 & 0 & 0 & 0 \\ 0 & 0 & 0 & 0 \end{pmatrix}, \quad V = \begin{pmatrix} \mu + \gamma_1 & 0 & 0 & 0 \\ 0 & \mu + \gamma_2 & 0 & 0 \\ 0 & 0 & \mu + \gamma_1 & 0 \\ 0 & 0 & 0 & \mu + \gamma_2 \end{pmatrix}.$$

The next generation matrix, $K = FV^{-1}$, and the Jacobian matrix, $(F - V)$, are reducible. The equations for the infected subpopulations decouple near the disease-free equilibrium, and K has two positive eigenvalues corresponding to the reproduction numbers for each strain:

$$\mathcal{R}_i = \frac{\beta_i S_o}{\mu + \gamma_i}, \quad i = 1, 2. \tag{6.17}$$

The basic reproduction number for the system is the maximum of the two. That is,

$$\mathcal{R}_o = \max_{i \in \{1,2\}} \mathcal{R}_i. \tag{6.18}$$

There is also a reproduction number associated with the strain one equilibrium, $S = \bar{S}$, $I_1 = \bar{I}_1$, $S_1 = \bar{S}_1$, $I_2 = S_2 = I_{21} = I_{12} = 0$, where

$$\bar{S} = \frac{\mu + \gamma_1}{\beta_1},$$

$$\bar{I}_1 = \frac{\Pi}{\mu + \gamma_1}\left(1 - \frac{1}{\mathcal{R}_1}\right),$$

$$\bar{S}_1 = \frac{\Pi\gamma_1}{\mu(\mu + \gamma_1)}\left(1 - \frac{1}{\mathcal{R}_1}\right).$$

Linearizing about the strain one equilibrium then gives

$$F = \begin{pmatrix} \beta_2\bar{S} & 0 & \beta_2\bar{S} \\ 0 & 0 & 0 \\ \sigma_1\beta_2\bar{S}_1 & 0 & \sigma_1\beta_2\bar{S}_1 \end{pmatrix}, \qquad V = \begin{pmatrix} \mu + \gamma_2 & 0 & 0 \\ 0 & \mu + \gamma_1 & 0 \\ 0 & 0 & \mu + \gamma_2 \end{pmatrix}.$$

These are found by considering I_2, I_{21} and I_{12} to be disease variables and S, I_1, S_1 and S_2 to be nondisease variables. Thus, the spectral radius of FV^{-1}, given by

$$\mathcal{R}_{12} = \frac{\mathcal{R}_2}{\mathcal{R}_1} + \frac{\sigma_1\gamma_1\mathcal{R}_2}{\mu + \gamma_1}\left(1 - \frac{1}{\mathcal{R}_1}\right), \qquad (6.19)$$

is the reproduction number for strain two near the strain one equilibrium. This, of course, is only valid if $\mathcal{R}_1 > 1$. It is reasonable to assume that all parameters are positive and $0 < \sigma_1 < 1$, so that $\mathcal{R}_{12} < \mathcal{R}_2$. This implies that there is a range of values for β_2 for which strain two can invade a disease-free population, but can not invade a population in which strain one is endemic. Strain one may protect the host population from strain two.

6.5 $\mathcal{R}_o$ and the Local Stability of the Disease-Free Equilibrium

The reproduction number for a disease is the number of secondary infections produced by an infected individual in a population of susceptible individuals. If the reproduction numbers, $\mathcal{R}_o = \rho(FV^{-1})$, computed in the previous examples are consistent with the differential equation model, then it should follow that the disease-free equilibrium is stable if $\mathcal{R}_o < 1$ and unstable if $\mathcal{R}_o > 1$. This is shown through a series of lemmas.

If each entry of a matrix T is nonnegative we write $T \geq 0$ and refer to T as a *nonnegative matrix*. A matrix of the form $A = sI - B$, with $B \geq 0$, is said to have the Z sign pattern. These are matrices whose offdiagonal entries are negative or zero. If in addition, $s \geq \rho(B)$, then A is called an M-matrix. Note that in this section, I denotes an identity matrix, not a population of infectious individuals. The following lemma is a standard result from [3].

Lemma 1. *If A has the Z sign pattern, then $A^{-1} \geq 0$ if and only if A is a nonsingular M-matrix.*

From Assumptions (A1) and (A2) it follows that each entry of F is nonnegative. From Assumptions (A1) and (A3) it follows that the offdiagonal entries of V are negative or zero. Thus V has the Z sign pattern. Assumption (A4) with Assumption (A1) ensures that the column sums of V are positive or zero, which, together with the Z sign pattern, implies that V is a (possibly singular) M-matrix [3, condition M_{35} of Theorem 6.2.3]. In what follows, it is assumed that V is nonsingular. In this case, $V^{-1} \geq 0$, by Lemma 1. Hence, $K = FV^{-1}$ is also nonnegative.

Lemma 2. *If F is nonnegative and V is a nonsingular M-matrix, then $\mathcal{R}_o = \rho(FV^{-1}) < 1$ if and only if all eigenvalues of $(F-V)$ have negative real parts.*

Proof. Suppose $F \geq 0$ and V is a nonsingular M-matrix. By Lemma 1, $V^{-1} \geq 0$. Thus, $(I - FV^{-1})$ has the Z sign pattern, and by Lemma 1, $(I - FV^{-1})^{-1} \geq 0$ if and only if $\rho(FV^{-1}) < 1$. From the equalities $(V - F)^{-1} = V^{-1}(I - FV^{-1})^{-1}$ and $V(V - F)^{-1} = I + F(V - F)^{-1}$, it follows that $(V - F)^{-1} \geq 0$ if and only if $(I - FV^{-1})^{-1} \geq 0$. Finally, $(V - F)$ has the Z sign pattern, so by Lemma 1, $(V - F)^{-1} \geq 0$ if and only if $(V - F)$ is a nonsingular M-matrix. Since the eigenvalues of a nonsingular M-matrix all have positive real parts, this completes the proof. $\square$

Theorem 1. *Consider the disease transmission model given by (6.1). The disease-free equilibrium of (6.1) is locally asymptotically stable if $\mathcal{R}_o < 1$, but unstable if $\mathcal{R}_o > 1$, where $\mathcal{R}_o$ is defined as in Sect. 6.3.*

Proof. Let F and V be as defined in Sect. 6.2, and let J_{21} and J_{22} be the matrices of partial derivatives of g with respect to x and y evaluated at the disease-free equilibrium. The Jacobian matrix for the linearization of the system about the disease-free equilibrium has the block structure

$$J = \begin{pmatrix} F - V & 0 \\ J_{21} & J_{22} \end{pmatrix}.$$

The disease-free equilibrium is locally asymptotically stable if the eigenvalues of the Jacobian matrix all have negative real parts. Since the eigenvalues of J are those of $(F - V)$ and J_{22}, and the latter all have negative real parts by Assumption (A5), the disease-free equilibrium is locally asymptotically stable if all eigenvalues of $(F-V)$ have negative real parts. By the assumptions on $\mathcal{F}$ and $\mathcal{V}$, F is nonnegative and V is a nonsingular M-matrix. Hence, by Lemma 2 all eigenvalues of $(F-V)$ have negative real parts if and only if $\rho(FV^{-1}) < 1$. It follows that the disease-free equilibrium is locally asymptotically if $\mathcal{R}_o = \rho(FV^{-1}) < 1$.

Instability for $\mathcal{R}_o > 1$ can be established by a continuity argument. If $\mathcal{R}_o \leq 1$, then for any $\epsilon > 0$, $((1 + \epsilon)I - FV^{-1})$ is a nonsingular M matrix and, by Lemma 1, $((1 + \epsilon)I - FV^{-1})^{-1} \geq 0$. By the proof of Lemma 2, all eigenvalues of $((1 + \epsilon)V - F)$ have positive real parts. Since $\epsilon > 0$ is arbitrary, and eigenvalues are continuous functions of the entries of the matrix, it follows that all eigenvalues of $(V - F)$ have nonnegative real parts. To reverse the argument, suppose all the eigenvalues of $(V - F)$ have nonnegative real parts. For any positive ϵ, $(V + \epsilon I - F)$ is a nonsingular M-matrix, and by the proof of Lemma 2, $\rho(F(V + \epsilon I)^{-1}) < 1$. Again, since $\epsilon > 0$ is arbitrary, it follows that $\rho(FV^{-1}) \leq 1$. Thus, $(F - V)$ has at least one eigenvalue with positive real part if and only if $\rho(FV^{-1}) > 1$, and the disease-free equilibrium is unstable whenever $\mathcal{R}_o > 1$. $\square$

6.6 $\mathcal{R}_o$ and Global Stability of the Disease-Free Equilibrium

The change of local stability at the threshold $\mathcal{R}_o = 1$, corresponds to a transcritical bifurcation in the solutions to (6.1). It can be shown that there is a branch of endemic equilibria emanating from the bifurcation point at $\mathcal{R}_o = 1$, $(x, y) = (x_o, y_o)$. For an introduction to the general theory of these bifurcations in the context of differential equations, see Wiggins [19].

For the simple example of Sect. 6.4.1, there is an equilibrium with

$$S_e = \frac{S_o}{\mathcal{R}_o},$$

$$I_e = \frac{\Pi \kappa}{(\mu + \alpha)(\mu + \kappa)} \left(1 - \frac{1}{\mathcal{R}_o} \right),$$

$$E_e = \frac{(\mu + \alpha)I_e}{\kappa},$$

$$R_e = \frac{\alpha I_e}{\mu},$$

defined for $\mathcal{R}_o > 1$. Since the endemic equilibria only exist for $\mathcal{R}_o > 1$, the bifurcation is said to be in the forward direction, and the disease-free equilibrium is the only equilibrium of the system when $\mathcal{R}_o < 1$. In models for which endemic equilibria exist near the disease-free equilibrium for $\mathcal{R}_o < 1$ the bifurcation is called a backward bifurcation.

Castillo-Chavez et al. [6] use a comparison theorem to derive sufficient conditions for the global asymptotic stability of the disease-free equilibrium of a general disease transmission model when $\mathcal{R}_o < 1$. Clearly, in the case of a backward bifurcation this equilibrium can not be globally asymptotically stable whenever $\mathcal{R}_o < 1$. In most models, however, one expects a second threshold for global stability. A slight change to the argument of [6] gives a sufficient condition for global stability in this case as well.

Consider the disease transmission model (6.1) written in the form

$$x' = -Ax - \hat{f}(x, y), \qquad (6.20a)$$
$$y' = g(x, y). \qquad (6.20b)$$

Theorem 2. *If A is a nonsingular M-matrix and $\hat{f} \geq 0$, then the disease-free equilibrium of (6.20) is globally asymptotically stable.*

Proof. Integrating (6.20a) leads to

$$x(t) = e^{-tA}x(0) - \int_0^t e^{-(t-s)A}\hat{f}(x(s), y(s))\,ds. \qquad (6.21)$$

It can be shown that $e^{-tA} \geq 0$ whenever A is an M-matrix. Since solutions of (6.20) remain nonnegative, it follows that

$$0 \leq x(t) \leq e^{-tA}x(0). \qquad (6.22)$$

Finally, since $e^{-tA} \to 0$ as $t \to \infty$, it follows that $x(t) \to 0$ as $t \to \infty$. $\square$

For the SEIR model of Sect. 6.4.1, take $A = V - F$ and write (6.4) as follows

$$x' = -(V - F)x - \begin{pmatrix} \beta(S_o - S)I \\ 0 \end{pmatrix}, \qquad (6.23a)$$

$$S' = \Pi - \mu S - \beta SI, \qquad (6.23b)$$

$$R' = \alpha I - \mu R. \qquad (6.23c)$$

From the previous section, we know that $(V - F)$ is a nonsingular M-matrix whenever $\mathcal{R}_o < 1$. Hence, to show that the disease-free equilibrium is globally asymptotically stable for $\mathcal{R}_o < 1$, it is sufficient to show that $S \leq S_o$. The total population $N = S + E + I + R$ satisfies $N' = \Pi - \mu N$, so that $N(t) = S_o - (S_o - N(0))e^{-\mu t}$, with $S_o = \Pi/\mu$. If $N(0) \leq S_o$, then $S(t) \leq N(t) \leq S_o$ for all time. If, on the other hand, $N(0) > S_o$, then $N(t)$ decays exponentially to S_o, and either $S(t) \to S_o$, or there is some time T after which $S(t) < S_o$. Thus, from time T onward, $x(t)$ is bounded above, in each component, by $e^{-(t-T)(V-F)}x(T)$, which decays exponentially to zero. Note that for the argument of global stability we are not concerned with the size of $x(T)$. In fact, if $N(0) > S_o$, $x(T)$ may be much larger than $x(0)$. In this case the exponential bound on $x(t)$ concerns a decay following an epidemic, not an immediate elimination of the disease. In contrast, if $N(0) < S_o$, then the bound on $x(t)$ is $e^{-(t-T)(V-F)}x(0)$, and no epidemic occurs.

A simple model with a backward bifurcation is the vaccination model proposed by Kribs-Zaleta and Velasco-Hernández [15].

$$S' = \Pi - (\mu + \xi)S + \theta S_v - \beta SI + \gamma I, \qquad (6.24a)$$

$$S'_v = \xi S - (\mu + \theta)S_v - \beta(1 - r)S_v I, \qquad (6.24b)$$

$$I' = \beta(S + (1 - r)S_v)I - (\gamma + \mu)I. \qquad (6.24c)$$

As with the model of Sect. 6.4.4, vaccination reduces the force of infection by a factor r. However, in this model, susceptible individuals are vaccinated continuously at a rate ξ, and the protection acquired from vaccination wanes at a rate θ. Additionally, individuals recover from infection with no immunity, regardless of their vaccination history. The model has a unique disease-free equilibrium, where $S_o = (1 - p)N_o$ and $S_{vo} = pN_o$ with $N_o = \Pi/\mu$ and $p = \xi/(\mu+\theta+\xi)$. In keeping with the conventions of the literature, we denote by $\mathcal{R}_o$, the basic reproduction number in the absence of vaccination, and by $\mathcal{R}_c$, the control reproduction number, or the basic reproduction number in the presence of vaccination. For the model above we have

$$\mathcal{R}_o = \frac{\beta N_o}{\gamma + \mu},$$

$$\mathcal{R}_c = \frac{\beta\left(S_o + (1 - r)S_{vo}\right)}{\gamma + \mu} < \mathcal{R}_o.$$

The disease-free equilibrium is locally asymptotically stable if $\mathcal{R}_c < 1$ and unstable if $\mathcal{R}_c > 1$.

Equation (6.24c) can be written as

$$I' = (\gamma + \mu)(\mathcal{R}_o - 1)I - \beta(N_o - S - (1 - r)S_v)I. \qquad (6.25)$$

The matrix A in this example is simply $(\gamma+\mu)(1-\mathcal{R}_o)$. As with our previous example, we can assume that $S+S_v \le N_o$, since N approaches N_o asymptotically. Hence, the second term in the above equation eventually becomes and remains negative and the disease-free equilibrium is globally asymptotically stable if $\mathcal{R}_o < 1$.

The disease-free equilibrium may be locally asymptotically stable, but not globally asymptotically stable if $\mathcal{R}_c < 1 < \mathcal{R}_o$. It is known that there may be multiple endemic equilibria for parameter values in this range; further details can be found in Kribs-Zaleta and Velasco-Hernández [15].

References

1. R. M. Anderson and R. M. May, *Infectious Diseases of Humans*, Oxford University Press, Oxford, 1991.
2. V. Andreasen, J. Lin, and S. A. Levin, *The dynamics of cocirculating influenza strains conferring partial cross-immunity*, J. Math. Biol., 35 (1997), pp. 825–842.
3. A. Berman and R. J. Plemmons, *Nonnegative Matrices in the Mathematical Sciences*, Academic, New York, 1970.

4. S. M. Blower, P. M. Small, and P. C. Hopewell, *Control strategies for tuberculosis epidemics: new models for old problems*, Science, 273 (1996), pp. 497–500.
5. C. Castillo-Chavez and Z. Feng, *To treat or not to treat: the case of tuberculosis*, J. Math. Biol., 35 (1997), pp. 629–656.
6. C. Castillo-Chavez, Z. Feng, and W. Huang, *On the computation of $\mathcal{R}_o$ and its role on global stability*, in Mathematical approaches for emerging and reemerging infectious diseases: models, methods and theory, C. Castillo-Chavez, S. Blower, P. van den Driessche, D. Kirschner, and A.-A. Yakubu, eds., Springer, Berlin Heidelberg New York, 2002, pp. 229–250.
7. B. R. Cherry, M. J. Reeves, and G. Smith, *Evaluation of bovine viral diarrhea virus control using a mathematical model of infection dynamics*, Prev. Vet. Med., 33 (1998), pp. 91–108.
8. O. Diekmann and J. A. P. Heesterbeek, *Mathematical epidemiology of infectious diseases*, Wiley series in mathematical and computational biology, Wiley, West Sussex, England, 2000.
9. Z. Feng and J. X. Velasco-Hernández, *Competitive exclusion in a vector-host model for the Dengue fever*, J. Math. Biol., 35 (1997), pp. 523–544.
10. S. Gandon, M. Mackinnon, S. Nee, and A. Read, *Imperfect vaccination: some epidemiological and evolutionary consequences*, Proc. R. Soc. Lond. B., 270 (2003), pp. 1129–1136.
11. J. M. Heffernan, R. J. Smith, and L. M. Wahl, *Perspectives on the basic reproductive ratio*, J. R. Soc. Interface, 2 (2005), pp. 281–293.
12. H. W. Hethcote, *The mathematics of infectious diseases*, SIAM Rev., 42 (2000), pp. 599–653.
13. M. W. Hirsch and S. Smale, *Differential Equations, Dynamical Systems, and Linear Algebra*, Academic, Orlando, 1974.
14. Y.-H. Hsieh and C. H. Chen, *Modelling the social dynamics of a sex industry: its implications for spread of HIV/AIDS*, Bull. Math. Biol., 66 (2004), pp. 143–166.
15. C. M. Kribs-Zaleta and J. X. Velasco-Hernández, *A simple vaccination model with multiple endemic states*, Math. Biosci., 164 (2000), pp. 183–201.
16. M. Nuño, Z. Feng, M. Martcheva, and C. Castillo-Chavez, *Dynamics of two-strain influenza with isolation and partial cross-immunity*, SIAM J. Appl. Math., 65 (2005), pp. 964–982.
17. M. G. Roberts and J. A. P. Heesterbeek, *A new method for estimating the effort required to control an infectious disease*, Proc. R. Soc. Lond., 270 (2003), pp. 1359–1364.
18. P. van den Driessche and J. Watmough, *Reproduction numbers and sub-threshold endemic equilibria for compartmental models of disease transmission*, Math. Biosci., 180 (2002), pp. 29–48.
19. S. Wiggins, *Introduction to Applied Nonlinear Dynamical Systems and Chaos*, Springer, Berlin Heidelberg New York, 1990.
20. M. J. Wonham, T. de Camino-Beck, and M. Lewis, *An epidemiological model for West Nile virus: Invasion, analysis and control applications*, Proc. R. Soc. Lond. B, 271 (2004), pp. 501–507.

Chapter 7
Spatial Structure: Patch Models

P. van den Driessche

Abstract Discrete spatial heterogenity is introduced into disease transmission models, resulting in large systems of ordinary differential equations. Such metapopulation models describe disease spread on a number of spatial patches. In the first model considered, there is no explicit movement of individuals; rather infectives can pass the disease to susceptibles in other patches. The second type of model explicitly includes rates of travel between patches and also takes account of the resident patch as well as the current patch of individuals. A formula for and useful bounds on the basic reproduction number of the system are determined. Brief descriptions of application of this type of metapopulation model are given to investigate the spread of bovine tuberculosis and the effect of quarantine on the spread of influenza.

7.1 Introduction

Basic deterministic models assume no spatial variation. However, since both the environment and any population are spatially heterogeneous, it is obviously desirable to include spatial structure into an epidemic model. Demographic and disease parameters may vary spatially, and a human population may live in cities or be scattered in rural areas. Populations travel, animals and people by foot, birds and mosquitoes (reservoir and vector for West Nile virus) by wing. In addition, people travel by air between cities, so diseases can be spread quickly between very distant places (as was the case with the SARS outbreak in 2003).

Department of Mathematics and Statistics, University of Victoria, P.O. BOX 3060 STN CSC, Victoria, BC, Canada V8W 3R4
pvdd@math.uvic.ca

Spatial structure can be included in either a continuous or discrete way. If time is assumed continuous, then continuous space yields reaction-diffusion equations, which are discussed in the chapter by Wu [15], whereas discrete space yields coupled patch models. These models, which are the focus of this chapter, are called metapopulation models . They usually consist of a system (often a large system) of ordinary differential equations with the dynamics of each patch coupled to that of other patches by travel. A patch can be a city, community, or some other geographical region. If time is assumed discrete, then continuous space yields integrodifference equation models, whereas discrete space yields coupled lattice or cellular automata models [8, page 268].

Four different types of metapopulation models from the literature are considered in this chapter. The first two models are fairly general and are formulated and discussed in detail, whereas the last two, which deal with influenza and with tuberculosis in possums, are described in less detail. But since these metapopulation models can be complicated, readers are asked to consult the references for more background and details.

7.2 Spatial Heterogeneity

Consider a basic susceptible, exposed, infective, recovered (SEIR) compartmental model such as is frequently used for childhood diseases; see, for example, [1], [13, Sect. 2.2 with $p = q = 0$]. To incorporate spatial effects, Lloyd and May [9] divide the population into connected subpopulations. Let S_i, E_i, I_i, R_i denote respectively the number of susceptible, exposed, infective and recovered individuals in patch i for $i = 1, ..., n$. The total population of patch i is $N_i = S_i + E_i + I_i + R_i$. The birth and natural death rate constant d is assumed to be the same in each patch, so that the total population of each patch remains constant. The average latent period $1/\epsilon$ and the average infectious period $1/\gamma$ are assumed to be the same in each patch. This spatial model can be written for $i = 1, ..., n$ as

$$
\begin{aligned}
S_i' &= dN_i - dS_i - \lambda_i S_i \\
E_i' &= \lambda_i S_i - (d + \epsilon)E_i \\
I_i' &= \epsilon E_i - (d + \gamma)I_i \\
R_i' &= \gamma I_i - dR_i;
\end{aligned}
\tag{7.1}
$$

with the force of infection in patch i given by a mass action type of incidence

$$
\lambda_i = \sum_{j=1}^{n} \beta_{ij} I_j.
$$

Thus infective individuals in one patch can infect susceptible individuals in another patch, but there is no explicit movement of individuals in this model. If the exposed period tends to zero, corresponding to $\epsilon \to \infty$, then this reduces to an SIR model; which is now analyzed. Please consult [9] for analysis of the SEIR model.

For the SIR model, the equations are

$$S_i' = dN_i - dS_i - \lambda_i S_i$$
$$I_i' = \sum_{j=1}^{n} \beta_{ij} I_j S_i - (d+\gamma)I_i \qquad (7.2)$$

and the disease-free equilibrium is $S_i = N_i$, $I_i = R_i = 0$. Using the next generation matrix method [14], the basic reproduction number $\mathcal{R}_0$ can be calculated from (7.2) as $\mathcal{R}_0 = \rho(FV^{-1})$ where the i,j entry of FV^{-1} is $\beta_{ij} N_i/(d+\gamma)$.

For the case that each patch has the same population (i.e., $N_i = N$) and β_{ij} are such that the endemic equilibrium values of S_i, I_i, and λ_i are independent of i, then the endemic equilibrium is given explicitly for $\mathcal{R}_0 > 1$ by

$$S_{i\infty} = S_\infty = \frac{N}{\mathcal{R}_0}, \quad I_{i\infty} = I_\infty = \frac{dN}{d+\gamma}(1 - \frac{1}{\mathcal{R}_0}), \quad R_{i\infty} = R_\infty = \frac{\gamma I_\infty}{d},$$
$$(7.3)$$

with $\lambda_{i\infty} = \lambda_\infty = d(\mathcal{R}_0 - 1)$.

As an example of a symmetric situation that satisfies the above requirements, assume that $\beta_{ij} = \beta$ if $i = j$ and $\beta_{ij} = p\beta$ with $p < 1$ if $i \neq j$. Thus the contact rate is the same within each patch and has a smaller value between each pair of different patches. Then matrix $B = [\beta_{ij}] = \beta(pJ_{n \times n} + (1-p)I_{n \times n})$ where $J_{n \times n}$ is the matrix of all ones and $I_{n \times n}$ is the identity matrix. The eigenvalues of B are $\beta[pn + (1-p)]$, which is simple, and $\beta(1-p)$ with multiplicity $n - 1$. Thus

$$\mathcal{R}_0 = \frac{\beta N[pn + (1-p)]}{(d+\gamma)},$$

which depends on the number of patches n and the coupling strength p.

Linearizing about the endemic equilibrium and assuming solutions are proportional to $\exp(zt)$ yields a characteristic equation that can be written in the form $\det(B - \Gamma I_{n \times n}) = 0$ with

$$\Gamma = (z + d + \lambda_\infty)\frac{(z + d + \gamma)}{(z+d)S_\infty}.$$

Thus Γ takes values that are the eigenvalues of B. The simple eigenvalue of B gives rise to the quadratic

$$z^2 + d\mathcal{R}_0 z + d(d+\gamma)(\mathcal{R}_0 - 1) = 0.$$

Following [1], the authors [9] set $d\mathcal{R}_0 = 1/A$ where A is the average age of first infection, and $1/\tau = d + \gamma \approx \gamma$ where τ is the average infective period. As $A >> \tau$, the quadratic can be approximated by

$$z^2 + z/A + 1/(A\tau) = 0,$$

which gives $z \approx -1/(2A) \pm i/\sqrt{A\tau}$. This represents a weakly damped oscillation about and towards the endemic equilibrium, with the period of oscillation shorter than the damping time. The number of individuals in each compartment in all patches oscillate in phase. The repeated eigenvalue (of multiplicity $n-1$) of B gives rise to internal modes that are strongly damped. Thus, for all but the smallest values of p, the oscillations quickly become phase locked. This result is based on a linear stability analysis, but the authors [9] believe that the endemic equilibrium is globally attracting for $\mathcal{R}_0 > 1$.

Simulation results for a two patch model ($n = 2$) for the above example are presented [9, Sect. 4]. Parameters are chosen to model measles epidemics in a population of $N_1 = N_2 = 10^6$, with $d = 0.02$ year^{-1}. The average infective period is taken as five days ($\tau = 5$) giving $\gamma = 73.0$ year^{-1}, with $\beta = 0.0010107$ year^{-1} infective^{-1}. When $p = 0$, these parameters give $\mathcal{R}_0 \approx 13.8$, with $A \approx 3.6$ years. Rapid phase locking occurs for p larger than about 0.002. For $p = 0.01$, numerical simulations show I_1 and I_2 synchronized by about five years as they approach I_∞ given by (7.3) with damped oscillations. A stochastic formulation of this SIR model is also shown numerically to give synchronization, although a slightly larger value of p is required. Assuming that the within-patch contact rate is seasonally forced, in-phase and out of phase biennial oscillations are seen with chaotic solutions possible for some parameter values (illustrated in [9, Fig. 3] with $p = 10^{-3}$). As found in other metapopulation models, with larger p values (i.e., stronger between patch coupling) the system effectively behaves more like that of a single patch.

7.3 Geographic Spread

Sattenspiel and Dietz [11] introduced a metapopulation epidemic model in which individuals are labeled with their city of residence as well as the city in which they are present at a given time. This model explicitly includes rates of travel between n patches, which can be cities or geographical regions. A susceptible-infective-recovered (SIR) model incorporating this spatial heterogeneity is formulated [11, Sect. 2], and the same spatial heterogeneity is incorporated in a susceptible- infective-susceptible (SIS) model formulated by Arino and van den Driessche [2]. The notation of [2] is used here.

To formulate the demographic model with travel, let $N_{ij}(t)$ be the number of residents of patch i who are present in patch j at a time t. Residents of patch i leave this patch at a per capita rate $g_i \geq 0$ per unit time, with a fraction $m_{ji} \geq 0$ going to patch j, thus $g_i m_{ji}$ is the travel rate from patch i to patch j. Here $m_{ii} = 0$ and $\sum_{j=1}^{n} m_{ij} = 1$. Residents of patch i who are in patch j return home to patch i with a per capita rate of $r_{ij} \geq 0$ with $r_{ii} = 0$. It is natural to assume that $g_i m_{ji} > 0$ if and only if $r_{ij} > 0$. These travel rates determine a directed graph with patches as vertices and arcs between vertices if the travel rates between them are positive. It is assumed that the travel rates are such that this directed graph is strongly connected.

Assume that birth occurs in the home patch at a per capita rate $d > 0$, and death occurs in any patch with this same rate. Then the population numbers satisfy the equations

$$N'_{ii} = \sum_{k=1}^{n} r_{ik} N_{ik} - g_i N_{ii} + d\left(\sum_{k=1}^{n} N_{ik} - N_{ii}\right) \tag{7.4}$$

$$N'_{ij} = g_i m_{ji} N_{ii} - r_{ij} N_{ij} - d N_{ij} \text{ for } i \neq j. \tag{7.5}$$

These equations describe the evolution of the number of residents in patch i who are currently in patch i (7.4) and those who are currently in patch $j \neq i$ (7.5). The number of residents of patch i, namely $N_i^r = \sum_{j=1}^{n} N_{ij}$ is constant, as is the total population of the n patch system. With initial conditions $N_{ij}(0) > 0$, the system (7.4)–(7.5) has an asymptotically stable equilibrium $\hat{N}_{ij}$.

An epidemic model is now formulated in each of the n patches, with $S_{ij}(t)$ and $I_{ij}(t)$ denoting the number of susceptible and infective individuals resident in patch i who are present in patch j at time t. Taking an SIS model with standard incidence [2], equations for the evolution of the number of susceptibles and infectives resident in patch i (with $i = 1, ..., n$) are

$$S'_{ii} = \sum_{k=1}^{n} r_{ik} S_{ik} - g_i S_{ii} - \sum_{k=1}^{n} \kappa_i \beta_{iki} \frac{S_{ii} I_{ki}}{N_i^p} + d\left(\sum_{k=1}^{n} N_{ik} - S_{ii}\right) + \gamma I_{ii} \tag{7.6}$$

$$I'_{ii} = \sum_{k=1}^{n} r_{ik} I_{ik} - g_i I_{ii} + \sum_{k=1}^{n} \kappa_i \beta_{iki} \frac{S_{ii} I_{ki}}{N_i^p} - (\gamma + d) I_{ii} \tag{7.7}$$

and for $j \neq i$

$$S'_{ij} = g_i m_{ji} S_{ii} - r_{ij} S_{ij} - \sum_{k=1}^{n} \kappa_j \beta_{ikj} \frac{S_{ij} I_{kj}}{N_j^p} - d S_{ij} + \gamma I_{ij} \tag{7.8}$$

$$I'_{ij} = g_i m_{ji} I_{ii} - r_{ij} I_{ij} + \sum_{k=1}^{n} \kappa_j \beta_{ikj} \frac{S_{ij} I_{kj}}{N_j^p} - (\gamma + d) I_{ij} \tag{7.9}$$

with $N_i^p = \sum_{j=1}^n N_{ji}$, the number present in patch i. Here $\beta_{ikj} > 0$ is the proportion of adequate contacts in patch j between a susceptible from patch i and an infective from patch k that results in disease transmission, $\kappa_j > 0$ is the average number of such contacts in patch j per unit time, and $\gamma > 0$ is the recovery rate of infectives (assumed the same in each patch). Note that the disease is assumed to be sufficiently mild so that it does not cause death and does not inhibit travel, and individuals do not change disease status during travel. Equations (7.6)–(7.9) together with nonnegative initial conditions constitute the SIS metapopulation model. It can be shown that the nonnegative orthant $R_+^{2n^2}$ is positively invariant under the flow and solutions are bounded.

The disease-free equilibrium is given by $S_{ij} = \hat{N}_{ij}, I_{ij} = 0$ for all $i, j = 1, ..., n$. If the system is at an equilibrium and one patch is at the disease-free equilibrium, then all patches are at the disease-free equilibrium; whereas if one patch is at an endemic disease level, then all patches are at an endemic level. These results hold based on the assumption that the directed graph determined by the travel rates is strongly connected. If this is not the case, then the results apply to patches within a strongly connected component.

For the n-patch connected model, the next generation matrix [6,14] can be determined from (7.7) and (7.9), leading to a formula for the basic reproduction number $\mathcal{R}_0$. To keep the notation simple, the formula for $n = 2$ patches is explicitly given here, thus $m_{12} = m_{21} = 1$, and g_1, g_2, r_{12}, r_{21} are assumed positive. For the general n case see [2, page 185]. Ordering the infective variables as $I_{11}, I_{12}, I_{21}, I_{22}$, the matrix of new infections at the disease-free equilibrium F is a block matrix with 4 blocks, in which each block F_{ij} is the 2×2 diagonal matrix

$$F_{ij} = diag\left(\kappa_1 \beta_{ij1} \frac{\hat{N}_{i1}}{\hat{N}_1^p}, \quad \kappa_2 \beta_{ij2} \frac{\hat{N}_{i2}}{\hat{N}_2^p}\right).$$

Matrix V, accounting for transfer between infective compartments, can be written as $V = V_1 \oplus V_2$, where $\oplus$ denotes the direct sum, and

$$V_1 = \begin{bmatrix} g_1 + \gamma + d & -r_{12} \\ -g_1 & r_{12} + \gamma + d \end{bmatrix}, \quad V_2 = \begin{bmatrix} r_{21} + \gamma + d & -g_2 \\ -r_{21} & g_2 + \gamma + d \end{bmatrix}.$$

Matrices V_1 and V_2 are irreducible nonsingular M-matrices (for definition and properties of M-matrices see [4]) thus their inverses are positive, and $V^{-1} = V_1^{-1} \oplus V_2^{-1}$. Using these blocks, $\mathcal{R}_0$ can easily be computed for a given set of parameter values as $\mathcal{R}_0 = \rho(FV^{-1})$, where ρ denotes the spectral radius. For n patches, a similar formula is obtained, with FV^{-1} being an $n^2 \times n^2$ positive matrix. It is apparent that $\mathcal{R}_0$ depends on the travel rates as well as the epidemic parameters. If $\mathcal{R}_0 < 1$, then the disease-free equilibrium is locally asymptotically stable; whereas if $\mathcal{R}_0 > 1$, then it is unstable.

Assume that the disease transmission coefficients are equal for all populations present in a patch, i.e., $\beta_{ijk} = \beta_k$ for $i, j = 1, ..., n$. With this assumption, the following bounds can be found for $\mathcal{R}_0$ for n patches:

$$\min_{i=1,...,n} \mathcal{R}_0^{(i)} \leq \mathcal{R}_0 \leq \max_{i=1,...,n} \mathcal{R}_0^{(i)} \tag{7.10}$$

where $\mathcal{R}_0^{(i)} = \kappa_i \beta_i / (d + \gamma)$ is the basic reproduction number of patch i in isolation. Thus if $\mathcal{R}_0^{(i)} < 1$ for all i, the disease dies out; whereas if $\mathcal{R}_0^{(i)} > 1$ for all i, then the disease-free equilibrium is unstable.

Numerical simulations presented in [2] with $n = 3$ patches show that (for parameter values relevant for gonorrhea) when $\mathcal{R}_0 > 1$, the number of infectives in each subpopulation goes to an endemic value. Further numerical investigations for $n = 2$ patches focus on a case in which in isolation the disease would be absent in patch 1 but endemic in patch 2 (i.e., $\mathcal{R}_0^{(1)} < 1$, $\mathcal{R}_0^{(2)} > 1$). The bounds in (7.10) give $\mathcal{R}_0 \in [\mathcal{R}_0^{(1)}, \mathcal{R}_0^{(2)}]$. Parameter values are chosen to be relevant for a disease like gonorrhea: $\gamma = 1/25$, $d = 1/(75 \times 365)$ with the time unit of one day. Suppose that $N_1^r = N_2^r = 1500$, $\kappa_1 = \kappa_2 = 1$, $r_{12} = r_{21} = 0.05$, $\beta_1 = 0.016$ giving $\mathcal{R}_0^{(1)} = 0.4$, and $\beta_2 = 0.048$ giving $\mathcal{R}_0^{(2)} = 1.2$. A change in travel rates g_1, g_2 can induce a bifurcation from $\mathcal{R}_0 < 1$ to $\mathcal{R}_0 > 1$ or vice versa, see [2, Fig. 3a]. Another view of this case is presented in Fig. 7.1 in which $g_1 = g_2$ and $\mathcal{R}_0$ is plotted as a function of $g_1 = g_2$, with $\mathcal{R}_0 = 1$ shown as a broken horizontal line. Thus travel can stabilize (small travel rates) or destabilize (larger travel rates) the disease-free equilibrium. These numerical results support the claim that for $\mathcal{R}_0 > 1$, the endemic equilibrium is unique and that $\mathcal{R}_0$ acts as a sharp threshold between extinction and invasion of the disease.

Similar conclusions are drawn for the more general SEIRS model [3], for which an explicit formula for $\mathcal{R}_0$ is derived. Sattenspiel and Dietz [11] suggested an application of their metapopulation SIR model to the spread of measles in the 1984 epidemic in Dominica. Travel rates of infants, school-age children and adults are assumed to be different, thus making the model system highly complex and requiring knowledge of more data for simulation. Sattenspiel and coworkers; see [10] and references therein, have since used this modeling approach for studying other infectious diseases in the historical archives; one such example is discussed further in the next section.

7.4 Effect of Quarantine on Spread of 1918–1919 Influenza in Central Canada

Work by Sattenspiel and Herring focuses on the spread of the 1918–1919 influenza epidemic in three communities in central Manitoba, Canada. The effect of quarantine on the spread of this epidemic is discussed by Sattenspiel

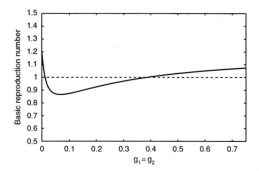

Fig. 7.1 The basic reproduction number as a function of travel rates for the two patch model. For parameter values, see text. Figure by Mahin Salmani

and Herring [12], which is a good source of references to their work. The three communities, all of which are former Hudson's Bay Company fur trade posts, are Norway House (population 746), Oxford House (population 322) and God's Lake (population 299). Many of the inhabitants were fur trappers and Norway House was on a main trading route. Oxford House was in direct contact with Norway House, and God's Lake was connected to the other posts by less travelled routes.

The influenza epidemic occurred in this region during the late fall of 1918 and the following winter. There were 107 deaths among residents of Norway House in a period of a year beginning in July 1918, with most of these probably from influenza. The time taken in winter to travel between Norway House and Oxford House was six days, a factor that slowed the spread of the disease, especially since influenza has a period of communicability of 3–5 days from clinical onset in adults [5, page 272]. Sattenspiel and Herring [12] used a fascinating mix of mathematical modeling and estimation of model parameters, both epidemiological parameters for influenza and anthropological parameters relating to population numbers and travel, to investigate the impact of attempts to limit travel during this influenza epidemic. Norway House was quarantined during December 1918 and January 1919, and this control measure is investigated. Quarantining here means the limit of travel of any individual regardless of disease status. The SIR model formulated for n patches [12, Sect. 3] is similar to the SIS model described in Sect. 7.3

above, except that birth and death are ignored (i.e., $d = 0$). Using the notation of Sect. 7.3, and letting $R_{ij}(t)$ denote the number of recovered individuals resident in patch i who are present in patch j at time t, the system of equations for individuals residing in patch i (with $i = 1, ...n$) is

$$S'_{ii} = \sum_{k=1}^{n} r_{ik}S_{ik} - g_iS_{ii} - \sum_{k=1}^{n} \kappa_i\beta_{iki}\frac{S_{ii}I_{ki}}{N_i^p}$$

$$I'_{ii} = \sum_{k=1}^{n} r_{ik}I_{ik} - g_iI_{ii} + \sum_{k=1}^{n} \kappa_i\beta_{iki}\frac{S_{ii}I_{ki}}{N_i^p} - \gamma I_{ii}$$

$$R'_{ii} = \sum_{k=1}^{n} r_{ik}R_{ik} - g_iR_{ii} + \gamma I_{ii}$$

and for $j \neq i$

$$S'_{ij} = g_im_{ji}S_{ii} - r_{ij}S_{ij} - \sum_{k=1}^{n} \kappa_j\beta_{ikj}\frac{S_{ij}I_{kj}}{N_j^p}$$

$$I'_{ij} = g_im_{ji}I_{ii} - r_{ij}I_{ij} + \sum_{k=1}^{n} \kappa_j\beta_{ikj}\frac{S_{ij}I_{kj}}{N_j^p} - \gamma I_{ij}$$

$$R'_{ij} = g_im_{ji}R_{ii} - r_{ij}R_{ij} + \gamma I_{ij}.$$

The basic reproduction number $\mathcal{R}_0$ is calculated as in Sect. 5.3 with $d = 0$.

Quarantine is incorporated by adjusting the rates of travel between patches, thus g_i and r_{ij} are multiplied by a factor q_i when patch i is quarantined. The parameter q_i lies in the range $0 < q_i < 1$, where 1 would correspond to no quarantine and 0 would correspond to perfect quarantine. The quarantine (control) reproduction number is calculated as for $\mathcal{R}_0$, but including the factor q_i.

Sattenspiel and Herring [12, Sect. 4] estimated model parameters needed for the above system with $n = 3$, for Norway House (NH), Oxford House (OH) and God's Lake (GL). They set $\gamma = 0.2$ (average influenza infective period 5 days), $\beta_{ikj} = \beta = 0.5$ (50% of all contacts result in infection) for all communities, and $\kappa_{NH} = 1$, $\kappa_{OH} = \kappa_{GL} = 0.5$ (twice as many contacts at NH). Estimates of g_i, m_{ji} and r_{ij} were made from records kept by the Hudson Bay Company of arrivals at and departures from each community, and population numbers were determined from census data. Quarantine at NH was incorporated by multiplying g_{NH}, $r_{NH,OH}$ and $r_{NH,GL}$ by a factor q_{NH}.

Simulation results of the above system [12, Sect. 5] show that quarantine causes a slight delay in the arrival of the epidemic peak, with an increase in the peak number of infectives at NH and a decrease in those at OH and GL. Quarantine is found to be most effective if started well before an epidemic peaks, but not right at the start of an epidemic. However, starting quarantining after an epidemic peaks, has little effect. The authors [12] conclude that when travel rates are low (as between these communities), quarantine must be highly effective before it significantly alters disease patterns.

7.5 Tuberculosis in Possums

An SEI metapopulation model for the spread of bovine tuberculosis (*Mycobacterium bovis*) in the common brushtail possum (*Trichosurus vulpecula*) in New Zealand is formulated and analyzed by Fulford et al. [7]. This is now described very briefly to demonstrate the use of a complex metapopulation model; please consult the original paper for full details.

Tuberculosis has a significant latent period and is a fatal disease in possums, thus an SEI model is appropriate. A two-age class model is formulated with juveniles and adults, with susceptible and exposed (but not infective) juveniles migrating as they mature (1 to 2 years old). In addition pseudo-vertical transmission is included, accounting for disease transmission between mothers to young in their pouch. For n patches, a system of $6n$ equations describes the dynamics. For a two patch model ($n = 2$), the authors determine the disease-free equilibrium numerically, provide a methodology for computing $\mathcal{R}_0$ by using the next generation matrix [6], and show how this can be generalized to n patches.

Parameters appropriate for tuberculosis in possums are taken for simulations and for a two patch model prevalence shows damped oscillations towards an endemic state with $\mathcal{R}_0 \in [1.55, 1.67]$. Possums are thought to spread tuberculosis to farmland, thus the model is applied to evaluate control (culling) strategies to reduce the control reproduction number below one [7, Sect. 6]. Several spatial configurations are considered and critical culling rates (giving the control reproduction number equal to one) are calculated.

7.6 Concluding Remarks

The metapopulation models discussed in the previous sections demonstrate that such spatial models are usually high dimensional and contain many parameters. However, extended models can be formulated, including biological realism such as age structure and control measures such as restriction of travel. Using the next generation matrix, the basic reproduction number $\mathcal{R}_0$ can be computed for estimated parameters. Simulations can easily be performed with parameters relevant for a particular disease with given demography and spatial structure. These models assume that the population of each patch is sufficiently large so that a deterministic model is appropriate and there is homogeneous mixing within each patch. As noted in [12], stochastic effects may be significant when patch populations are small, such as in the three communities modeled by [12] and discussed in Sect. 7.4 above.

References

1. Anderson, R.M. and May, R.M.: Infectious Diseases of Humans. Oxford University Press, Oxford (1991)
2. Arino, J. and van den Driessche, P.: A multi-city epidemic model. Math. Popul. Stud. **10**, 175–193 (2003)
3. Arino, J. and van den Driessche, P.: The basic reproducton number in a multi-city compartment model. LNCIS. **294**, 135–142 (2003)
4. Berman, A. and Plemmons, R.J.: Nonnegative Matrices in the Mathematical Sciences. Academic, New York (1979)
5. Chin, J.: Control of Communicable Diseases Manual. 17th Edition. American Public Health Association, Washington (2000)
6. Diekmann, O., and Heesterbeek, J.A.P.: Mathematical Epidemiology of Infectious Diseases. Model Building, Analysis and Interpretation. Wiley, New York (2000)
7. Fulford, G.R., Roberts, M.G., and Heesterbeek, J.A.P.: The metapopulation dynamics of an infectious disease: tuberculosis in possums. Theor. Pop. Biol. **61**, 15–29 (2002)
8. Kot, M.: Elements of Mathematical Ecology. Cambridge University Press, Cambridge (2001)
9. Lloyd, A., and May, R.M.: Spatial heterogeneity in epidemic models. J. Theor. Biol. **179**, 1–11 (1996)
10. Sattenspiel, L.: Infectious diseases in the historical archives: a modeling approach. In: Herring, D.A. and Swedlund, A.C. (eds) Human Biologists in the Archives. Cambridge University Press, Cambridge 234–265 (2003)
11. Sattenspiel, L. and Dietz, K.: A structured epidemic model incorporating geographic mobility among regions. Math. Bios. **128**, 71–91 (1995)
12. Sattenspiel, L. and Herring, D.A.: Simulating the effect of quarantine on the spread of the 1918–1919 flu in central Canada. Bull. Math. Biol. **65**, 1–26 (2003)
13. van den Driessche, P.: Deterministic compartmental models: extensions of basic models. Chapter 5 of Mathematical Epidemiology (this volume).
14. van den Driessche, P., and Watmough, J.: Reproduction numbers and sub-threshold endemic equilibria for compartmental models of disease transmission. Math. Bios. **180**, 29–48 (2002)
15. Wu, J.: Spatial Structure: Partial differential equation models. Chapter 8 of Mathematical Epidemiology (this volume)

Chapter 8
Spatial Structure: Partial Differential Equations Models

Jianhong Wu

Abstract This chapter introduces some basic concepts and techniques in modeling spatial spread of diseases involving hosts moving randomly during certain stages of the disease progression. First we derive some reaction diffusion models using the conservation law and Fick's law of diffusion. We then discuss the usefulness of these models in describing disease spread rates and evaluating the effectiveness of some spatially relevant disease control strategies. We illustrate the general theory via two case studies, one about the spread of rabies in continental Europe during the period 1945–1985 and another about spread rates of West Nile virus in North America.

8.1 Introduction

As discussed in [19], spatial structures play an important role in describing the spreading of communicable diseases, not only because the environment is heterogeneous but also because individuals move around in space. Many prevention and control strategies involve spatial aspects such as immigration, vaccination, border control and restriction of individual movements.

In this chapter, we introduce an approach, based on reaction diffusion equations, to describe the spread of communicable diseases in spatially structured populations. We shall also illustrate this approach and demonstrate its applications via two case studies; one is about the spread of rabies in continental Europe during the period 1945–1985 and another is about spread rates of West Nile virus.

Center for Disease Modeling, Department of Mathematics and Statistics, York University, Toronto, ON, Canada M3J 1P3
wujh@mathstat.yorku.ca

We have no intention here to give a comprehensive introduction to a subject that has been intensively studied, and we refer to [10] that has great influence on the next two sections of this chapter, and [5–7, 15, 18] for relevant literature and detailed discussions.

8.2 Model Derivation

We first consider the case in which a collection of individuals moves randomly in one dimensional space, with an average step length Δx in every time unit Δt. Assume the growth rate (with respect to time) at spatial location x and time t is given by $f(t, x)$ (this term, in most cases, also depends explicitly on the numbers of the individuals) and assume the probability of moving to the left and to the right are both equal, and hence are 0.5.

Let $u(t, x)$ be the number of individuals within the spatial segment $[x, x + \Delta x]$ at the time t. Then

$$u(t + \Delta t, x) - u(t, x) = \frac{1}{2}u(t, x - \Delta x) + \frac{1}{2}u(t, x + \Delta x) - u(t, x) + f(t, x)\Delta t.$$

Using the Taylor series expansions for $u(t, x \pm \Delta x)$ and $u(t + \Delta t, x)$ as follows

$$u(t, x \pm \Delta x) = u(t, x) \pm \frac{\partial}{\partial x}u(t, x)\Delta x + \frac{1}{2}\frac{\partial^2}{\partial x^2}u(t, x,)(\Delta x)^2 + \cdots ,$$
$$u(t + \Delta t, x) = u(t, x) + \frac{\partial}{\partial t}u(t, x)\Delta t + \frac{1}{2}\frac{\partial^2}{\partial t^2}u(t, x)(\Delta t)^2 + \cdots ,$$

we obtain

$$\frac{\partial}{\partial t}u(t, x)\Delta t + \frac{1}{2}\frac{\partial^2}{\partial t^2}u(t, x)(\Delta t)^2 + \cdots = \frac{1}{2}\frac{\partial^2}{\partial x^2}u(t, x)(\Delta x)^2 + \cdots + f(t, x)\Delta t,$$

$$(8.1)$$

where $\cdots$ denotes higher order terms. Assume that the temporal and spatial scales are chosen appropriately so that

$$\frac{(\Delta x)^2}{2\Delta t} = D \tag{8.2}$$

remains to be a given constant, called the *diffusion coefficient*. Dividing (8.1) by Δt and then letting $\Delta t \to 0$ and $\Delta x \to 0$, we get

$$\frac{\partial}{\partial t}u(t, x) = D\frac{\partial^2}{\partial x^2}u(t, x) + f(t, x). \tag{8.3}$$

Note the above derivation also suggests how D can be estimated from field data.

The above reaction diffusion equation can also be established through the *conservation equation* based on a certain *balance law* [5]. To describe the model derivation, we consider the spatial segment $[x, x + \Delta x]$ and note that the change of the total number of individuals in this segment is due to the flow into and out of the interval through its boundary points; and due to the (local) growth process that reproduces the individuals within the segment. In other words, if we denote by $J(t, x)$ the number of individuals crossing at x in the positive direction per unit time, then we have the following balance law:

$$\frac{\partial}{\partial t} u(t, x) \Delta x = J(t, x) - J(t, x + \Delta x) + f(t, x) \Delta x.$$

Again, we use the Taylor series expansion for $J(t, x + \Delta x)$ to obtain that

$$\frac{\partial}{\partial t} u(t, x) \Delta x = -\frac{\partial}{\partial x} J(t, x) \Delta x - \frac{1}{2} \frac{\partial^2}{\partial x^2} J(t, x)(\Delta x)^2 + \cdots + f(t, x) \Delta x.$$

Dividing by Δx and then letting $\Delta x \to 0$ gives

$$\frac{\partial}{\partial t} u(t, x) = -\frac{\partial}{\partial x} J(t, x) + f(t, x). \tag{8.4}$$

It remains to specify $J(t, x)$, the flux of individuals at (t, x). A popular choice of such a flux is based on the so-called *Fick's law*, which states that the flux due to random motion is approximately proportional to the local gradient of the number of individuals. This yields

$$J(t, x) = -D \frac{\partial}{\partial x} u(t, x). \tag{8.5}$$

Combining (8.4) and (8.5) gives the reaction diffusion equation (8.3).

In population ecology, we can translate Fick's law of diffusion into the statement that the individuals move from a region of high concentration to a region of low concentration in search for limited resources. We must, however, use this law with caution when modeling spatial spread of infectious diseases since the individual movement behaviors may be altered during the course of outbreaks of diseases.

To determine the value of $u(t, x)$ in space and for all future time $t \geq 0$, we need to specify the initial distribution of the population $u(0, x)$ (*initial condition*). Also, when the space is bounded, say $x \in (0, L)$, we need to specify the boundary conditions. Typical boundary conditions include the homogeneous *Dirichlet condition*

$$u(t, 0) = u(t, L) = 0$$

when the boundary is uninhabitable, or the homogeneous *Neumann condition*

$$\frac{\partial}{\partial x} u(t, 0) = \frac{\partial}{\partial x} u(t, L) = 0$$

when there is no flux through boundaries.

8.3 Case Study I: Spatial Spread of Rabies in Continental Europe

We now demonstrate the usefulness of the partial differential equation approach for the study of spatial spread of rabies in continental Europe during the period roughly 1945–1985.

Starting on the edge of the German/Polish border, the front of the epizootic moved westward at an average speed of about 30–60 km a year. The spread of the epizootic was essentially determined by the ecology of the fox population as foxes are the main carrier of the rabies under consideration. If the fox population density is estimated at different times as the rabies epizootic passes, the wave is seen to consist of two main parts: the front through which the population is rapidly decreasing in magnitude and the much longer tail where there are essentially periodic outbreaks of the disease.

A model was formulated in [9] in order to describe the front of the wave, its speed and the total number of foxes infected after the front passes, and the connection of the wave speed to the so-called propagation speed of the disease. We shall also use this case study to illustrate how the partial differential equation model can help us in designing some spatial intervention strategies by considering the minimal length of protective zones. Various extensions of this basic model were also proposed to discover the mechanism for the periodic outbreaks and to estimate the periods and amplitudes, and we shall briefly discuss these extensions at the end of this section.

We start with a list of basic facts and some standing assumptions that we will use in our modeling and analysis.

- Foxes are the main carriers of rabies in the rabies epizootic considered.
- The rabies virus is contained in the saliva of the rabid fox and is normally transmitted by bite.
- Rabies is invariably fatal in foxes.
- Foxes are territorial and seem to divide the countryside into non-overlapping home ranges which are marked out by scent.
- The rabies virus enters the central nervous system and induces behavioral changes in its host. If the spinal cord is involved it often causes paralysis. However, if it enters the limbic system the foxes become aggressive, lose their sense of direction and territorial behavior and wander about in a more or less random way.

The last observation is the basis on which a reaction diffusion equation can be used to model the dynamics of rabid foxes. To formulate a deterministic model, at time t and spatial location x, let

$$S(t, x) = \text{the total number of susceptible foxes,}$$
$$I(t, x) = \text{the total number of infective foxes.}$$

Note that for the sake of simplicity in the above definition of $I(t, x)$, we do not distinguish rabid foxes and those in the incubation period, although it should be pointed out that only a fraction of infective foxes, namely, rabid foxes, transmit the disease.

A salient feature of rabies is the rather lengthy incubation period of between 12 and 150 days from the time of an infected bite to the onset of the clinical infectious stage. This feature was taken into account in the ordinary differential equation model [1] and its reaction diffusion analogue was developed in [16].

Ignoring this lengthy incubation period, then the model formulated in [9] in a one-dimensional unbounded domain takes the following form

$$\begin{cases} \frac{\partial}{\partial t} S(t, x) = -KS(t, x)I(t, x), \\ \frac{\partial}{\partial t} I(t, x) = D\frac{\partial^2}{\partial x^2} I(t, x) + KS(t, x)I(t, x) - \mu I(t, x), \end{cases}$$

where

K = the transmission coefficient,
μ^{-1} = life expectancy of an infective fox,
D = diffusion coefficient = A/k,
k = the average time until a fox leaves its territory,
A = the average area of a typical fox's territory.

In [16], other approaches, based on field observations of net distances traveled by infectives during observation periods, were developed for estimating the parameter D.

Rescaling the variables by

$$u(t, x) = I(t, x)/S_0, \quad v(t, x) = S(t, x)/S_0,$$
$$x^* = (KS_0/D)^{1/2} x, \quad t^* = KS_0 t,$$
$$r = \mu/(KS_0),$$

where S_0 is the initial susceptible density that is assumed to be uniform in space, and dropping the asterisks for convenience, we obtain the non-dimensional system

$$\begin{cases} \frac{\partial}{\partial t} u(t, x) = \frac{\partial^2}{\partial x^2} u(t, x) + u(t, x)(v(t, x) - r), \\ \frac{\partial}{\partial t} v(t, x) = -u(t, x)v(t, x). \end{cases} \tag{8.6}$$

Observe that r^{-1} is in fact the basic reproduction number of the corresponding ODE model. Therefore, if $r > 1$ the infection dies out quickly. Epidemiologically, this is reasonable since $r > 1$ if and only if $\mu > KS_0$. That is, $r > 1$ (or equivalently, the basic reproduction number is less than 1) if the mortality rate is greater than the rate of recruitment of new infectives. In this case, rabies cannot persist.

The above discussion also gives the minimum fox density $S_c := \mu/K$ below which rabies cannot persist. Mathematically, in [8], it was proven that if $r \geq 1$,

$u(0, x) \geq 0$ for all $x \in R$ and u has bounded support, and if $v(0, x) = 1$ for $x \in R$, then $u(t, x) \to 0$ as $t \to \infty$ uniformly on R.

A natural question is what happens if $r < 1$. In what follows, we will illustrate that if the initial distribution of susceptibles is uniformly equal to 1 (that is, $S = S_0$ everywhere), then a small localized introduction of rabies evolves into a traveling wave with a certain wave speed.

A solution of (8.6) is called a *traveling wave* (or *traveling wavefront*) at speed c if

$$u(t, x) = f(z), \ v(t, x) = g(z)$$

where

$$z = x - ct$$

is the *wave variable* and f and g are *waveforms* (or *wave profiles*).

Intuitively speaking, a traveling wave is a solution that moves in space with a constant speed c and without changing shape. In other words, if a fox or an observer moves at the same speed of the wave, the fox will not notice the change of the wave.

Substituting the above special form into system (8.6), we obtain the equations for the waveforms:

$$\begin{cases} f'' + cf' + fg - rf = 0, \\ cg' - fg = 0. \end{cases} \tag{8.7}$$

This is a system of ordinary differential equations, where primes denote differentiation with respect to z.

To solve the above system for the waveforms, we need to specify the asymptotic boundary conditions that are given naturally by

$$f(\pm\infty) = 0, \ g(+\infty) = 1, g(-\infty) = a, \tag{8.8}$$

where a is an important parameter to be found, that tells us the proportion of susceptible foxes that remain after the infective wave has passed, and this number is given by

$$a - r \ln a = 1. \tag{8.9}$$

To obtain the above formula (8.9), we rewrite the system for the waveforms as

$$\frac{1}{c}f'' + f' + g' - r\frac{g'}{g} = 0$$

that gives, after integration, the following

$$\frac{1}{c}f' + f + g - r \ln g = B \tag{8.10}$$

for a constant B that can be found, using the boundary condition at $z = \infty$, as $B = 1$. Therefore, using the boundary condition at $-\infty$, we get $a - r \ln a = 1$.

This is a very useful relation in order to obtain an estimation of r (and hence K). For example, in [14], it is suggested that the mortality rate is about 65–80% during the height of the epizootic. Therefore, if we take the fraction a of surviving foxes to be 0.2, we obtain approximately $r = 0.5$.

It follows also from (8.10) that system (8.7) is equivalent to the following planar system

$$\begin{cases} f' = c[r \ln g - f - g + 1], \\ g' = fg/c. \end{cases} \tag{8.11}$$

The linearization around the stationary point $(0, 1)$ has eigenvalues

$$\lambda_\pm = -\frac{1}{2}[c \pm \sqrt{c^2 - 4(1 - r)}].$$

Hence, if $c^2 < 4(1 - r)$ we have complex eigenvalues and all of the trajectories cannot stay in the positive quadrant near $(0, 1)$.

If $c \geq 2\sqrt{1 - r}$, $(0, 1)$ is a stable node and $(0, a)$ is a saddle point with the unstable trajectory entering the positive quadrant that, using some phase-plane analysis (see [8]), converges to $(0, 1)$ as $z \to \infty$. Therefore, in [9] it is shown that if $r < 1$ there exists a traveling wave of system (8.6) subject to boundary condition (8.8) with the speed

$$c \geq c_0 = 2\sqrt{1 - r}. \tag{8.12}$$

The traveling wave with the minimal wave speed c_0 is of paramount importance since any initial function $u(0, x)$ of compact support splits up into two traveling waves going in opposite directions with the same speed. More precisely, it was proved in [8] that if $u(0, \cdot)$ has compact support, then for every $\delta > 0$ there exists N so that

$$u(t, x + c_0 t - \ln t/c_0) \leq \delta$$

for every $t > 0$ and for all $x > N$. Therefore, if a fox travels with speed $c(t) = c_0 - (c_0 t)^{-1} \ln t$ towards $+\infty$ (in space) to the right of the support of $u(0, \cdot)$, the infection will never overtake the fox (hence the title "Run for your life, a note on the asymptotic speed of propagation of an epidemic" of the paper [3]). In other words, the asymptotic speed of the infection must be less than $c(t)$. As a consequence, if $u(t, x)$ takes the form of a traveling wave for large t, it must do so for the one with the minimal speed c_0.

A key issue for potential applications of the model is to identify all parameters involved. The parameter r is related to the transmission coefficient K which can hardly be estimated directly. Fortunately, as discussed above, formula (8.9) enables us to calculate r indirectly by considering the mortality as the epizootic front passes. We have $r = 0.5$, hence the disease reduces the fox population by about 50%.

The next parameter is μ. Recall that $1/\mu$ is the life expectancy of an infective fox. An infective fox first goes through an incubation period that can vary from 12 to 150 days, and then a rabid state lasting from 3 to 10 days. Thus, a life expectancy of about 35 days give μ as approximately $(1/10\,\text{yr})^{-1}$.

The diffusion coefficient D is the one related to the spatial spread and, as shown in [9], this can be calculated by the formula $D = A/k$. Fox territories can range from $2.5\,\text{km}^2$ to 16km^2 depending on the habitat, food availability and fox density. If we assume an average territory to be about $5\,\text{km}^2$ and that an infective fox leaves its territory about the time it becomes rabid, that is after about a month, then k is approximately 12yr^{-1}. Thus we get D as approximately $60\,\text{km}^2\text{yr}^{-1}$.

Putting this together, we then obtain the minimal wave speed of about 50 km per year, which seems to be in good agreement with the empirical data from Europe.

The diffusion model provides a useful framework to evaluate some spatially related control measures. For example, in [9], some estimate about a protective barrier is given. The mathematical formulation can be stated as follows: Let $0 \leq x \leq L$ be a protective barrier between a rabies free region $x > L$ and an infected region $x < 0$. How large should $L > 0$ be in order for $I(t, x) < \epsilon$ for all $t \geq 0$ and for all $x > L$, where $\epsilon > 0$ is a parameter, below which the infection is regarded as dying out?

This problem was investigated in [9] via numerical simulations, and the result of course also depends on the susceptible fox density in the protective zone and the parameter r. Reduction of the susceptible fox population in the protective zone can be achieved be shooting, gassing, vaccination, etc. Admittedly, the above model is only an approximation, but such a relatively simple model that captures some basic features of the disease spread requires very fewer parameters to estimate.

In the model considered so far, the natural birth and death are assumed to be balanced. Using a classical logistic model for the growth of susceptible foxes, we can explain the tail part of the wave, and in particular, the oscillatory behavior. Indeed, Anderson et al. [1] speculated that the periodic outbreak is primarily an effect of the incubation period, and Dunbar [4] and Murray et al. [16] obtained some qualitative results that show sustained oscillations if the classical logistic model is used and the carrying capacity of the environment is sufficiently large.

The model also ignores the fact that juvenile foxes leave their home territory in the fall, traveling distances that typically may be 10 times a territory size in search of a new territory. If a fox happens to have contracted rabies around the time of such long-distance movement, it could certainly increase the rate of spread of the disease into uninfected areas. This factor was pointed out in [16], and the impact of the age-dependent diffusion of susceptible foxes was recently considered in [17] by using structured population models.

8.4 Case Study II: Spread Rates of West Nile Virus

Although West Nile virus (WNv) was isolated in the West Nile district of Uganda in 1937, and WNv in the Eastern Hemisphere has been maintained in an enzootic cycle involving culicine mosquitoes (vectors) and birds (reservoirs), WNv activities in North America were not recorded until August of 1999 in the borough of Queens, New York City [2]. In the subsequent five years the epidemic has spread spatially to most of the west coast of North America, as a consequence of the interplay of disease dynamics and bird and mosquito movement. We refer to [21] for a detailed discussion of the ecological and epidemiological aspects of the disease spread and the recent modeling efforts of the WNv transmission dynamics.

Here we present the work [11], where the spread of WNv is investigated by spatially extending the non-spatial dynamical model [20] to include diffusive movements of birds and mosquitoes. The simplified spatial model that is analyzed takes the form:

$$\begin{cases} \frac{\partial I_V}{\partial t} = \alpha_V \beta_R \frac{I_R}{N_R}(A_V - I_V) - d_V I_V + \epsilon \frac{\partial^2 I_V}{\partial x^2}, \\ \frac{\partial I_R}{\partial t} = \alpha_R \beta_R \frac{N_R - I_R}{N_R} I_V - \gamma_R I_R + D \frac{\partial^2 I_R}{\partial x^2}, \end{cases} \tag{8.13}$$

where the parameters and variables are defined below:

d_V: adult female mosquito death rate,
γ_R: bird recovery rate from WNv,
β_R: biting rate of mosquitoes on birds,
α_V, α_R: WNv transmission probability per bite to mosquitoes and birds, respectively,
ϵ, D: diffusion coefficients for mosquitoes and birds respectively,
$I_V(t, x), I_R(t, x)$: numbers of infectious (infective) female mosquitoes and birds at time t and spatial location $x \in R$,
N_R: number of live birds,
A_V: number of adult mosquitoes.

The initial model is much more complicated and system (8.13) is obtained after a sequential procedure of simplification. Indeed, in the work [11] the female mosquito population is divided into larval, susceptible, exposed and infectious classes, and the bird population consists of compartments for susceptible, infectious, removed and dead birds. Under the assumption that the death rate of birds due to WNv can be ignored and the removed birds become immediately susceptible (no temporary immunity arising from WNv), it is shown that N_R remains a constant and the number of removed birds tend to zero. Hence the spatially homogeneous model for infectious birds is given by

$$\frac{dI_R}{dt} = \alpha_R \beta_R \frac{N_R - I_R}{N_R} I_V - \gamma_R I_R. \tag{8.14}$$

Furthermore, if exposed mosquitoes are immediately infective, then the exposed class (of mosquitoes) can be ignored and the dynamical system for the adult and larval mosquitoes is a simple planar linear system, solutions of which approach constants. Therefore, the spatially homogeneous version for the infectious mosquitoes becomes

$$\frac{dI_V}{dt} = \alpha_V \beta_R \frac{I_R}{N_R}(A_V - I_V) - d_V I_V. \tag{8.15}$$

Phase-plane analysis of the spatially homogeneous coupled system (8.14)-(8.15) shows that a nontrivial (endemic) equilibrium (I_V^*, I_R^*) exists if and only if the basic reproduction number $\mathcal{R}_0$ is large than 1, where

$$\mathcal{R}_0 = \sqrt{\frac{\alpha_V \alpha_R \beta_R^2 A_V}{d_V \gamma_R N_R}}.$$

Moreover, this endemic equilibrium, if it exists, is globally asymptotically stable in the positive quadrant.

For the spatially varying model (8.13), the vector field is cooperative, therefore an application of the general result in [12] ensures that there exists a minimal speed of traveling fronts c_0 such that for every $c \geq c_0$, the nonlinear system (8.13) has a nonincreasing traveling wave solution $(I_V(x - ct), I_R(x - ct))$ with speed c so that

$$\lim_{(x-ct)\to-\infty} (I_V, I_R) = (I_V^*, I_R^*), \qquad \lim_{(x-ct)\to\infty} (I_V, I_R) = (0,0).$$

Here a traveling wavefront with speed c for system (8.13) is a solution that has the form $(I_V(x - ct), I_R(x - ct))$ and connects the disease-free and endemic equilibria so that the above boundary conditions are satisfied. Note that the traveling wave solution is then given by

$$\begin{cases} -cI_V' = \epsilon I_V'' + \alpha_V \beta_R \frac{I_R}{N_R}(A_V - I_V) - d_V I_V, \\ -cI_R' = D I_R'' + \alpha_R \beta_R \frac{N_R - I_R}{N_R} I_V - \gamma_R I_R. \end{cases}$$

What makes the minimal wave speed c_0 so important epidemiologically is the following mathematical relation: the minimal wave speed c_0 coincides with the spread rate c^* defined as follows: if the initial values of $(I_V(\cdot, 0), I_R(\cdot, 0))$ have compact support and are not identical to either equilibrium, then for small $\epsilon > 0$,

$$\lim_{t\to\infty} \left\{ \sup_{|x|\geq(c^*+\epsilon)t} \|(I_V(t,x), I_R(t,x))\| \right\} = 0,$$
$$\lim_{t\to\infty} \left\{ \sup_{|x|\leq(c^*-\epsilon)t} \|(I_V(t,x), I_R(t,x)) - (I_V^*, I_R^*)\| \right\} = 0.$$

This relation holds due to the cooperative nature of the vector field. It is also due to this nature that the spread speed c^* is linearly determined: namely, the spread speed is the same as the number $\tilde{c}$ so that the solution $(\tilde{I}_V, \tilde{I}_R)$ with nontrivial initial values of compact support of the linearized system of (8.13) about the disease endemic equilibrium satisfies, for small ϵ,

$$\lim_{t\to\infty} \left\{ \sup_{|x|\geq(\tilde{c}+\epsilon)t} \|(\tilde{I}_V(t,x), \tilde{I}_R(t,x))\| \right\} = 0,$$
$$\lim_{t\to\infty} \left\{ \sup_{|x|\leq(\tilde{c}-\epsilon)t} \|(\tilde{I}_V(t,x), \tilde{I}_R(t,x))\| \right\} > 0.$$

Consequently, it is shown in [11] that

$$c_0 = c^* = \tilde{c} = \inf_{\lambda>0} \sigma_1(\lambda)$$

where $\sigma_1(\lambda)$ is the largest eigenvalue of the matrix

$$B_\lambda = \lambda \begin{pmatrix} \epsilon & 0 \\ 0 & D \end{pmatrix} + \lambda^{-1} \begin{pmatrix} -d_V & \alpha_V \beta_R \frac{A_V}{N_R} \\ \alpha_R \beta_R & -\gamma_R \end{pmatrix}.$$

The characteristic equation of B_λ is given by

$$p(\sigma; \lambda, \epsilon) = \sigma^2 - \sigma\lambda^{-1}[\theta + (D+\epsilon)\lambda^2]$$
$$+ \lambda^{-2}[d_V\gamma_R - \alpha_V\alpha_R\beta_R^2 \frac{A_V}{N_R}] - Dd_V - \epsilon\gamma_R + \epsilon D\lambda^2 = 0.$$

In the general case $\epsilon > 0$, the larger root $\sigma_1(\lambda, \epsilon)$ can have more than one extremum and hence it is difficult to obtain a result for the minimal spread rate by examining roots of $p(\sigma; \lambda, \epsilon)$. However, the case with $\epsilon = 0$ is sufficiently simple and due to the continuous dependence of eigenvalues on parameters, it is shown in [11] that as $\epsilon \to 0$, the spread speed rate approaches the positive square root of the largest zero of an explicitly defined cubic.

In [11], an example is provided to show how the spread rate varies as a function of the bird diffusion coefficient D, in the range $D \in [0, 14] \, km^2/day$ as estimated in [20]. This example is based on the assumption that $A_V/N_R = 20$ and $\gamma_R = 0.01/day$ and using the parameters $d_V = 0.029, \alpha_V = 0.16, \alpha_R = 0.88, \beta_R = 0.3/day$ estimated in [20]. Since WNv has spread across North America in about five years, the observed spread rate is about $1,000 \, km/year$. This, together with the aforementioned functional relation between the spread rate and the diffusion rate of birds, shows that a diffusion coefficient of about 5.94 is needed in the model to achieve the observed spread rate.

Needless to say, the reaction-diffusion system (8.13) is a first approximation for the spatial spread of WNv, and it is based on the assumption of random flight of birds and mosquitoes. In reality, as pointed out in [11], flight is influenced by topographical, environmental and other factors. The work in [13] based on a patchy model seems to indicate the spread speed may be different if the movement of birds has preference to direction. Certainly, to incorporate more ecology and epidemiology, models should contain more realistic bird and mosquito movements.

8.5 Remarks

We conclude this chapter with a few remarks. First of all, we note that reaction diffusion equations arise naturally from modeling spatial spread of infectious diseases when a subpopulation moves randomly in space, and use of such a model is appropriate when transmission mechanisms and control measures involve spatial movements.

We must, however, be extremely cautious when modeling spatial spread when individual movement is not so obviously random. Other type of models will be needed, and in particular, the discrete space models considered in [19] seem to be more appropriate. When other factors such as disease age and social structures are considered, model systems could be much more complicated.

We have shown that when the space is large in scale, traveling waves of the reaction diffusion equations are important as they describe the progress of the disease to uninfected regions. The wave speed is obviously important to understand the speed of propagation: in some cases it coincides with the spread rate and can be determined by considering the linearization of the nonlinear system at a disease endemic equilibrium.

References

1. R. M. Anderson, H. C. Jackson, R. M. May and A. M. Smith, Population dynamics of fox rabies in Europe, Nature, 289, 765–771 (1981).
2. C. Bowman, A. Gumel, P. van den Driessche, J. Wu and H. Zhu, A mathematical model for assessing control strategies against West Nile virus, Bull. Math. Biol., 67, 1107–1133 (2005).
3. O. Diekmann, Run for your life, a note on the asymptotic speed of propagation of an epidemic, J. Differ. Equ., 33, 58–73 (1979).
4. S. R. Dunbar, Travelling wave solutions of diffusive Lotka-Volterra equations, J. Math. Biol. 17, 11–32 (1983).
5. L. Edelstein Keshet, Mathematical Models in Biology, Birkhäuser Mathematics Series, McGraw-Hill, Toronto, 1988.
6. K. P. Hadeler, The role of migration and contact distributions in epidemic spread, in Bioterrorisum: Mathematical Modeling Applications in Homeland Security (H. T. Banks and C. Castillo-Chavez eds.), pp. 199–210, 2003. Siam, Philadelphia.
7. D. S. Johns and B. D. Sleeman, Differential Equations and Mathematical Biology, Chapman & Hall/CRC Mathematical Biology and Medicine Series, Florida, 2003.
8. A. Källén, Thresholds and travelling waves in an epidemic model for rabies, Nonlinear Anal. 8, 651–856 (1984).
9. A. Källén, P. Arcuri and J. D. Murray, A simple model for the spatial spread and control of rabies, J. Theor. Biol., 116, 377–393 (1985).
10. M. Kot, Elements of Mathematical Ecology, Cambridge University Press, Cambridge, 2001.
11. M. Lewis, J. Rencławowicz and P. van den Driessche, Traveling waves and spread rates for a West Nile virus model, Bull. Math. Biol., 68, 3–23 (2006).
12. B. Li, H. Weinberger and M. Lewis, Spreading speed as slowest wave speeds for cooperative systems, Math. Biosci., 196, 82–98 (2005).

13. R. Liu, J. Shuai, J. Wu and X. Zou, Modelling spatial spread of West Nile virus and impact of directional dispersal of birds, Math. Biosci. Eng., 3, 145–160 (2006).

14. D. W. MacDonald, Rabies and Wildlife. A Biologist's Prespective, Oxford, Oxford University Press, 1980.

15. J. D. Murray, Mathematical Biology, Springer, Berlin Heidelberg New York, 1989.

16. J. D. Murray, E. A. Stanley and D. L. Brown, On the spatial spread of rabies among foxes, Proc. R. Soc. Lond., B229, 111–150 (1986).

17. C. Ou and J. Wu, Spatial spread of rabies revisited: influence of age-dependent diffusion on nonlinear dynamics, SIAM J. Appl. Math., 67, 138–164, (2006).

18. S. I. Rubinow, Introduction to Mathematical Biology, Wiley, New York, 1975.

19. P. van den Driessche, Spatial Structure: Patch Models, Chapter 7 of Mathematical Epidemiology (this volume).

20. M. J. Wonham, T. de-Camino-Beck and M. Lewis, An epidemiological model for West Nile virus: invasion analysis and control applications, Proc. R. Soc. Lond. B, 271, 501–507, (2004).

21. M. J. Wonham and M. A. Lewis, A Comparative Analysis of Models for West Nile Virus, Chapter 14 of Mathematical Epidemiology (this volume).

Chapter 9
Continuous-Time Age-Structured Models in Population Dynamics and Epidemiology

Jia Li and Fred Brauer

Abstract We present continuous-time models for age-structured populations and disease transmission. We show how to use the method of characteristic lines to analyze the model dynamics and to write an age-structured population model as an integral equation model. We then extend to an age-structured SIR epidemic model. As an example we describe an age-structured model for AIDS, derive a formula for the reproductive number of infection, and show how important a role pair-formation plays in the modeling process. In particular, we outline the semi-group method used in an age-structured AIDS model with non-random mixing. We also discuss models for populations and disease spread with discrete age structure.

9.1 Why Age-Structured Models?

In the simplest models for a single population all members are assumed to be interchangeable. However, even the simplest models for disease transmission include structuring the population by disease state (susceptible, exposed, infective, or removed).

More advanced population models add some structure to the population such as specification of spatial location or age. Age is one of the most important characteristics in the modeling of populations and infectious diseases. Individuals with different ages may have different reproduction and survival capacities. Diseases may have different infection rates and mortality rates for different age groups [1].

Department of Mathematical Sciences, University of Alabama in Huntsville, 301, Sparkman Dr., Huntsville, AL 35899, USA li@math.uah.edu

Department of Mathematics, University of British Columbia, 1984, Mathematics Road, Vancouver BC, Canada V6T 1Z2 brauer@math.ubc.ca

Individuals of different ages may also have different behaviours, and behavioural changes are crucial in control and prevention of many infectious diseases. Young individuals tend to be more active in interactions with or between populations, and in disease transmissions.

Sexually-transmitted diseases (STDs) are spread through partner interactions with pair-formations, and the pair-formations are clearly age-dependent in most cases. For example, most AIDS cases occur in the group of young adults.

Childhood diseases, such as measles, chicken pox, and rubella, are spread mainly by contacts between children of similar ages. More than half of the deaths attributed to malaria are in children under five years of age due to their weaker immune systems. This suggests that in models for disease transmission in an age structured population it is necessary to allow the contact rates between two members of the population to depend on the ages of both members.

In order to describe age-structured models for disease transmission we must first develop the theory of age-structured populations. In fact, the first models for age-structured populations [34] were designed for the study of disease transmission in such populations.

9.2 Modeling Populations with Age Structure

Let $\rho(t, a)$ be the age-density function at time t with $a \in [0, a_+]$, where $a_+ < \infty$ is the maximum age of individuals, or with $a \in [0, \infty)$ for convenience of mathematical description. Then $\int_{a_1}^{a_2} \rho(t, a) da$ is the number of individuals having ages in the interval $[a_1, a_2]$ at time t, and $\int_0^\infty \rho(t, a) da = P(t)$ is the total population size at t. Let β be the age specific fertility rate, or birth rate, so that $\int_{a_1}^{a_2} \beta \rho(t, a) da$ is the number of offspring produced by individuals with ages in $[a_1, a_2]$ in unit time at time t. Then $\int_0^\infty \beta \rho(t, a) da = B(t)$ is the total number of newborns, at time t. The age specific fertility may depend on the population density so that $\beta = \beta(a, \rho(t, a))$, or may depend on the total population so that $\beta = \beta(a, P)$. The reader should note that here β is not related to the contact rate for disease transmission in compartmental models introduced in earlier chapters. Here we assume the fertility to be time-independent. Let μ be the age specific mortality, or death rate, so that $\int_0^\infty \mu \rho(t, a) da$ is the total number of deaths at time t, occurring in one unit time. Similarly, the age specific mortality may depend on the population density so that $\mu = \mu(a, \rho(t, a))$, or may depend on the total population size so that $\mu = \mu(a, P)$. Again we assume the mortality to be time-independent. In this chapter we consider the case in which both the fertility and mortality depend on the total population size rather than on the age-specific population density.

Suppose that the population changes from time t to $t + h$, with $h > 0$. The number of newborns in the time interval $[t, t + h]$ is $\int_t^{t+h} B(s)ds = \int_t^{t+h} \int_0^\infty \beta(\sigma, P)\rho(s, \sigma)d\sigma ds$. Note that the number of individuals who die at time $t + s$, having age less than or equal to $a + s$, is $\int_0^{a+s} \mu(\sigma, P)\rho(t + s, \sigma)d\sigma$. Then the total number of deaths in the time interval $[t, t + h]$ is $\int_0^h \int_0^{a+s} \mu(\sigma, P)\rho(t + s, \sigma)d\sigma ds$.

Let $N(t, a) = \int_0^a \rho(t, \sigma)d\sigma$ be the number of individuals having ages less than or equal to a at time t, and assume that there is no migration. Then the change in the population size from time t to $t + h$ is the total number of births minus the total number of deaths during the time interval $[t, t + h]$, that is,

$$N(t + h, a + h) - N(t, a) = \int_t^{t+h} B(s)ds - \int_0^h \int_0^{a+s} \mu(\sigma, P)\rho(t + s, \sigma)d\sigma ds. \tag{9.1}$$

The instantaneous rate of change of the population size is

$$\lim_{h \to 0} \frac{N(t + h, a + h) - N(t, a)}{h} = N_t(t, a) + N_a(t, a) = \int_0^a \rho_t(t, \sigma)d\sigma + \rho(t, a).$$

Dividing (9.1) by h and then letting $h \to 0$ yields

$$\int_0^a \rho_t(t, \sigma)d\sigma + \rho(t, a) = B(t) - \int_0^a \mu(\sigma, P)\rho(t, \sigma)d\sigma. \tag{9.2}$$

Setting $a = 0$ in (9.2), we have $\rho(t, 0) = B(t)$. Differentiating equation (9.2) with respect to a, we have

$$\rho_t(t, a) + \rho_a(t, a) = -\mu(a, P)\rho(t, a). \tag{9.3}$$

Then we arrive at the following system of a first order partial differential equation with corresponding initial and boundary conditions:

$$\begin{aligned} \rho_t(t, a) + \rho_a(t, a) &= -\mu(a, P)\rho(t, a), \\ \rho(t, 0) &= \int_0^\infty \beta(a, P)\rho(t, a)da = B(t), \\ \rho(0, a) &= \phi(a), \end{aligned} \tag{9.4}$$

where $\phi(a)$ is the initial age distribution. For continuity at $(0, 0)$ it would be necessary to require that

$$\phi(0) = \int_0^\infty \beta(a, P)\phi(a)da,$$

but because it is possible to allow discontinuous solutions of (9.4) this requirement is usually ignored.

The partial differential equation in (9.4) is commonly called the Lotka–McKendrick equation [26, 42].

9.2.1 Solutions along Characteristic Lines

Fix t_0 and a_0 and consider the functions $\bar\rho(h) := \rho(t_0 + h, a_0 + h)$ and $\bar\mu(h) := \mu(a_0 + h, P(t_0 + h))$. This amounts to following the age cohort of members of the population with age a_0 at time t_0. Then equation (9.3) is equivalent to

$$\frac{d\bar\rho}{dh} + \bar\mu(h)\bar\rho = 0. \tag{9.5}$$

Solving (9.5) yields

$$\bar\rho(h) = \bar\rho(0)e^{-\int_0^h \bar\mu(\tau)d\tau}, \tag{9.6}$$

that is,

$$\rho(t_0 + h, a_0 + h) = \rho(t_0, a_0)e^{-\int_0^h \bar\mu(\tau)d\tau}. \tag{9.7}$$

For $a > t$, setting $(t_0, a_0) = (0, a - t)$ and $h = t$, we have

$$\rho(t, a) = \rho(0, a - t)e^{-\int_0^t \bar\mu(\tau)d\tau} = \phi(a - t)e^{-\int_0^t \mu(a-t+\tau, P(\tau))d\tau}, \tag{9.8}$$

and for $t > a$, setting $(t_0, a_0) = (t - a, 0)$ and $h = a$, we have

$$\rho(t, a) = \rho(t - a, 0)e^{-\int_0^a \bar\mu(\tau)d\tau} = B(t - a)e^{-\int_0^a \mu(\tau, P(t-a+\tau))d\tau}, \tag{9.9}$$

[17, 38, 42]. Then, we obtain the following expressions for solutions along the lines of characteristics for system (9.4):

$$\rho(t, a) = \begin{cases} \phi(a - t)e^{-\int_0^t \mu(a-t+\tau, P(\tau))d\tau}, & a > t, \\ B(t - a)e^{-\int_0^a \mu(\tau, P(t-a+\tau))d\tau}, & t > a. \end{cases} \tag{9.10}$$

Thus we have obtained an expression for the population density function for all (t, a) by following each age cohort along a characteristic line. Notice, however, that the solutions in (9.10) involve the total population size P which depends on $\rho(t, a)$.

9.2.2 Equilibria and the Characteristic Equation

One of the important properties in the study of population dynamics is the asymptotic behavior of the steady states or equilibria of the populations. For system (9.4), a steady state, or an equilibrium distribution, $\rho^*(a)$, satisfies the equations

$$
\begin{aligned}
\frac{d\rho^*(a)}{da} &= -\mu(a, P^*)\rho^*(a), \\
\rho^*(0) &= \int_0^\infty \beta(a, P^*)\rho^*(a)da, \\
P^* &= \int_0^\infty \rho^*(a)da.
\end{aligned}
\tag{9.11}
$$

Suppose that system (9.11) has a solution $\rho^*(a)$. Then we can investigate the local stability of this steady state or equilibrium by linearization of system (9.4) about $\rho^*(a)$ as follows.

Let $y(t, a) = \rho(t, a) - \rho^*(a)$, and write $Y(t) = \int_0^\infty y(t, a)da$. Then substitution into (9.4) yields

$$
y_t + y_a = \rho_t + \rho_a - \rho_a^* = -\mu(a, Y + P^*)(y + \rho^*) - \rho_a^*,
$$

and

$$
y(t, 0) = \rho(t, 0) - \rho^*(0) = \int_0^\infty \beta(a, Y + P^*)(y + \rho^*)\, da - \rho^*(0),
$$

where $P^* = \int_0^\infty \rho^*(a)da$. For $\rho(t, a)$ near ρ^*, we have, using (9.11),

$$
\begin{aligned}
y_t + y_a &\approx -\mu(a, P^*)y - \mu(a, P^*)\rho^* - \rho^*\mu_P(a, P^*)Y - \rho_a^* \\
&= -\mu(a, P^*)y - \rho^*\mu_P(a, P^*)Y,
\end{aligned}
\tag{9.12}
$$

and

$$
\begin{aligned}
y(t, 0) &\approx \int_0^\infty (\beta(a, P^*)y + \rho^*(a)\beta_P(a, P^*)Y)\, da \\
&= \int_0^\infty \beta(a, P^*)y(t, a)da + \int_0^\infty \rho^*(a)\beta_P(a, P^*)daY(t) \\
&= \int_0^\infty K(a, \rho^*, P^*)y(t, a)da,
\end{aligned}
\tag{9.13}
$$

where

$$
K(a, \rho^*, P^*) = \beta(a, P^*) + \int_0^\infty \rho^*(\sigma)\beta_P(\sigma, P^*)d\sigma.
\tag{9.14}
$$

Hence, for $\rho(t, a)$ near ρ^*, we arrive at the linearized equation

$$
y_t + y_a = -\mu(a, P^*)y - \rho^*\mu_P(a, P^*)\int_0^\infty y(t, a)da,
\tag{9.15}
$$

with the linearized integral boundary condition

$$
y(t, 0) = \int_0^\infty K(a, \rho^*, P^*)y(t, a)da.
\tag{9.16}
$$

Suppose further that $y(t, a) = u(a)e^{\xi(t-a)}$, and write $w = \int_0^\infty u(a)e^{-\xi a}da$. By substituting them into (9.15) and (9.16), respectively, we have

$$\begin{aligned}
\frac{du(a)}{da} &= -\mu(a, P^*)u(a) - \rho^*\mu_P(a, P^*)e^{\xi a}\int_0^\infty u(a)e^{-\xi a}da \\
&= -\mu(a, P^*)u(a) - \rho^*\mu_P(a, P^*)e^{\xi a}w,
\end{aligned} \tag{9.17}$$

and

$$u(0) = \int_0^\infty K(a, \rho^*, P^*)u(a)e^{-\xi a}da. \tag{9.18}$$

Solving (9.17), we have

$$u(a) = e^{-\int_0^a \mu(\alpha, P^*)d\alpha}\left(u(0) - E(\xi, a)\, w\right), \tag{9.19}$$

where

$$E(\xi, a) = \int_0^a e^{\left(\int_0^s \mu(\alpha, P^*)d\alpha + \xi s\right)}\rho^*\mu_P(s, P^*)ds.$$

Then substituting (9.19) into (9.18) and w, we obtain the following linear system

$$\begin{aligned}
u(0) &= \int_0^\infty Ke^{-\left(\int_0^a \mu(\alpha, P^*)d\alpha + \xi a\right)}da\, u(0) - \int_0^\infty Ke^{-\xi a}E(\xi, a)da\, w, \\
w &= \int_0^\infty e^{-\left(\int_0^a \mu(\alpha, P^*)d\alpha + \xi a\right)}da\, u(0) - \int_0^\infty e^{-\left(\int_0^a \mu(\alpha, P^*)d\alpha + \xi a\right)}E(\xi, a)da\, w,
\end{aligned}$$

or equivalently, the linear system

$$\begin{aligned}
\left(1 - \int_0^\infty Ke^{-\left(\int_0^a \mu(\alpha, P^* d\alpha + \xi a\right)}da\right)u(0) + \int_0^\infty Ke^{-\xi a}E(\xi, a)da\, w &= 0, \\
\int_0^\infty e^{-\left(\int_0^a \mu(\alpha, P^*)d\alpha + \xi a\right)}da\, u(0) - \left(1 + \int_0^\infty e^{-\left(\int_0^a \mu(\alpha, P^*)d\alpha + \xi a\right)}E(\xi, a)da\right)w &= 0,
\end{aligned} \tag{9.20}$$

in the unknowns $u(0)$ and w. Hence, there exists a non-zero solution $(u(0), w)$ to system (9.20) if and only if

$$\begin{aligned}
&\left(1 - \int_0^\infty Ke^{-\left(\int_0^a \mu(\alpha, P^*)d\alpha + \xi a\right)}da\right)\left(1 + \int_0^\infty e^{-\left(\int_0^a \mu(\alpha, P^*)d\alpha + \xi a\right)}E(\xi, a)da\right) \\
&+ \int_0^\infty Ke^{-\xi a}E(\xi, a)da\int_0^\infty e^{-(\mu+\xi)a}da = 0.
\end{aligned} \tag{9.21}$$

Equation (9.21) is an equation in ξ. There exists a solution of the form $y(t, a) = u(a)e^{\xi(t-a)}$ of the linearization (9.15) and (9.16) if and only if there exists a solution ξ to equation (9.21). Equation (9.21) is called the characteristic equation of system (9.4) as in [9–11].

9.3 Age-Structured Integral Equations Models

Integral equations have also been used for modeling of age-structured populations. These integral equations can be derived from system (9.4), or more specifically from (9.10).

Write $\Pi(a, P) = e^{-\int_0^a \mu(\tau, P(t-a+\tau))d\tau}$. Then it follows from (9.10) that

$$
\begin{aligned}
B(t) &= \int_0^t \beta(a, P)\rho(t, a)da + \int_t^\infty \beta(a, P)\rho(t, a)da \\
&= \int_0^t \beta(a, P)\Pi(a, P)B(t-a)da + \int_t^\infty \beta(a, P)\frac{\Pi(a, P)}{\Pi(a-t, P)}\phi(a-t)da,
\end{aligned}
\tag{9.22}
$$

and

$$
\begin{aligned}
P(t) &= \int_0^t \rho(t, a)da + \int_t^\infty \rho(t, a)da \\
&= \int_0^t \Pi(a, P)B(t-a)da + \int_t^\infty \frac{\Pi(a, P)}{\Pi(a-t, P)}\phi(a-t)da.
\end{aligned}
\tag{9.23}
$$

The coupled equations (9.22) and (9.23) are a system of nonlinear integral equations. In general, it cannot be solved analytically. We consider two special cases as follows.

If the birth rate is age-independent and density-dependent, that is, if $\beta = \beta(P)$, then the equation for the total number of newborns becomes

$$
B(t) = \int_0^\infty \beta(P(t))\rho(t, a)da = P(t)\beta(P(t)).
\tag{9.24}
$$

Substituting (9.24) into (9.23), we have

$$
P(t) = \int_t^\infty \frac{\Pi(a, P(t))}{\Pi(a-t, P(t))}\phi(a-t)da + \int_0^t \Pi(a, P(t))P(t-a)\beta(P(t-a))da.
\tag{9.25}
$$

Equation (9.25) is a delayed integral equation.

If the death rate is age-independent and density-dependent, that is, if $\mu = \mu(P)$, then integration of the partial differential equation in (9.4) yields the nonlinear ordinary differential equation

$$
P'(t) + P(t)\mu(P(t)) = B(t),
\tag{9.26}
$$

where $'$ denotes d/dt, which is coupled with the following integral equation for B derived from (9.22):

$$
B(t) = \int_t^\infty \beta(a, P)\frac{\Pi(a, P)}{\Pi(a-t, P)}\phi(a-t)da + \int_0^t \beta(a, P)\Pi(a, P)B(t-a)da.
\tag{9.27}
$$

Under further assumptions, systems (9.24) and (9.25), or (9.26) and (9.27) may become analytically solvable. For example, if β and μ are both functions of total population size only, then $P(t)$ is obtained by solving the ordinary differential equation (9.26) with $B(t)$ given by (9.24).

Whereas system (9.22) and (9.23) cannot in general be solved analytically, the equilibrium age distributions provide useful information for the population dynamics.

At an equilibrium age distribution, $\rho^*(a)$, we write

$$P^* = \int_0^\infty \rho^*(a)da, \quad B^* = \int_0^\infty \beta(a, P^*)\rho^*(a)da, \tag{9.28}$$

which are constant. It follows from (9.11) that

$$\rho^*(a) = \rho^*(0)e^{-\int_0^a \mu(a,P^*)da} = B^*\Pi(a, P^*). \tag{9.29}$$

Substituting (9.29) into B^* in (9.28), we can solve for B^* to get

$$B^* = \int_0^\infty \beta(a, P^*)\rho^*(a)da = B^* \int_0^\infty \beta(a, P^*)\Pi(a, P^*)da. \tag{9.30}$$

Then there exists a positive solution B^* to equation (9.30) if and only if there exists positive P^* such that

$$\int_0^\infty \beta(a, P^*)\Pi(a, P^*)da = 1. \tag{9.31}$$

Define $R(P) = \int_0^\infty \beta(a, P)\Pi(a, P)da$, which is called the reproduction number and is an expected number of newborns that an individual produces over its lifetime when the total population size is P. At an equilibrium age distribution P^*, the reproduction number is equal to one.

Substituting (9.29) into P^* in (9.28), we have

$$P^* = B^* \int_0^\infty \Pi(a, P^*)da.$$

Notice that $\int_0^\infty \Pi(a, P^*)da$ is the average life expectancy of individuals, when the population is at the equilibrium P^*. Then the total population size P^* equals the total number of surviving newborns at the equilibrium.

9.3.1 The Renewal Equation

We consider a special case where the birth and death rates are density-independent such that $\beta = \beta(a)$ and $\mu = \mu(a)$. In this case

$$\Pi(a, P) = e^{-\int_0^a \mu(\sigma)d\sigma}$$

is a function of a only. Then equation (9.22) becomes the linear integral equation

$$B(t) = F(t) + \int_0^t L(t - a)B(a)da, \qquad (9.32)$$

where

$$F(t) = \int_t^\infty \beta(a)\frac{\Pi(a)}{\Pi(a - t)}\phi(a - t)da, \quad L(t) = \beta(t)\Pi(t).$$

Equation (9.32) is a linear Volterra integral equation of the second kind. It is called the renewal equation or Lotka equation for the population [4, 26].

Because of the linearity of equation (9.32), we can use Laplace transformation techniques to investigate the properties of the dynamics of the population. Let $\widehat{B}(s)$, $\widehat{F}(s)$, and $\widehat{L}(s)$ be the Laplace transforms of $B(t)$, $F(t)$, and $L(t)$, respectively. Notice that the integral in (9.32) is the convolution of K and B. Then taking the Laplace transform of each term in (9.22), we have

$$\widehat{B}(s) = \widehat{F}(s) + \widehat{L}(s)\widehat{B}(s).$$

Solving for $\widehat{B}(s)$, we obtain

$$\widehat{B}(s) = \frac{\widehat{F}(s)}{1 - \widehat{L}(s)} = \widehat{F}(s) + \frac{\widehat{F}(s)\widehat{L}(s)}{1 - \widehat{L}(s)}. \qquad (9.33)$$

Since $\widehat{F}(s)$ and $\widehat{L}(s)$ are analytic functions, the properties of $\widehat{B}(s)$ are determined by the property of $1 - \widehat{L}(s)$.

It follows from

$$\widehat{L}(s) = \int_0^\infty \beta(a)\Pi(a)e^{-sa}da$$

that

$$\widehat{L}(0) = \int_0^\infty \beta(a)\Pi(a)da = R(0),$$

$$\frac{d\widehat{L}(s)}{ds} = -\int_0^\infty a\beta(a)\Pi(a)da < 0,$$

and

$$\lim_{s \to -\infty} \widehat{L}(s) = +\infty, \qquad \lim_{s \to +\infty} \widehat{L}(s) = 0.$$

Hence there exists a unique $s_0 \in \mathbb{R}$ such that $\widehat{L}(s_0) = 1$. Whether s_0 is positive, zero, or negative, depends on whether $R(0)$ is greater than, equal to, or less than one.

Moreover, it is easy to check that if there exits a complex number $s = \alpha + i\gamma$ such that $\widehat{L}(s) = 1$, then it follows from the real part of $\widehat{L}(s) = 1$,

$$\int_0^\infty \beta(a)\Pi(a)e^{-\alpha a}\cos\gamma a\,da = 1 = \int_0^\infty \beta(a)\Pi(a)e^{-s_0 a}da,$$

that $\alpha \leq s_0$.

Hence s_0 is a dominant root of $\widehat{L}(s) = 1$. With this dominant root, s_0, it can be shown that

$$B(t) = b_0 e^{s_0 t}\left(1 + \Omega_1(t)\right), \quad P(t) = p_0 e^{s_0 t}\left(1 + \Omega_2(t)\right),$$

where $b_0 \geq 0$ and $p_0 \geq 0$ are real numbers, and $\lim_{t\to\infty}\Omega_k(t) = 0$, $k = 1, 2$. (See, e.g., [26, Sect. I, 5].)

Therefore, the location of s_0 determines the asymptotic behavior of the population. Equation $\widehat{K}(s) = 1$ is called the Lotka characteristic equation for the renewal equation (9.32).

Now that we have an understanding of age-structured population models, we can begin to study age-structured disease transmission models.

9.4 Age-Structured Epidemic Models

Suppose that we have an age-structured population described by (9.4) in which there is an infectious disease of SIR type. We introduce functions $S(t, a), I(t, a), R(t, a)$ representing the age distribution at time t of susceptible, infective, and removed members, respectively, so that

$$S(t, a) + I(t, a) + R(t, a) = \rho(t, a).$$

As we have seen, the rate of change in time of a function $X(t, a)$ of time and age is

$$X_t(t, a) + X_a(t, a).$$

Thus we may write a system of equations

$$\begin{aligned}
S_t(t, a) + S_a(t, a) &= -\left(\mu(a) + \lambda(t, a)\right)S(t, a),\\
I_t(t, a) + I_a(t, a) &= \lambda(t, a)S(t, a) - \left(\mu(a) + \gamma(a) + \delta(a)\right)I(t, a),\\
R_t(t, a) + R_a(t, a) &= -\mu(a)R(t, a) + \gamma(a)I(t, a),
\end{aligned}$$

to describe the transmission dynamics of the disease in the age-structured population. Here $\mu(a)$ is the natural death rate in each class, $\gamma(a)$ is the recovery rate, $\delta(a)$ is the disease death rate, and $\lambda(t, a)$ represents the infection rate. To this system of partial differential equations we must add the initial conditions

$$S(0, a) = \Phi(a), \quad I(0, a) = \Psi(a), \quad R(0, a) = 0, \tag{9.34}$$

where Φ and Ψ are the initial distributions of susceptibles and infectives, respectively. In addition there is the birth or renewal condition (assuming that the age-dependent birth rate does not depend on disease status and that all newborns are in the susceptible class).

$$S(t,0) = \int_0^\infty \beta(a)\rho(t,a)da. \tag{9.35}$$

Further analysis requires some assumption on the nature of the infection term $\lambda(t,a)$. One possibility is *intracohort* mixing,

$$\lambda(t,a) = f(a)I(t,a),$$

corresponding to the assumption that infection can be transmitted only between individuals of the same age. Another possibility is *intercohort* mixing,

$$\lambda(t,a) = \int_0^\infty b(a,\alpha)I(t,\alpha)d\alpha,$$

with $b(a,\alpha)$ giving the rate of infection from contacts between an infective of age α with a susceptible of age a. For intercohort mixing it is necessary to make further assumptions on the mixing, that is, on the nature of the function $b(a,\alpha)$. One possibility here would be separable pair formation,

$$b(a,\alpha) = b_1(a)b_2(\alpha).$$

Rather than pursuing the general analysis further here, we refer the reader to more advanced references such as [26], and turn to an example that will illustrate the main ideas.

9.5 A Simple Age-Structured AIDS Model

Consider a simple age-structured epidemic model in which HIV/AIDS is spread in a homosexual population of ages in $[a_0, \infty]$, where a_0 is the minimal sexually active age. We divide the population into the groups of susceptible individuals, infective individuals, and AIDS cases, denoted by S, I, and A, respectively.

Assume that there is an input flow, $\Lambda(a)$ for all ages a, entering only the susceptible group. For this simple model, we further assume that the number of susceptible individuals of age a_0 is a constant B, and that no individuals with age a_0 are infected yet. Let $\mu(a)$ be the natural death rate of all individuals in the population, $\gamma(a)$ the HIV developing rate for infective individuals, and $\delta(a)$ the AIDS induced death rate of AIDS cases. Then the transmission dynamics of the disease are governed by the following system of equations [24]:

$$S_t(t,a) + S_a(t,a) = \Lambda(a) - (\mu(a) + \lambda(t,a)) S(t,a), \tag{9.36a}$$
$$S(t,a_0) = B, \tag{9.36b}$$
$$S(0,a) = \Phi(a), \tag{9.36c}$$

$$I_t(t,a) + I_a(t,a) = -(\mu(a) + \gamma(a)) I(t,a) + \lambda(t,a)S(t,a), \tag{9.36d}$$
$$I(t,a_0) = 0, \tag{9.36e}$$
$$I(0,a) = \Psi(a), \tag{9.36f}$$

$$A_t(t,a) + A_a(t,a) = -\delta(a)A(t,a) + \gamma(a)I(t,a), \tag{9.36g}$$
$$A(t,a_0) = 0, \tag{9.36h}$$
$$A(0,a) = 0, \tag{9.36i}$$

where Φ and Ψ are the initial distributions of susceptibles and infectives, respectively.

The infection rate, λ, is determined by

$$\lambda(t,a) = r(a) \int_{a_0}^{\infty} \beta(a,a')\rho(t,a,a')\frac{I(t,a')}{T(t,a')}da', \tag{9.37}$$

where $T(t,a) = S(t,a) + I(t,a)$ is the total number of sexually active individuals, $r(a)$ the number of partners that an individual of age a has per unit time, $\beta(a,a')$ the transmission probability of a susceptible individual of age a infected by an infected partner of age a', and $\rho(a,a',t)$ the rate of pair-formation between individuals of ages a and a'.

The transmission probability can be further described by

$$\beta(a,a') = f(a)g(a'),$$

where $f(a)$ is the susceptibility of individuals of age a, and $g(a')$ is the infectiousness of individuals of age a'. Then

$$\lambda(t,a) = r(a)f(a) \int_{a_0}^{\infty} g(a')\rho(t,a,a')\frac{I(t,a')}{T(t,a')}da'. \tag{9.38}$$

9.5.1 The Reproduction Number

One of the fundamental questions of mathematical epidemiology is to find the reproduction number, which determines whether an infectious disease spreads in a susceptible population when the disease is introduced into the population [1,13–15,19,21,23,37,41]. A possible formula for the reproduction number can be derived by determination of the condition for local stability of the infection-free equilibrium [4,25,28].

Model (9.36) has an infection-free equilibrium, $(S, I, A) = (S^0(a), 0, 0)$, where

$$S^0(a) = Be^{-M(a)} + e^{-M(a)} \int_{a_0}^a e^{M(x)} \Lambda(x) dx$$

with $M(a) = \int_{a_0}^a \mu(s) ds$.

We assume a separable pair-formation such that

$$\rho(t, a, a') = p_1(a) p_2(a').$$

Then we perturb the infection-free equilibrium by letting $u(t, a) = S(t, a) - S^0(a)$. Substitution into (9.37) leads to

$$\lambda(t, a) = r(a) f(a) p_1(a) \int_{a_0}^\infty g(a') p_2(a') \frac{I(t, a')}{T(t, a')} da'$$
$$\approx r(a) f(a) p_1(a) \int_{a_0}^\infty \frac{g(a') p_2(a')}{S^0(a')} I(t, a') da' := \tilde\lambda(t, a).$$

Then linearizing system (9.36) yields the linear system:

$$\begin{aligned}
u_t + u_a &= -\mu(a) u - \tilde\lambda(t, a) S^0(a), \\
I_t + I_a &= -(\mu(a) + \gamma(a)) I + \tilde\lambda(t, a) S^0(a), \\
A_t + A_a &= -\delta(a) A + \gamma(a) I.
\end{aligned} \quad (9.39)$$

Assume

$$u(t, a) = \tilde u(a) e^{c(t-a)}, \quad I(t, a) = \tilde I(a) e^{c(t-a)}.$$

Then $\tilde u(a)$ and $\tilde I(a)$ satisfy the following system of ordinary differential equations:

$$\frac{d\tilde u(a)}{da} = -\mu(a) \tilde u(a) - b(a) e^{ca} W, \quad (9.40)$$

$$\frac{d\tilde I(a)}{da} = -(\mu(a) + \gamma(a)) \tilde I(a) + b(a) e^{ca} W, \quad (9.41)$$

where $b(a) = S^0(a) r(a) f(a) p_1(a)$, and

$$W = \int_{a_0}^\infty \frac{g(a') p_2(a')}{S^0(a')} e^{-ca'} \tilde I(a') da'. \quad (9.42)$$

Solving (9.41), we have

$$\tilde I(a) = W e^{-M(a) - \Gamma(a)} \int_{a_0}^a e^{M(s) + \Gamma(s)} b(s) e^{cs} ds, \quad (9.43)$$

with $\Gamma(a) = \int_{a_0}^{a} \gamma(v)dv$. Substituting (9.43) into (9.42), we obtain

$$W = W \int_{a_0}^{\infty} \frac{g(a')p_2(a')}{S^0(a')} e^{-M(a')-\Gamma(a')} \int_{a_0}^{a'} e^{M(s)+\Gamma(s)} b(s) e^{-c(a'-s)} ds da'.$$
(9.44)

Define

$$H(c) = \int_{a_0}^{\infty} \frac{g(a')p_2(a')}{S^0(a')} e^{-M(a')-\Gamma(a')} \int_{a_0}^{a'} e^{M(s)+\Gamma(s)} b(s) e^{-c(a'-s)} ds da'.$$
(9.45)

Then there exists a nonzero solution W to equation (9.44) if and only if there exists a real or complex number c such that

$$H(c) = 1.$$
(9.46)

For all real numbers c, we have $\lim_{c\to\infty} H(c) = 0$. Then it follows from

$$\frac{dH(c)}{dc} = -\int_{a_0}^{\infty} \frac{g(a')p_2(a')}{S^0(a')} e^{-M(a')-\Gamma(a')}$$
$$\cdot \int_{a_0}^{a'} (a'-s) e^{M(s)+\Gamma(s)} b(s) e^{-c(a'-s)} ds da' < 0,$$

that $H(c)$ is a decreasing function. Hence, if c is a real solution of equation (9.46), then $c > 0$, provided $H(0) > 1$, and $c < 0$, provided $H(0) < 1$.

Suppose $c = \alpha + i\gamma$ is a complex solution of equation (9.46). Then substituting it into (9.46) and separating the real and imaginary parts yields

$$1 = \mathrm{Re}H(c) = \int_{a_0}^{\infty} \frac{g(a')p_2(a')}{S^0(a')} e^{-M(a')-\Gamma(a')}$$
$$\cdot \int_{a_0}^{a'} e^{M(s)+\Gamma(s)} b(s) e^{-\alpha(a'-s)} \cos\gamma(a'-s) ds da' \leq H(\alpha).$$

If $H(0) < 1$, then $\alpha < 0$. Hence $H(0) = 1$ is a threshold for the stability of the infection-free equilibrium. Define $R_0 = H(0)$. Then R_0 is the reproduction number of infection for system (9.36). Equation (9.46) is called the characteristic equation.

9.5.2 Pair-Formation in Age-Structured Epidemic Models

Sexually transmitted diseases (STDs) spread through sexual activities between partners. The pair-formation, or mixing, is one of the key terms in modeling of STDs [18, 23]. In Sect. 9.5.1, we assume the function describing pair-formation in model (9.36) to be separable, which makes the mathematical analysis more tractable. However, it has been shown that the assumption of a separable pair-formation function is equivalent to assuming a total proportionate or random partnership formation [2, 3, 5–7]. We briefly explain it as follows.

Let $\rho(t, a, a')$ be the pair-formation or mixing, which is the proportion of partners with age a' that an individual of age a has at time t. Let $r(t, a)$ be the average number of partners that an individual of age a has per unit of time, and let $T(t, a)$ be the total number of individuals of age a at time t. Then the function $\rho(t, a, a')$ has the properties

1. $0 \leq \rho(t, a, a') \leq 1$,
2. $\int_0^\infty \rho(t, a, a')da' = 1$,
3. $\rho(t, a, a')r(t, a)T(t, a) = \rho(t, a', a)r(t, a')T(t, a')$,
4. $r(t, a)T(t, a)r(t, a')T(t, a') = 0 \Longrightarrow \rho(t, a, a') = 0$.

Properties (1) and (2) follow from the fact that $\rho(t, a, a')$ is a proportion so that it is always between zero and one, and its total sum equals one. Property (3) comes from the fact that the total number of pairs of individuals of age a with individuals of age a' needs to be equal to the total number of pairs of individuals of age a' with individuals of age a. Moreover, if there are no active individuals, then there is no pair-formation, which leads to property (4).

9.5.2.1 Total Proportionate Mixing

Suppose that the pair-formation is a separable function such that

$$\rho(t, a, a') = \rho_1(t, a)\rho_2(t, a'). \tag{9.47}$$

It follows from property 2) that

$$\int_0^\infty \rho(t, a, a')da' = \int_0^\infty \rho_1(t, a)\rho_2(t, a')da' = \rho_1(t, a) \int_0^\infty \rho_2(t, a')da' = 1,$$

for all t. Hence

$$\rho_1(t, a) = \frac{1}{\int_0^\infty \rho_2(t, a')da'}$$

is independent of a. Denote it by $L(t)$. Then

$$\rho(t, a, a') = L(t)\rho_2(t, a'). \tag{9.48}$$

It follows from property 3) and (9.48) that

$$L(t)\rho_2(t, a')r(t, a)T(t, a) = \rho(t, a', a)r(t, a')T(t, a'). \tag{9.49}$$

Integrating (9.49) with respect to a from 0 to ∞ yields

$$L(t)\rho_2(t,a') \int_0^\infty r(t,a)T(t,a)da = r(t,a')T(t,a'). \tag{9.50}$$

Hence

$$L(t)\rho_2(t,a') = \frac{r(t,a')T(t,a')}{\int_0^\infty r(t,a)T(t,a)da}, \tag{9.51}$$

which implies that $\rho(t,a,a')$ satisfies

$$\rho(t,a,a') = \frac{r(t,a')T(t,a')}{\int_0^\infty r(t,a)T(t,a)da}. \tag{9.52}$$

Notice that the right-hand side in (9.52) is the fraction of the total partners of age a' in the population, or the availability of partners of age a'. A pair-formation or mixing function satisfying (9.49) is called a total proportionate mixing. Such a mixing depends completely on the availability of partners and is a kind of random mixing. While it may be appropriate to assume a proportionate mixing or random mixing in special cases such as modeling of HIV/AIDS for homosexual men, in general, it is necessary to assume the pair-formation or mixing function to be non-separable.

9.5.3 The Semigroup Method

As we have suggested in Sect. 9.5.2.1, in general the mixing function should be assumed non-separable. The mathematical analysis then becomes more difficult. A possible way to investigate dynamical behavior of models with non-separable mixing is to utilize the semigroup method. We outline the method for a simplified age-structured HIV/AIDS model with non-separable mixing as follows.

Consider ages in a finite interval $[0,\omega]$, where ω is the maximal sexually active age and assume the infection rate has the form

$$\lambda(t,a) = h(a) \int_0^\omega \rho(a,a') \frac{I(t,a')}{T(t,a')} da'.$$

Let $x = S/T$ and $y = I/T$. Then the dynamics of the age-structured epidemic model can be determined by the equation

$$y_t(t,a) + y_a(t,a) = (-\gamma(a)y + \lambda(t,a)) \tag{9.53}$$

with infection rate

$$\lambda(t,a) = h(a) \int_0^\omega \rho(a,a')y(t,a')da'. \tag{9.54}$$

Linearizing equation (9.53) with (9.54), we have

$$y_t + y_a = -\gamma(a)y + h(a)\int_0^\omega \rho(a, a')y(t, a')da'. \qquad (9.55)$$

Define linear operators $\mathcal{B}$ and $\mathcal{P}$ by

$$(\mathcal{B}f)(a) = -\frac{df(a)}{da} - \gamma(a)f(a),$$
$$(\mathcal{P}f)(a) = h(a)\int_0^\omega \rho(a, a')f(a')da'.$$

Then equation (9.55) can be written as

$$\frac{dy}{dt} = (\mathcal{B} + \mathcal{P})\, y. \qquad (9.56)$$

The operator $\mathcal{B} + \mathcal{P}$ generates a C_0 semigroup $T(t)$, for $t \geq 0$, and the semigroup $T(t)$ is eventually uniformly continuous. The growth bound of $T(t)$ is the spectral bound of $\mathcal{B} + \mathcal{P}$. It can be shown [27, 29, 33] that the resolvent of $\mathcal{B} + \mathcal{P}$, denoted by $R(\lambda; \mathcal{B} + \mathcal{P})$, is equal to $(S_\lambda - I)^{-1}G$ where

$$(Gf)(a) = \int_0^a e^{-\lambda(a-\sigma)}\Gamma(a)\Gamma^{-1}(\sigma)f(\sigma)d\sigma,$$

$$(S_\lambda f)(a) = \int_0^\omega \int_0^a e^{-\lambda(a-\sigma)}\Gamma(a)\Gamma^{-1}(\sigma)h(\sigma)p(\sigma, \xi)d\sigma f(\xi)d\xi.$$

Here we write $\Gamma(a) = e^{-\int_0^a \gamma(s)ds}$. Therefore, we can define the reproduction number of the epidemic, R_0, as the spectral radius of the operator

$$(Sf)(a) = \int_0^\omega \int_0^a \Gamma(a)\Gamma^{-1}(\sigma)h(\sigma)p(\sigma, \xi)d\sigma f(\xi)d\xi.$$

Consider a special case where the pair-formation is a finite sum of separable functions given by

$$\rho(a, a') = \sum_{j=1}^n p_j(a)q_j(a').$$

Then the reproduction number, R_0, is the largest positive eigenvalue λ_1 of the nonnegative matrix

$$\hat{K} = \begin{pmatrix} \int_0^\omega q_1(a)H_1(a)da & \cdots & \int_0^\omega q_1(a)H_n(a)da \\ & \vdots & \\ \int_0^\omega q_n(a)H_1(a)da & \cdots & \int_0^\omega q_n(a)H_n(a)da \end{pmatrix},$$

where

$$H_j(a) = \Gamma(a) \int_0^a h(\sigma)\Gamma^{-1}(\sigma)p_i(\sigma)d\sigma.$$

In particular, for $n = 2$, we have the explicit expression

$$R_0 = \frac{1}{2} \int_0^\omega (q_1 H_1 + q_2 H_2)\,da \\ + \frac{1}{2}\sqrt{\left(\int_0^\omega (q_1 H_1 - q_2 H_2)\,da\right)^2 + 4\int_0^\omega q_1 H_2 da \int_0^\omega q_2 H_1 da}$$

for the reproduction number of infection [33].

9.6 Modeling with Discrete Age Groups

Under certain conditions, the age-structured partial differential equation model (9.4) can be reduced to a system of ordinary differential equations [22, 32, 40].

Partition the age interval into a finite number n of subintervals $[a_0, a_1)$, $[a_1, a_2), \cdots, [a_{n-1}, a_n)$, where $a_0 = 0$ and $a_n \leq \infty$. Denote the number of individuals with ages in interval $[a_{i-1}, a_i]$ by $H_i(t)$, so that $H_i(t) = \int_{a_{i-1}}^{a_i} \rho(t, a)da$, $i = 1, \cdots, n$. Then integrating the partial differential equation in (9.3) from a_0 to a_1, we have

$$\frac{dH_1(t)}{dt} + \rho(t, a_1) - \rho(t, a_0) + \int_{a_0}^{a_1} \mu(a, P)\rho(t, a)da = 0. \qquad (9.57)$$

Assume that individuals with ages in each interval have the same vital rates such that $\beta(a, P) = \beta_i$, $\mu(a, P) = \mu_i$, for a in $[a_{i-1}, a_i]$, $i = 1, \cdots, n$. Here β_i and μ_i are age-independent, but may be density-dependent. Then, in the age interval $[0, a_1]$, we have

$$\rho(t, 0) = \sum_1^n \beta_i\, H_i(t), \quad \int_{a_0}^{a_1} \mu\, \rho(t, a)da = \mu_1\, H_1(t),$$

which leads to

$$\frac{dH_1}{dt} = \sum_1^n \beta_i\, H_i - (m_1 + \mu_1)H_1. \qquad (9.58)$$

Here m_1 is the progression rate from groups 1 to 2, defined by $m_1 = \rho(t, a_1)/H_1(t)$, and we assume it is time-independent.

Integrating (9.3) from a_{i-1} to a_i for $2 \leq i \leq \infty$, we have

$$\frac{dH_i}{dt} = m_{i-1}\, H_{i-1} - (m_i + \mu_i)\, H_i, \quad i = 2, \cdots, n, \qquad (9.59)$$

where m_i is the age progression rate from groups i to $i+1$, and we let $m_n = 0$. Then the system in (9.4) is reduced into a system of n ordinary differential equations.

9.6.1 Examples

We provide two simple examples to demonstrate how the discrete age group model described by equations (9.58) and (9.59) can be applied to populations and infectious diseases.

9.6.1.1 A Two-Age-Group Population Model

There are many means by which individuals of a species might compete for resources and by which intra-specific competition might express itself. Organisms which do not undergo such radical changes during their life cycles (e.g., birds, mammals, most reptiles, fishes, and hemimetabolous insects such as aphids, true bugs and grasshoppers) can experience considerable competition between juveniles and adults for common resources. Intra-specific competition can also occur to organisms with simple life cycles. The well studied flour beetles of genus *Tribolium* whose adult and larval stages utilize food resources in common provide a case in point [8].

Let $J(t)$ and $A(t)$ denote the densities of juveniles and adults at time t, respectively. Using the model in (9.58) and (9.59), with $n = 2$, we have the two age-group model

$$\begin{aligned} J'(t) &= \beta(J, A)A - (m(J, A) + \mu_1(J, A))\, J, \\ A'(t) &= m(J, A)J - \mu_2(J, A)A, \end{aligned} \tag{9.60}$$

where $'$ denotes d/dt, β is the birth rate of adults, μ_i, $i = 1, 2$, are death rates for juveniles and adults, respectively, and m is the age progression rate.

Models similar to (9.60) have been studied intensively. Readers are referred to [12, 16, 20, 30, 31, 35, 38, 39].

9.6.1.2 A Multi-Age-Group Malaria Model

Malaria is by far the world's most important tropical parasitic disease, which kills more people than any other communicable disease with the exception of tuberculosis. Approximately 10.5% (1,098,000) of deaths in children in developing countries in 2002 were due to malaria. Generally children have weaker immune systems, having not been as exposed to as much illness as adults.

It is known that there is acquired immunity in humans, even though the mechanisms of immunity to malaria are not fully understood. The acquired immunity appears to depend on both the duration and the intensity of past exposure to infection. Recovery from a primary infection with malaria does not imply fully protective immunity against reinfection. Immunity against malaria evidently influences the production of gametocytes. Frequency and intensity of gametocytemia decrease with increasing age until they reach a minimum among adults [43].

Therefore, in modeling of malaria transmission, age-structured models are more appropriate, and this can provide insight into the spread of malaria among different age groups, and can help identify efficient disease control strategies, for example, by targeting certain age-groups for vaccination.

Consider a human population in which malaria spreads. Divide the human population into four classes: susceptibles, exposed who are the individuals infected but not yet transmitting the disease, infectives, and recovereds who are recovered and also immune from re-infection. Denote them as $S(a,t)$, $E(a,t)$, $I(a,t)$, and $R(a,t)$, respectively. We further divide the human population into n age groups such that S_i, E_i, I_i, and R_i, $i = 1,\ldots,n$, are the susceptible, exposed, infective, and recovered individuals in age group i. Then the malaria transmission dynamics in the human population are governed by the system of ordinary differential equations

$$
\begin{aligned}
S_1'(t) &= B(t) - (\mu_1 + \eta_1)S_1 - \lambda_1(t)S_1, \\
S_j'(t) &= \eta_{j-1}S_{j-1} - \lambda_j(t)S_j - (\mu_j + \eta_j)S_j, \quad j = 2,\ldots,n, \\
E_1'(t) &= \lambda_1(t)S_1 - (\mu_1 + \epsilon_1 + \eta_1)E_1, \\
E_j'(t) &= \lambda_j(t)S_j + \eta_{j-1}E_{j-1} - (\mu_j + \epsilon_j + \eta_j)E_j \quad j = 2,\ldots,n, \\
I_1'(t) &= \epsilon_1 E_1 - (\mu_1 + \gamma_1 + \omega_1 + \eta_1)I_1, \\
I_j'(t) &= \epsilon_j E_j + \eta_{j-1}I_{j-1} - (\mu_j + \gamma_j + \omega_j + \eta_j)I_j \quad j = 2,\ldots,n, \\
R_1'(t) &= \gamma_1 I_1 - (\mu_1 + \eta_1)R_1, \\
R_j'(t) &= \eta_{j-1}R_{j-1} + \gamma_j I_j - (\mu_j + \eta_j)R_j \quad j = 2,\ldots,n,
\end{aligned}
\tag{9.61}
$$

where $B(t)$ is a input flow into the susceptible class, μ_i the age specific natural death rates, ω_i the age specific disease induced death rates, η_i the age progression rate, ε_i the age specific disease progression rates, and γ_i the age specific recovery rates.

The infection rates $\lambda_j(t)$ for humans are related to the vector (mosquito) population and are given by

$$
\lambda_j(t) = \frac{bN_v(t)}{N(t)}\beta_j \frac{I_v(t)}{N_v(t)} = \frac{b\beta_j I_v(t)}{N(t)}, \quad j = 1,\ldots,n,
\tag{9.62}
$$

where b is the number of bites on humans taken per mosquito in unit time, N_v the total mosquito population, $N = \sum_{j=1}^{n}(S_j + E_j + I_j + R_j)$ the total human population, I_v the number of infective mosquitoes, and β_j the probability of infection for humans in group j.

Due to the short life span of the mosquito populations, age structure is not incorporated into the mosquitoes. It is also assumed that all mosquitoes will die before recovering from infection. Then $N_v = S_v + E_v + I_v$, where S_v and E_v are the numbers of susceptible and exposed mosquitoes. The dynamics of the mosquito population are described by the equations

$$
\begin{aligned}
S_v'(t) &= M_v - \lambda_v S_v - \mu_v S_v, \\
E_v'(t) &= \lambda_v(N_v - E_v - I_v) - (\mu_v + \varepsilon_v), \\
I_v'(t) &= \varepsilon_v E_v - \mu_v I_v,
\end{aligned}
\tag{9.63}
$$

where M_v is an input flow of susceptible mosquitoes, μ_v is the natural death rate of mosquitoes, ε_v is the disease progression rate for exposed mosquitoes, and λ_v is the infection rate for mosquitoes given by

$$
\lambda_v(t) = b \sum_{j=1}^{n} \left(\frac{\beta_{v_j} I_j(t)}{N(t)} \right).
\tag{9.64}
$$

Here β_{v_j} are the infection rate of mosquitoes by infected humans in group j.

System (9.61) is strongly coupled, which increases the difficulty of mathematical analysis. Readers are referred to [36] for preliminary studies.

References

1. R. M. Anderson and R. M. May, Infectious Diseases of Humans, Dynamics and Control, Oxford University Press, Oxford, 1991.
2. S. P. Blythe and C. Castillo-Chavez, Like-with-like preference and sexual mixing models, Math. Biosci., 96 (1989), 221–238.
3. S. P. Blythe, C. Castillo-Chavez, J. S. Palmer and M. Cheng, Toward a unified theory of sexual mixing and pair formation, Math. Biosci., 107 (1991), 379–405.
4. F. Brauer and C. Castillo-Chavez, Mathematical Models in Population Biology and Epidemiology, Springer, Berlin Heidelberg New York, 2001.
5. S. Busenberg and C. Castillo-Chavez, Interaction, pair formation and force of infection terms in sexually transmitted diseases, in: C. Castillo-Chavez, (ed.), Mathematical and Statistical Approaches to AIDS Epidemiology, Lecture Notes in Biomathematics, Vol. 83, Springer, Berlin Heidelberg New York, (1989), 289–300.
6. S. Busenberg and C. Castillo-Chavez, A general solution of the problem of mixing of subpopulations and its application to risk- and age-structured epidemic models for the spread of AIDS, IMA J. Math. Appl. Med. Biol., 8 (1991), 1–29.
7. C. Castillo-Chavez and S. P. Blythe, Mixing Framework for Social/Sexual Behavior, in: C. Castillo-Chavez, (ed.), Mathematical and Statistical Approaches to AIDS Epidemiology, Lecture Notes in Biomathematics, Vol. 83, Springer, Berlin Heidelberg New York, (1989), 275–288.
8. R. F. Costantino and R. A. Desharnais, Population Dynamics and the 'Tribolium' Model: Genetics and Demography, Mono. Theor. Appl. Gen., 3 (1991), Springer, Berlin Heidelberg New York.

9. J. M. Cushing, Existence and stability of equilibria in age-structured population dynamics, J. Math. Biol., 20 (1984), 259–276.

10. J.M. Cushing, Equilibria in structured populations, J. Math. Biol., 23 (1985), 15–39.

11. J. M. Cushing, An Introduction to Structured Population Dynamics, SIAM, Philadelphia, 1998.

12. J. M. Cushing and Jia Li, Juvenile versus adult competition, J. Math. Biol., 29 (1991), 457–473.

13. O. Diekmann, J. A. P. Heesterbeek and J. A. J. Metz, On the definition and computation of the basic reproduction ratio R_0 in models for infectious diseases in heterogeneous populations, J. Math. Biol., 28 (1990), 365–382.

14. O. Diekmann, K. Dietz and J.A.P. Heesterbeek, The basic reproduction ratio for sexually transmitted diseases, Part 1: Theoretical considerations, Math. Biosci., 107 (1991), 325–339.

15. O. Diekmann and J. A. P. Heesterbeek, Mathematical Epidemiology of Infectious Diseases, Wiley, New York, 2000.

16. W. S. Gurney and R. M. Nisbet, Ecological Dynamics, Oxford University Press, Oxford, 1998.

17. M. E. Gurtin and R. C. MacCamy, Non-linear age-dependent population dynamics, Archive for Rational Mechanics and Analysis, 54 (1985), 281–300.

18. K. P. Hadeler, R. Waldstatter and A. Worz-Busekros, Models for pair-formation in bisexual populations, J. Math. Biol., 26 (1988), 635–649.

19. K. P. Hadeler and J. Müller, Vaccination in age structured populations I: The reproductive number, in: V. Isham and G. Medley, (eds.), Models for Infectious Human Diseases: Their Structure and Relation to Data, Cambridge University Press, Combridge, (1995), 90–101.

20. A. Hastings, Age-dependent predation is not a simple process, I. Continuous models, Theor. Popul. Biol., 23 (1983), 347–362.

21. J. A. P. Heesterbeek, R_0, Thesis, Centre for Mathematics and Computer Science, Amsterdam, (1991).

22. H. W. Hethcote, An age-structured model for pertussis transmission, Math. Biosci., 145 (1997), 89–136.

23. H. W. Hethcote, The mathematics of infectious diseases, SIAM Rev, 42 (2000), 599–653.

24. J. M. Hyman, Jia Li and E. A. Stanley, Threshold conditions for the spread of the HIV infection in age-structured populations of homosexual men, J. Theor. Biol., 166 (1994), 9–31.

25. J. M. Hyman and Jia Li, An intuitive formulation for the reproductive number for the spread of diseases in heterogeneous populations, Math. Biosci., 167 (2000), 65–86.

26. M. Iannelli, Mathematical Theory of Age-Structured Population Dynamics, Appl. Math. Monogr. C.N.R., 7 (1995).

27. H. Inaba, Threshold and stability for an age-structured epidemic model, J. Math. Biol., 28 (1990), 411–434.

28. J. A. Jacquez, C. P. Simon, and J. Koopman, The reproductive number in deterministic models of contagious diseases, Comm. Theor. Biol., 2 (1991), 159–209.

29. K. J. Engel and R. Nagel, One-Parameter Semigroups for Linear Evolution Equations, Springer, Berlin Heidelberg New York, 1999.

30. T. Kostova and Jia Li, Oscillations and stability due to juvenile versus adult competition, Int. J. Comput. Math. Appl., 32 (1996), 57–70.

31. T. Kostova, Jia Li and M. Friedman, Two models for competition between age classes, Math. Biosci., 157 (1999), 65–89.

32. Jia Li and T. G. Hallam, Survival in continuous structured population models, J. Math. Biol., 26 (1988), 421–433.

33. Jia Li, Threshold conditions in age-structured AIDS models with biased mixing, CNLS Newsletter, Los Alamos National Laboratory, 58 (1990), 1–10.

34. A. G. McKendrick, Applications of mathematics to medical problems, Proc. Edinb. Math. Soc. 44 (1926), 98–130.
35. R. M. Nisbet and L. Onyiah, Population dynamic consequences of competition within and between age classes. J. Math. Bio., 32 (1994), 329–344.
36. T.R. Park, Age-Dependence in Epidemic Models of Vector-Borne Infections, Ph.D. Dissertation, University of Alabama in Huntsville, 2004.
37. R. Ross, The Prevention of Malaria, Murray, London, 1909.
38. H. R. Thieme, Mathematics in Population Biology, Princeton University Press, Princeton, 2003.
39. W. O. Tschumy, Competition between juveniles and adults in age-structured populations, Theor. Popul. Biol., 21 (1982), 255–268.
40. D. W. Tudor, An age-dependent epidemic model with application to measles, Math. Biosci., 73 (1985), 131–147.
41. P. Waltman, Deterministic Threshold Models in the Theory of Epidemics, Lect. Notes Biomath., 1 (1974), Springer, Berlin Heidelberg New York.
42. G. F. Webb, Theory of Age Dependent Population Dynamics, Marcel Dekker, New York, 1985.
43. W. H. Wernsdorfer, The importance of malaria in the world, in: Kreier, J. P. (ed.), Malaria, Epidemology, Chemotherapy, Morphology, and Metabolism, Vol. 1, Academic, New York, 1980.

Chapter 10
Distribution Theory, Stochastic Processes and Infectious Disease Modelling

Ping Yan

Abstract The occurrence of a major outbreak, the shape of the epidemic curves, as well as the final sizes of outbreaks, are realizations of some stochastic events with some *probability distributions*. These distributions are manifested through some *stochastic mechanisms*. This chapter divides a typical outbreak in a closed population into two phases, the initial phase and beyond the initial phase. For the initial phase, this chapter addresses several aspects: the invasion (i.e. the risk of a large outbreak); quantities associated with a small outbreak; and characteristics of a large outbreak. In a large outbreak beyond the initial phase, the focus is on its final size. After a review of distribution theories and stochastic processes, this chapter separately addresses each of these issues by asking questions such as: Are the latent period and/or the infectious period distributions playing any role? What is the role of the contact process for this issue? Is the basic reproduction number R_0 sufficient to address this issue? How many *stochastic mechanisms* may manifest observations that may resemble a power-law distribution, and how much detail is really needed to address this specific issue? etc. This chapter uses distribution theory and stochastic processes to capture the agent–host–environment interface during an outbreak of an infectious disease. With different phases of an outbreak and special issues in mind, modellers need to choose which detailed aspects of the distributions and the stochastic mechanisms need to be included, and which detailed aspects need to be ignored. With these discussions, this chapter provides some syntheses for the concepts and models discussed in some proceeding chapters, as well as some food for thought for following chapters on case studies.

Centre for Communicable Diseases and Infection Control, Infectious Diseases and Emergency Preparedness Branch, Public Health Agency of Canada, 100 Elangtine Drive, AL0602-B, Tunney's Pasture, Ottawa, ON, Canada K1A 0K9
Ping_Yan@phac-aspc.gc.ca

10.1 Introduction

This chapter limits the discussions to human–human transmission through direct contacts involving an agent (e.g. virus, bacteria, etc.) in a closed population. The *agent* has biological characteristics. The human *hosts* may differ in susceptibility. The *environment* is where contacts and transmissions take place. One wishes to control the outbreak aimed at preventing a large outbreak from happening. Should it happen, the number of individuals who are infectious at time t can be approximated by a curve, either symmetrical or slightly negatively skewed, like that illustrated in Fig. 10.1.

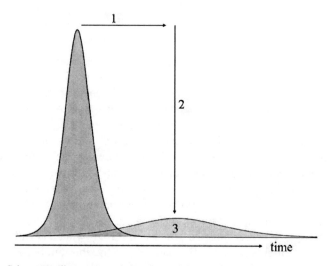

Fig. 10.1 Schematic illustration of the three objectives

Further public health objectives include:

1. Reducing the initial growth of the curve (and delaying the peak)
2. Reducing the peak burden
3. Reducing the final size, defined as the total number of individuals or the proportion of individuals in the population that will be eventually infected by the end of the outbreak.

Brauer [1] (Chap. 2 of this book) discussed the initial growth rate and the maximum value of $I(t)$ for the number of infectious individuals at time t in a compartment model. The final size of an outbreak was not only discussed in deterministic compartment models, but also by Allen [2] (Chap. 3 of this book) with respect to stochastic models. There concepts are also embedded in network models introduced by Brauer [3] in Chap. 4 of this book.

10.2 A Review of Some Probability Theory and Stochastic Processes

10.2.1 Non-negative Random Variables and Their Distributions

10.2.1.1 The Distribution Functions, the Expectation (Mean) and the Variance

Most random variables in this chapter, discrete or continuous, take non-negative values $X \geq 0$. The expected (mean) value and variance are denoted by $E[X]$ and $var[X]$. The cumulative distribution function (c.d.f.) is denoted by $F_X(x) = \Pr\{X < x\}$ satisfying $F_X(0) = 0$, monotonically increasing, and $F_X(\infty) = 1$. The survivor function is $\overline{F}_X(x) = 1 - F_X(x) = \Pr\{X \geq x\}$. The lower case $f_X(x)$ is used for the probability density function (p.d.f.) $f_X(x) = -\frac{d}{dx}\overline{F}_X(x) = \lim_{\delta \to 0} \frac{\Pr\{x \leq X < x+\delta\}}{\delta}$ if X is absolutely continuous; and for the probability mass function (p.m.f.) $f_X(x) = \overline{F}_X(x) - \overline{F}_X(x+1) = \Pr\{X = x\}$ if X is discrete taking values $x = 0, 1, 2, \cdots$. $f_X(x)$ satisfies (1) $f_X(x) > 0$; (2) $\sum_{x=0}^{\infty} f_X(x) = 1$ for discrete and $\int_0^{\infty} f_X(x)dx = 1$ for continuous random variable X.

It can be shown that when $X \geq 0$,

$$E[X] = \begin{cases} \sum_{x=0}^{\infty} \overline{F}_X(x), & \text{if } X \text{ discrete,} \\ \int_0^{\infty} \overline{F}_X(x)dx, & \text{if } X \text{ continuous.} \end{cases} \tag{10.1}$$

10.2.1.2 The Probability Generating Function (p.g.f.)

If X is discrete taking values $x = 0, 1, 2, \cdots$, the probability generating function, as previously introduced in Allen [2] and Brauer [3], is a mathematical tool to study its distribution. It is defined as

$$G_X(s) = E\left(s^X\right) = \sum_{x=0}^{\infty} s^x \Pr\{X = x\}, \tag{10.2}$$

satisfying

$$G_X(0) = \Pr\{X = 0\}, G_X(1) = 1, G_X'(s) > 0, G_X''(s) > 0. \tag{10.3}$$

If the p.m.f. $f(x) = \Pr\{X = x\}$ is given, $G_X(s)$ is uniquely defined through (10.2). If $G_X(s)$ is given, provided that it is a smooth function of s with higher order of derivatives, then

$$\Pr\{X = x\} = \frac{1}{x!}G_X^{(x)}(0), \ x = 0, 1, 2, \cdots \tag{10.4}$$

where $G_X^{(x)}(0) = \frac{d^x}{ds^x}G_X(s)\big|_{s=0}$, so that the p.m.f. $f(x)$ can be uniquely generated through $G_X(s)$. Calculations of moments and of some probabilities are very easy. The mean and variance of X are

$$E[X] = G_X'(1), \tag{10.5}$$
$$var[X] = G_X''(1) + G_X'(1) - (G_X'(1))^2.$$

10.2.1.3 The Hazard Function

If X is continuous, the hazard function, $h_X(x) \overset{\text{def.}}{=} \lim_{\delta \to 0} \frac{\Pr\{x \le X < x + \delta | X \ge x\}}{\delta}$ uniquely determines the distribution of X, through the relationships $f_X(x) = h_X(x)\overline{F}_X(x)$ and $\overline{F}_X(x) = \exp\left(-\int_0^x h_X(u)du\right)$. The shape of $h_X(x)$ can be used to define and classify families of continuous distributions for X. Some commonly considered candidates are:

1. Constant: $h_X(x) = \alpha$
2. $\lim_{x \to \infty} h_X(x) = \alpha$, but $h_X(x)$ is monotone, either increasing or decreasing
3. Monotone with $\lim_{x \to \infty} h_X(x) = \begin{cases} \infty, & \text{if increasing} \\ 0, & \text{if decreasing} \end{cases}$
4. Non-monotone: $h_X(x)$ initially increases to a maximum value and then decreases with $\lim_{x \to \infty} h_X(x) = 0$.

In the first case when $h_X(x) = \alpha$, $\overline{F}_X(x) = \exp(-\alpha x)$ and $f_X(x) = \alpha \exp(-\alpha x)$. This gives the exponential distribution. α is often called *rate*. If X represents the infectious period in a Susceptible-Infective-Removed (SIR) model, α is often referred to as the removal rate. The second equation in (2.1) of Brauer [1] is a special case of a more general equation

$$\frac{d}{dt}I(t) = i(t) - \int_0^t i(s) f_X(t - s)ds \tag{10.6}$$

$$= i(t) - \int_0^t i(s) h_X(t - s)\overline{F}_X(t - s)ds$$

where the special case is $h_X(x) = \alpha$ and $\overline{F}_X(x) = \exp(-\alpha x)$ such that $\frac{d}{dt}I(t) = i(t) - \alpha \int_0^t i(s)e^{-\alpha(t-s)}ds$. Note that $I(t) = \int_0^t i(s)e^{-\alpha(t-s)}ds$, because $\int_0^t i(s)e^{-\alpha(t-s)}ds = \int_0^t i(s) \Pr\{X > t - s\}ds$ is the number of individuals infected before t who have not yet been removed. If one further assumes $i(t) = \beta I(t)S(t)$, then (10.6) reduces to $I' = (\beta S - \alpha) I$ which is (2.1) of Brauer [1]. In addition, the exponential distribution has memoryless property: $\Pr\{X > x + y | X > x\} = \frac{\exp(-\alpha(x+y))}{\exp(-\alpha x)} = \exp(-\alpha y)$. For X with exponential distribution, the rate is reciprocal to the mean duration: $E[X] = \frac{1}{\alpha}$.

The exponentially distributed infectious period is the underlying assumption in the continuous time Markov chain model in Allen [2].

In the second case $\lim_{x\to\infty} h_X(x) = \alpha$, a commonly used model is the gamma distribution with p.d.f.

$$f_X(x) = \frac{\alpha(\alpha x)^{\kappa-1}}{\Gamma(\kappa)}e^{-\alpha x} \tag{10.7}$$

and survivor function $\overline{F}_X(x) = \frac{\alpha^{\kappa}}{\Gamma(\kappa)}\int_x^\infty u^{\kappa-1}e^{-\alpha u}du$, where α is a scale parameter. κ is the shape parameter that determines the shape of the hazard function. When $\kappa > 1$, the hazard is an increasing function of x and when $\kappa < 1$, a decreasing function of x. When $\kappa = 1$, it is the exponential distribution. It can be shown that for any $\kappa > 0$, $h_X(x) = \frac{f_X(x)}{\overline{F}_X(x)} \to \alpha$ as $x \to \infty$.

In the third case, the monotone hazard function satisfies $\lim_{x\to\infty} h_X(x) = \infty$, if increasing, and $\lim_{x\to\infty} h_X(x) = 0$, if decreasing. A common choice is $h_X(x) = \beta\alpha^{\beta}x^{\beta-1}$, a power function of x such that when $\beta = 1$, $h_X(x) = \alpha$; when $\beta > 1$, $h_X(x)$ is an increasing function of x and when $\beta < 1$, $h_X(x)$ is a decreasing function of x. β is the shape parameter that determines the shape of the hazard function. The corresponding p.d.f. and the survivor function are $f_X(x) = \beta\alpha(\alpha x)^{\beta-1}e^{-(x\alpha)^\beta}$ and $\overline{F}_X(x) = e^{-(\alpha x)^\beta}$. α is a scale parameter. This is the Weibull distribution.

Of non-monotone hazard functions that initially increase to a maximum value and then decrease with $\lim_{x\to\infty} h(x) = 0$, there are two commonly used distributions in the literature: the log-normal distribution with p.d.f. $f_X(x) = \frac{\theta}{x\sqrt{2\pi}}e^{\frac{-\theta^2(\log\alpha x)^2}{2}}$ and the log-logistic distribution with p.d.f. $f_X(x) = \frac{\alpha\theta(\alpha x)^{\theta-1}}{\left(1+(\alpha x)^\theta\right)^2}$. In both distributions α is a scale parameter and θ is the shape parameter.

Another continuous distribution for X used in this chapter is the Pareto distribution with hazard function $h_X(x) = \frac{\alpha\kappa}{1+\alpha x}$ which is monotonically decreasing $\lim_{x\to\infty} h_X(x) = 0$. It has p.d.f. $f_X(x) = \frac{\kappa\alpha}{(1+\alpha x)^{\kappa+1}}$ and survivor function $\overline{F}_X(x) = \frac{1}{(1+\alpha x)^\kappa}$.

10.2.1.4 The Laplace Transform

For a non-negative continuous random variable X, it is sometimes convenient to work with the Laplace transforms

$$\mathcal{L}_X(r) = \int_0^\infty e^{-rx}f_X(x)dx,$$

and its related function $\mathcal{L}_X^*(r) = \int_0^\infty e^{-rx}\overline{F}_X(x)dx = \frac{1-\mathcal{L}(r)}{r}$, provided that they exist. As a function of r, the following statements hold:

- $\lim_{r\to 0}\mathcal{L}_X^*(r) = \int_0^\infty \overline{F}_X(x)dx = E[X]$
- $\mathcal{L}_X^*(r)$ is monotonically decreasing
- $\lim_{r\to\infty}\mathcal{L}_X^*(r) = 0$

10.2.2 Some Important Discrete Random Variables Representing Count Numbers

10.2.2.1 The Random Number Associated with the *Basic Reproduction Number* R_0

Let us call an *infectious contact* a contact at which transmission takes place [4]. We use the notation N for a discrete valued random variable, defined as

N = the number of infectious contacts made by an infective individual, throughout its entire infectious period, in a wholly susceptible population.

The expected value of N is the basic reproduction number $R_0 = E[N]$. Uncertainties are captured by $var[N]$ as well as the probability distribution $\Pr\{N = x\}$, $x = 0, 1, 2, \cdots$. Some public health questions can be sufficiently addressed by $R_0 = E[N]$ alone. Some other public health questions can be addressed by the first two moments: $E[N]$ and $var[N]$. For some questions, one needs to know the precise distribution for N. However, there are many other important public health issues that are not addressed by the distribution N. The distribution for N may arise from a combination of the following stochastic mechanisms:

1. The contact network structure and stochastic features that give rise to the contact process
2. The individual (host) properties that determine transmissibility per contact, such as host susceptibility
3. The probability distributions of durations of time-to-events, such as the infectious period.

The same probability distribution $\Pr\{N = x\}$ can arise from different stochastic mechanisms. For some public health questions, different stochastic mechanisms may give different answers, even if the probability distribution $\Pr\{N = x\}$ is the same. For some other public health aspects, different distributions for $\Pr\{N = x\}$ may give the same answer as long as certain aspects of the underlying stochastic mechanisms remain the same.

10.2.2.2 Discrete Random Numbers Associated with the *Final Size* of an Outbreak

Calendar time is denoted by t. At $t = 0$, there are S_0 initial susceptible individuals and I_0 initial infectious individuals in a closed population. The population size $n = S_0 + I_0$. S_0 and I_0 are fixed integers. Under the assumption $\frac{S_0}{n} \approx 1$ (i.e. S_0 is very large) and $\frac{I_0}{n} = \varepsilon \approx 0$ (i.e. I_0 is very small), if infective cases are removed at the end of their infectious period by recovery with immunity, or by complete isolation through intervention, or death, there may be a proportion of susceptible individuals that eventually escape from infection when the outbreak is over. The following pair of quantities are random:

- S_∞ = the total number of susceptible individuals who have escaped from infection in the population
- $Z = n - S_\infty$ = the final size, with a probability distribution $\Pr\{Z = z\}$, $z = 1, 2, \cdots$

Diekmann and Heesterbeek [5] describe the distinction of a *small* outbreak from a *large* outbreak. Using the random variable Z,

Small outbreak. As $n \to \infty$, the infectious agent produces a handful of cases and the outbreak becomes extinct, such that the expected number of cases, $E[Z]$, remains finite. The expected outbreak size as a proportion, $\frac{E[Z]}{n}$, is concentrated at zero.

Large outbreak. As $n \to \infty$, the final outbreak size as a proportion is a positive quantity, such that the expected final outbreak size number scales linearly with the size of the susceptible population. In other words, $E[Z] \to \infty$ but $\frac{E[Z]}{n} \to \eta$ where η is a positive quantity, $0 < \eta < 1$.

There are occasions that $E[Z] \to \infty$ but $\frac{E[Z]}{n} \to 0$ (e.g., in the order $\sim n^{\frac{2}{3}}$). In other words, although the final size as a proportion (scaled to the population size n) is concentrated at zero, the final outbreak size as absolute numbers can be very large. In this case, the outbreak size is neither small, nor large.

10.2.2.3 The Time-to-Extinction of a Small Outbreak

A random variable T_g is defined as the time to extinction **at** generation $g = 1, 2, 3, \cdots$, where the event $\{T_g = g\}$ refers to {no infected case at generation g and at least one infected case at generation $g-1$}. The generation time g is different from the calendar time t. In certain situations, observations arising from an outbreak can be identified by generation time.

10.2.2.4 Three Commonly Used Discrete Distributions

Allen [2] used p.g.f. to determine the probability of extinction in the context of *branching processes*. Brauer [3] reviewed the use of p.g.f. to characterize and calculate *degree distributions* with respect to the spread of diseases over contact networks. In this chapter, p.g.f. will be used to generate the distributions for N which is related to the basic reproduction number R_0, to derive probability distributions for the final size $\Pr\{Z = z\}$ in small outbreaks, and where possible, to derive the probability $\Pr\{T_g = g\}$.

Table 10.1 Three discrete distributions frequently used in this chapter

	$f_X(x)$	$G_X(s)$	$E[X]$	$var[X]$
Poisson	$\frac{\lambda^x}{x!}e^{-\lambda}$	$e^{-\lambda(1-s)}$	λ	λ
Geometric	$\frac{\lambda^x}{(1+\lambda)^{x+1}}$	$\frac{1}{1+\lambda(1-s)}$	λ	$\lambda^2 + \lambda$
Neg. binomial	$\frac{\Gamma\left(x+\frac{1}{\phi}\right)}{\Gamma(x+1)\Gamma\left(\frac{1}{\phi}\right)}\frac{(\lambda\phi)^x}{(1+\lambda\phi)^{x+\frac{1}{\phi}}}$	$\frac{1}{[1+\lambda\phi(1-s)]^{\frac{1}{\phi}}}$	λ	$\phi\lambda^2 + \lambda$

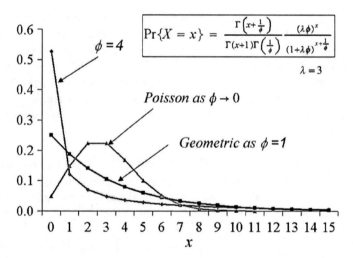

Fig. 10.2 Comparing the negative binomial distributions at $\lambda = 3$

We use the notation X which could be N, Z or T_g depending on context. Table 10.1 lists three commonly referred discrete distributions in this chapter. In discrete distributions, $f_X(x) = F_X(x+1) - F_X(x)$. The shapes of $f_X(x)$ for these distributions are compared in Fig. 10.2. The Poisson distribution is a special case of the negative binomial distribution as $\phi \to 0$; the geometric distribution is a special case of the negative binomial distribution as $\phi = 1$.

10.2.3 Continuous Random Variables Representing Time-to-Event Durations

In models involving a sequence of events, one may treat each pair of successive events as an *initiating event* that leads to a *subsequent event* over a random duration $X \geq 0$. Durations can be categorized by: (1) the natural history of infectiousness of an infected individual (e.g. latent and infectious periods); (2) the natural history of clinical manifestation (e.g. incubation period and duration of illness); and (3) the reaction time of the public health system (e.g. time to detect an infection or to isolate an infectious individual).

Table 10.2 Some common durations in infectious disease models

Initiating event	Subsequent event	Duration X
Infection of a virus	Becoming infectious	Latent period
Becoming infectious	Removal	Infectious period
Infection of a virus	Onset of symptoms	Incubation period
Infection of a virus	First positive test (diagnosis)	Time to testing
Diagnosis	Starting treatment	Time to treatment
Diagnosis	Isolation	Time to isolation

Models are often developed along the event history based on one or a combination of the three categories. For example, models based on the natural history of infectiousness of an infected host might be a deterministic or stochastic SIR or a Susceptible-Latent-Infective-Removed (SEIR) model, like those discussed in previous chapters.

10.2.3.1 Ordering and Tail Properties of Non-negative Continuous Random Variables

Very often one wants to compare two non-negative random variables X_1 and X_2, either through some overall summary of their distributions, or through their tail properties. For notational simplicity, we write $\overline{F}_1(x) = \overline{F}_{X_1}(x)$ and $\overline{F}_2(x) = \overline{F}_{X_2}(x)$ hereafter.

Definition 1. X_2 is longer than X_1 in stochastic order, denoted by $X_2 \geq_{st} X_1$, if corresponding survivor functions $\overline{F}_2(x) \geq \overline{F}_1(x)$ for all x.

In general, $\overline{F}_2(x) \geq \overline{F}_1(x) \Rightarrow E[X_2] = \int_0^\infty \overline{F}_2(x)dx \geq E[X_1] = \int_0^\infty \overline{F}_1(x)dx$. The reverse is not true for non-exponential distributions.

Definition 2. X_2 is longer than X_1 in hazard rate order, denoted by $X_2 \geq_{hr} X_1$, if corresponding hazard functions $h_2(x) \leq h_1(x)$, for all $x > 0$.

Note that $X_2 \geq_{hr} X_1 \Rightarrow X_2 \geq_{st} X_1$,

Definition 3. X_2 is longer than X_1 in Laplace transform order, denoted by $X_2 \geq_{\mathcal{L}} X_1$, if $\mathcal{L}_2^*(r) = \int_0^\infty e^{-rx}\overline{F}_2(x)dx \geq \mathcal{L}_1^*(r) = \int_0^\infty e^{-rx}\overline{F}_1(x)dx$.

Note that for the Laplace transforms $\mathcal{L}_2(r) = \int_0^\infty e^{-rx} f_2(x)dx$ and $\mathcal{L}_1(r) = \int_0^\infty e^{-rx} f_1(x)dx$, $X_2 \geq_{\mathcal{L}} X_1 \Leftrightarrow \mathcal{L}_2(r) \leq \mathcal{L}_1(r)$ since $\mathcal{L}_X^*(r) = \frac{1-\mathcal{L}_X(r)}{r}$.

In the special case when X is exponentially distributed, with $\mu_X = E[X]$, the hazard function is $h_X(x) = \frac{1}{\mu_X}$; the survivor function is $\overline{F}_X(x) = e^{-\frac{x}{\mu_X}}$; and $\mathcal{L}_X^*(r) = \frac{\mu_X}{1+r\mu_X}$. If X_2 and X_1 are both exponentially distributed, $X_2 \geq_{hr} X_1 \Leftrightarrow X_2 \geq_{st} X_1 \Leftrightarrow X_2 \geq_{\mathcal{L}} X_1 \Leftrightarrow E[X_2] \geq E[X_1]$. In general,

$$X_2 \geq_{hr} X_1 \Rightarrow X_2 \geq_{st} X_1 \Rightarrow X_2 \geq_{\mathcal{L}} X_1. \tag{10.8}$$

Another stochastic property used in this chapter is the relationship between the hazard function and the tail properties of randomly distributed durations. Denote $X_s = (X - s | X > s)$ as the *residual life* of a duration X conditioning on $X > s$. If X stands for the infectious period, then X_s stands for the time remaining to be infectious, after s amount of time since the beginning of the infectious period. Thus

$$\Pr\{X_s > x\} = \Pr\{X > s + x | X > s\} = \frac{\overline{F}_X(s+x)}{\overline{F}_X(s)}$$

is the survivor function of X_s. Given the hazard function $h_X(x)$, it can be shown that $\frac{\overline{F}_X(s+x)}{\overline{F}_X(s)} = \exp\left(-\int_s^{s+x} h_X(u)du\right)$. This leads to:

1. If $h_X(x)$ is strictly increasing with $h_X(x) \to \infty$, then $\frac{\overline{F}_X(s+x)}{\overline{F}_X(s)}$ is a decreasing function of s, $\lim_{s\to\infty} \frac{\overline{F}_X(s+x)}{\overline{F}_X(s)} = 0$, for any $x > 0$.
2. If there exists $x_0 \geq 0$ such that for $x > x_0$, $h_X(x)$ is a decreasing function of x with $\lim_{x\to\infty} h_X(x) = 0$, then

$$\lim_{s\to\infty} \frac{\overline{F}_X(s+x)}{\overline{F}_X(s)} = 1, \text{ for any } x > 0. \tag{10.9}$$

The distribution is said to be *heavy tailed* if it satisfies (10.9). Intuitively, if X ever exceeds a large value, then it is just as likely to exceed any larger value.
3. If $h_X(x) \to \alpha$, the distribution has *exponential tail*

$$\lim_{s\to\infty} \frac{\overline{F}_X(s+x)}{\overline{F}_X(s)} = e^{-\alpha x}, \text{ for any } x > 0. \tag{10.10}$$

4. A distribution is *sub-exponential*, if $\lim_{x\to\infty} \frac{\overline{F}_X(x)}{\exp(-\alpha x)} = \infty$, implying that it has a heavier tail (goes to zero more slowly) than an exponential tail. All heavy tailed distributions are also sub-exponential.

Another form of heavy tailed distribution is that of the *Pareto form*, such that for some $\theta > 0$, $A > 0$, $\lim_{x\to\infty} \frac{\overline{F}_X(x)}{x^{\theta+1}} = A$. They are heavy tailed with *power-law* property $\propto \frac{1}{x^{\theta+1}}$.

Table 10.3 Comparison of hazard functions and tail properties

	p.d.f. $f_X(x)$	Hazard $h_X(x)$	$\Pr\{X_s > x\}$, $x > 0$
Exponential	$\alpha e^{-\alpha x}$	Constant α	$\frac{\overline{F}_X(s+x)}{\overline{F}_X(s)} = e^{-\alpha x}$ (exponential)
Gamma	$\frac{\alpha(\alpha x)^{\kappa-1}}{\Gamma(\kappa)} e^{-\alpha x}$	Converge $h(x) \to \alpha$	$\lim_{s\to\infty} \frac{\overline{F}_X(s+x)}{\overline{F}_X(s)} = e^{-\alpha x}$ (exponential)
Weibull (if $\beta < 1$)	$\beta\alpha^\beta x^{\beta-1} e^{-(x\alpha)^\beta}$	$h(x) \downarrow 0$	$\lim_{s\to\infty} e^{\alpha^\theta \left(s^\theta - (s+x)^\theta\right)} = 1$ (heavy, not power-law)
Log-logistic	$\frac{\alpha\theta(\alpha x)^{\theta-1}}{(1+(\alpha x)^\theta)^2}$	$h(x) \downarrow 0$ After $x > x_0$	$\lim_{s\to\infty} \frac{1+(\alpha x)^\theta}{1+(\alpha(x+y))^\theta} = 1$ (heavy, not power-law)
Log-normal	$\frac{\theta}{x\sqrt{2\pi}} e^{\frac{-\theta^2(\log\alpha x)^2}{2}}$	$h(x) \downarrow 0$ After $x > x_0$	$\lim_{s\to\infty} \frac{\overline{F}_X(s+x)}{\overline{F}_X(s)} = 1$ (heavy, not power-law)
Pareto	$\frac{\kappa\alpha}{(1+\alpha x)^{\kappa+1}}$	$h(x) = \frac{\alpha\kappa}{1+x\alpha} \downarrow 0$	$\lim_{s\to\infty} \frac{(s\alpha+1)^\kappa}{(s\alpha+x\alpha+1)^\kappa} = 1$ (heavy and power-law)

10.2.4 Mixture of Distributions

In a heterogenous population an individual i is associated with a random variable X_i following a distribution with p.m.f. or p.d.f. $f(x|\theta_i)$ specified up to a parameter θ_i. If the heterogeneity can be observed through a vector of covariates $\underline{z}$, say, such as gender, birth date, height, etc., a common practice in statistics is to model θ_i as a function of $\underline{z}$ via a generalized linear model $\eta(\theta_i) = \beta_1 z_1 + \beta_2 z_2 + \cdots + \beta_q z_q$, where $\eta(\cdot)$ is a link function such that $-\infty < \eta(\theta_i) < \infty$.

If the heterogeneity is not observable, one assumes that θ_i varies among individuals as independently and identically distributed (i.i.d.) random variables with expectation $E_\theta[\cdot]$ such that at the population level, one may model X arising from a distribution given by

$$f_X(x) = E_\theta\left[f(x|\theta)\right] = \int_{\theta\in\Theta} f(x|\theta)dU(\theta),$$

where $U(\theta)$ is a c.d.f. of the mixing distribution.

Example 1. Each individual is associated with an infectious period with a constant removal rate λ_i, and λ_i itself varies across individuals. Then at the individual level, this leads to an exponentially distributed infectious period with p.d.f. $f(x|\lambda_i) = \lambda_i e^{-\lambda_i x}$. If λ_i has a gamma distribution with p.d.f. $u(\lambda) = \frac{\lambda^{\kappa-1}}{\alpha^\kappa \Gamma(\kappa)} e^{-\lambda/\alpha}$, then the resulting distribution is the Pareto distribution with p.d.f. $f_X(x) = \frac{\kappa\alpha}{(1+\alpha x)^{\kappa+1}}$, survivor function $\overline{F}_X(x) = \frac{1}{(1+\alpha x)^\kappa}$ and hazard function $h_X(x) = \frac{\alpha\kappa}{1+x\alpha}$.

There is often unobservable heterogeneity with respect to the infectious period, where *removal* can be due to deaths, recovery, or public health intervention such as isolation. By this example, even at the individual level it is justified to use an exponentially distributed infectious period with a removal rate λ_i, observed data from a population often arise as if the infectious period follows a distribution with a power-law heavy tail.

10.2.4.1 Mixed Poisson Distributions

Mixed Poisson distributions play a central role in the chapter. It is constructed by $f(x|\lambda) = \frac{\lambda^x}{x!} e^{-\lambda}$ and an arbitrary mixing distribution $U(\lambda)$. It is a discrete distribution with p.m.f. and p.g.f.

$$f_X(x) = \Pr\{X = x\} = \int_0^\infty \frac{\lambda^x}{x!} e^{-\lambda} dU(\lambda)$$

$$G_X(s) = \int_0^\infty e^{\lambda(s-1)} dU(\lambda) = E_\lambda \left[e^{\lambda(s-1)} \right].$$

Example 2. In 1920, Greenwood and Yule [6] derived the negative binomial distribution as a mixed Poisson distribution. The mixing distribution is a gamma distribution with p.d.f. $u(\lambda) = \frac{1}{\phi\mu\Gamma(\frac{1}{\phi})} \left(\frac{\lambda}{\phi\mu} \right)^{\frac{1}{\phi}-1} e^{-\frac{\lambda}{\phi\mu}}$ with $E[\lambda] = \mu$ and $var[\lambda] = \phi\mu^2$. The p.m.f. of the mixed Poisson distribution is

$$f_X(x) = \Pr\{X = x\} = \frac{\Gamma\left(x + \frac{1}{\phi}\right)}{\Gamma(1+x)\,\Gamma\left(\frac{1}{\phi}\right)} \frac{(\phi\mu)^x}{(1+\phi\mu)^{x+\frac{1}{\phi}}} \tag{10.11}$$

and p.g.f. $G_X(s) = \frac{1}{[1+\phi\mu(1-s)]^{\frac{1}{\phi}}}$. The mean and variance is $E[X] = \mu$ and $var[X] = \phi\mu^2 + \mu$. The special case when $\phi = 1$ gives the exponential distribution for $u(\lambda)$ and geometric distribution for $\Pr\{X = x\}$. The special case when $\phi \to 0$ degenerates $u(\lambda)$ to a fixed point and results in the Poisson distribution. Note that the p.d.f. of gamma distributions displayed in

Fig. 10.3 shows the remarkable resemblance of the p.m.f. to the negative binomial distributions in Fig. 10.2.

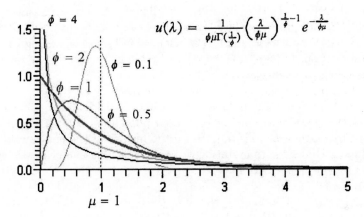

Fig. 10.3 P.d.f. for gamma distributions with $\mu = 1$ and different ϕ values

10.2.5 Stochastic Processes

A *stochastic process* $\{X(t),\ t \in T\}$ is a collection of random variables. For each t in the index set T, $X(t)$ is a random variable. If t is referred to as time, then $X(t)$ is the state of the process at time t. A realization of $\{X(t), t \in T\}$ is called a *sample path*. This chapter is restricted to discrete, non-negative sample paths, that is, for given t, $X(t)$ is a non-negative discrete random variable taking values $x = 0, 1, 2, \cdots$. The index t can be either discrete, or continuous. The distribution of $X(t)$ for given t often depends on the past history of the process $\mathcal{H}_t = \{X(u), 0 < u \le t^-\}$. The conditional probability $\Pr\{X(t) = x | \mathcal{H}_t\}$ is meaningful.

If the conditional distribution of the future state at time $t + s$, given the present state at time s and all past states, depends only on the present state and is independent from the past, that is, for all continuous time $s, t > 0$ and non-negative integers i, j

$$\Pr\{X(t + s) = i | X(s) = i, \mathcal{H}_s\} = \Pr\{X(t + s) = i | X(s) = i\}, \qquad (10.12)$$

we call it a *Markov process*. If, in addition, $\Pr\{X(t + s) = i | X(s) = i\} = \{X(t) = i | X(0) = i\}$ is independent of s, then the stochastic process $\{X(t)\}$ is a continuous time stationary Markov chain.

10.2.5.1 Continuous Time Markov Chain SIR Model

An SIR model can be expressed as a bivariate continuous time-homogeneous Markov chain $\{S(t), I(t)\}$ as described in Allen [2]. At any given time t, both $S(t)$ and $I(t)$ take integer values. Transitions can only occur from $\{S(t) = s, I(t) = i\}$ to $\{S(t + \Delta t) = s - 1, I(t + \Delta t) = i + 1\}$ corresponding to a new infection, or from $\{S(t) = s, I(t) = i\}$ to $\{S(t + \Delta t) = s, I(t + \Delta t) = i - 1\}$ corresponding to a removal.

Let $R(t)$ be the number of removed individuals at time t; $M(t) = I(t) + R(t)$ be the total number of infected individuals by time t; and $S(t) + I(t) + R(t) = n$, assuming a closed population. $\{R(t)\}$ and $\{M(t)\}$ are also stochastic processes. Let H_t be the history of the epidemic process up to time t. Under the time stationary Markov chain assumption, the model in Allen [2] can be re-written as

$$\begin{cases} \Pr\{dM(t) = 1, dR(t) = 0 | H_t\} \approx \beta \frac{S(t)}{n} I(t) dt, \\ \Pr\{dM(t) = 0, dR(t) = 1 | H_t\} \approx \frac{1}{\mu} I(t) dt, \\ \Pr\{dM(t) = 0, dR(t) = 0 | H_t\} \approx 1 - \beta \frac{S(t)}{n} I(t) dt - \frac{1}{\mu} I(t) dt. \end{cases} \tag{10.13}$$

Assuming the outbreak is originated from a single initial infective, the stochastic means follow the following renewal-type equations: $E[M(t)] = 1 + \beta \int_0^t e^{-\frac{1}{\mu} x} E[M(t - x)] dx$, $E[R(t)] = 1 - e^{-\frac{1}{\mu} t} + \beta \int_0^t e^{-\frac{1}{\mu} x} E[R(t - x)] dx$. Therefore,

$$E[I(t)] = e^{-\frac{1}{\mu} t} + \beta \int_0^t e^{-\frac{1}{\mu} x} E[I(t - x)] dx \tag{10.14}$$

where $I(t) = M(t) - R(t)$. The first term $e^{-\frac{1}{\mu} t}$ in (10.14) is the probability that the initial infective is still infectious at time t according to an exponentially distributed infectious period. With respect to the integration for the second term, during the interval $(0, t]$ the initial infective makes many infectious contacts with a constant rate β, so that the expected number of infectious contacts at an infinitesimal interval containing $x \in (0, t]$ is $\beta e^{-\frac{1}{\mu} x}$. The expected number of individuals who are still infectious at time t evolved from such contacts at $x \in (0, t]$ is $E[I(t - x)]$. It can be shown that [7]

$$\frac{d}{dt} E[I(t)] = \left\{ \frac{\beta E[S(t)]}{n} - \frac{1}{\mu} \right\} E[I(t)] + \frac{\beta}{n} cov\{S(t), I(t)\} \tag{10.15}$$

where $E[S(t)]$, $E[I(t)]$ and $cov\{S(t), I(t)\}$ are expected values for the random variables $S(t)$ and $I(t)$ as well as their covariance at fixed time t.

Comparison with the Deterministic SIR Model

The deterministic counterpart of (10.13) is

$$\begin{cases} \frac{d}{dt}S(t) = -\beta\frac{S(t)}{n}I(t), \\ \frac{d}{dt}I(t) = \left\{\beta\frac{S(t)}{n} - \frac{1}{\mu}\right\}I(t). \end{cases} \tag{10.16}$$

The solution of the deterministic equations is not simply the mean of the stochastic process since the covariance term in (10.15) is ignored. Nevertheless, for a broad class of processes, the deterministic solution is a good approximation to the stochastic mean of a major outbreak when n is large.

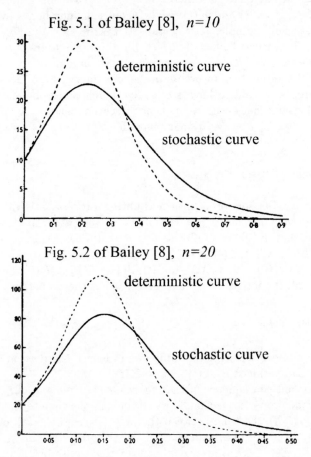

Fig. 10.4 $E[I(t)]$ by stochastic SIR model compared with its deterministic counterpart, from Bailey [8]

Bailey [8] gives a thorough discussion of the properties of deterministic and stochastic versions of the SIR model, along with comparisons of $E[I(t)]$ (stochastic model) versus $I(t)$ (deterministic model) for $n = 10$ and $n = 20$. These are reproduced as Fig. 10.4.

10.2.5.2 Counting Processes

A *counting process* $\{K(t),\ t \in [0, \infty)\}$ is a stochastic process with the meaning:

$$K(t) = \text{cumulative number of events by time } t.$$

$K(t)$ is a non-negative integer random variable for fixed time t, non-decreasing $(K(s) \le K(t)$, if $s < t)$ and $K(t) - K(s)$ equals the number of events in the interval $(s, t]$. For example, $\{R(t)\}$ and $\{M(t)\}$ in (10.13) are counting processes. The derivative $\beta(x) = \frac{d}{dx}E[K(x)]$ is the *instantaneous increment* of the counting process $\{K(x)\}$. A counting process is said to have *independent increment*, if the numbers of events in disjoint intervals are independent. A counting process is said to have *stationary increment*, if the distribution of numbers of events that occur in any time interval depends only on the length of the time interval. For a stationary increment process, $\beta(x) = $ constant β.

The Stationary Poisson Process

The stationary Poisson process is a counting process satisfying both the *independent increment* and the *stationary increment* properties. In addition, it satisfies $K(0) = 0$. It is also a stationary Markov chain with $\Pr\{K(t + \Delta t) = k + 1 | K(t) = k, \mathcal{H}_t\} = \Pr\{K(\Delta t) = 1 | K(0) = 0\}$ such that $\Pr\{K(\Delta t) = 1 | K(0) = 0\} = \lambda \Delta t + o(\Delta t)$, $\Pr\{K(\Delta t) \ge 2 | K(0) = 0\} = o(\Delta t)$, where λ is call the intensity of the process. It has the property

$$\Pr\{K(t + s) - K(s) = k\} = \Pr\{K(t) = k\} = \frac{(\lambda t)^k}{k!} e^{-\lambda t}$$

so that the number of events in an interval of length t is Poisson distributed with mean and variance $E[K(t)] = var[K(t)] = \lambda t$.

An important consequence of the *independent increment* and the *stationary increment* properties is that, if we denote X_1 the time to the first event, and for $k \ge 1$, X_k the time between the $(k-1)$th and the kth events, then $X_k : k = 1, 2, \cdots$ are i.i.d. exponentially distributed random variables with mean $\frac{1}{\lambda}$. $(X_1, X_2, \cdots, X_k)$ are called *inter-arrival times* between events.

Let $Y_1 = X_1, Y_2 = X_1 + X_2, \cdots, Y_k = X_1 + X_2 + \cdots + X_k$ denote the times that events occur, it can be further shown that, conditioning on $K(t) = k$, the k *arrival times* of events $(Y_1, Y_2, \cdots, Y_k)$ have the same distribution as the order statistics corresponding to k independent uniformly distributed random variables on the interval $(0, t)$.

Mixed Poisson Process

Analogue to the mixed Poisson distributions, if individual i is associated with a Poisson process with intensity λ_i which varies across individuals and if factors resulting in the heterogeneity cannot be observed or at present understood, one may consider a random variable $\xi > 0$ with $E[\xi] = \lambda$ and $var[\xi] = \sigma > 0$. This model is expressed by a conditional process $\{K(t)|\xi\}$ and a marginal process $\{K(t)\}$.

At the individual level, the cumulative number of infectious contacts is the conditional process $\{K(t)|\xi\}$ and is assumed to follow a Poisson process with intensity rate ξ. The mean and variance are

$$E[K(t)|\xi] = var[K(t)|\xi] = \xi t. \tag{10.17}$$

The process as seen at the cohort level is the marginal process $\{K(t)\}$ and is a mixed Poisson process, of which, for given time t, the probability $\Pr\{K(t) = k\}$ is

$$\Pr\{K(t) = k\} = \int_0^\infty \frac{(\xi t)^k}{k!} e^{-\xi t} dU(\xi). \tag{10.18}$$

For given t, one can write the p.g.f. for $K(t)$ as

$$G_K(s, t) = \int_0^\infty e^{\xi t(s-1)} dU(\xi) = E_\xi[(e^{t(s-1)})^\xi]. \tag{10.19}$$

Proposition 1. *The mixed Poisson process $\{K(t)\}$ is not a Poisson process. It preserves the stationary increment of the Poisson process, but loses the independent increment property.*

To show it is not a Poisson process, it is only necessary to notice that

$$E[K(t)] = E\{E[K(t)|\xi]\} = \lambda x \tag{10.20}$$
$$var[K(t)] = E\{var[K(t)|\xi]\} + var\{E[K(t)|\xi]\}$$
$$= \lambda x + \sigma(\lambda x)^2 > E[K(t)].$$

To prove that $\{K(t)\}$ does not have independent increment, one only needs to show that $\Pr\{K(s) = k, K(s+t) - K(s) = l\} \neq \Pr\{K(s) = k\} \Pr\{K(s+t) - K(s) = l\}$.

Unlike the Poisson process where $E[X_1] = \frac{1}{\lambda}$, in the mixed Poisson process, $E[X_1] > \frac{1}{\lambda}$ as explained by Jensen's inequality,

$$E[X_1] = E\{E[X_1|\xi]\} = E\left[\frac{1}{\xi}\right] > \frac{1}{E[\xi]} = \frac{1}{\lambda}.$$

10.2.5.3 Branching Processes

Consider a population consisting of individuals or particles able to produce offspring of the same kind. Each individual i is associated with a generation time T_G and by the end of T_G, it produces a random number N_i of new offsprings. We use $g = 0, 1, 2, \cdots$ as the discrete time unit to represent generations. Let X_0 be the number of individuals initially present at the zeroth generation. All offsprings of the zeroth generation constitute the first generation and their number is denoted by X_1. Let X_g denote the size of the gth generation. Given X_{g-1}, if the distribution of X_g only depends on X_{g-1}, then $\{X_g\}$ is a discrete time Markov chain. This process $\{X_g\}$ is called a branching process. Suppose that $X_0 = 1$, we can calculate

$$X_g = \sum_{i=0}^{X_{g-1}} N_i.$$

In most branching processes, it is assumed that N_i is i.i.d. according to a random variable N which does not vary over generations. We say a branching process becomes extinct at generation g, if $X_{g-1} > 0$ but $X_g = 0$. In such case, we denote $Z = \sum_g X_g$ as the final size upon extinction.

Branching processes are often used in infectious disease epidemiology to approximate the initial stage of an outbreak, where the depletion of the number of susceptible individuals is negligible. A discrete time branching process that is well studied in the literature is the Galton–Watson process.

Galton–Watson process.   Each infected individual is associated with a fixed length of generation time T_G. At its end, this individual produces a random number of N secondary infections.

There are various types of continuous time branching processes.

Bellman–Harris processes.   Infected individuals have independently and arbitrarily distributed generation time T_G. Each individual produces a random number of N secondary infections only at the end of the generation time. T_G and N are independent.

Crump–Mode–Jagers (CMJ) processes.   Infected individuals have independently and arbitrarily distributed generation time T_G. Throughout the generation time, each individual produces secondary infections according to a counting process $\{K(x)\}$. Different individuals follow the same counting process. It is assumed that the generation time T_G and $\{K(x)\}$ are independent. By the end of the generation time, an infected individual produces a random number of N secondary infections.

The CMJ processes have been used to approximate the early phase of the SIR models [9,10]), where the generation time T_G has the meaning of the infectious period.

In an SEIR model, let us denote T_E as the *latent period* during which an infected individual is unable to transmit the disease and T_I as the *infectious period* during which an infected individual is able to transmit the disease, and let $T_G = T_E + T_I$. If the infectious period degenerates to $\mu = E[T_I] \to 0$, such that by the end of the latent period, the infected individual instantly produces N new infections and is removed, then $T_G = T_E$ and the branching process approximations are Bellman–Harris processes, including special cases (1) the Markov branching process assuming T_E is exponentially distributed; (2) the Galton–Watson process assuming T_E is non-random. If there is no latent period, $T_G = T_I$, the branching process approximations are CMJ processes. Therefore, we get:

Combined Bellman–Harris–CMJ processes. T_E and T_I are independent. T_E is random with an arbitrary but specified distribution. Only after T_E amount of time has elapsed, an infected individual can produce secondary infections according to a CMJ process.

In these branching processes, there is an embedded Galton–Watson branching process to track the generations. The basic reproduction number corresponds to the mean value of this embedded Galton–Watson branching process with $R_0 = E[N]$. A schematic illustration of these branching processes is given in Fig. 10.5.

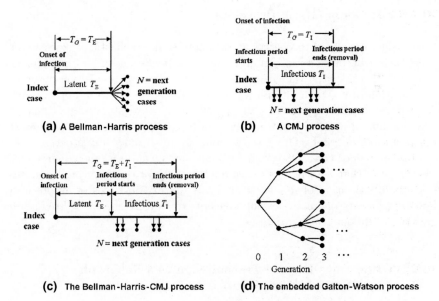

(a) A Bellman-Harris process

(b) A CMJ process

(c) The Bellman-Harris-CMJ process

(d) The embedded Galton-Watson process

Fig. 10.5 Illustrations of branching processes in the context of SEIR models

10.2.6 Random Graph and Random Graph Process

As introduced in Brauer [3], a *graph* $\mathcal{G}$ consists of a set of vertices $V = \{v_1, ..., v_n\}$ and a set of pairs of distinct vertices called edges. The *degree* of a vertex v_i is the number of vertices adjacent to it which is the number of edges attached to vertex v_i. A subgraph $\mathcal{G}'$ is a graph whose vertices and edges form subsets of the vertices and edges of a given graph $\mathcal{G}$. A *component* is a connected subgraph. A graph may consist of a number of disjoint components. The size of a graph is $n =$ the total number of vertices.

10.2.6.1 Random Graph

A *random graph* is obtained by starting with a set of n vertices and adding edges between them at random. The degrees in a random graph are random associated with the *degree distribution*. In addition, a random graph has several measures on its geometry, such as measures on its *connectivity*, *diameter*, sizes of its components, *clustering coefficient*, etc. All these measures are outcomes of random events. Different random graph models produce different probability distributions on graphs.

10.2.6.2 Random Graph Processes

If new edges are added to vertices according to some stochastic processes over time, then it becomes a random graph process. It has not only a time dimension, but also a spatial dimension. At any snapshot at time t, one observes a realization of a random graph $\mathcal{G}_t$. This is analogous to a random variable X versus a stochastic process $\{X(t)\}$. One may aggregate, or superimpose, these random graphs per unit of time over a fixed period of time. The result is still a random graph. This gives an analogy to the counting process.

Similar to such concepts as independent increment, stationary increment, Markov, etc. that determine the temporal growth of the counting processes (e.g. the distribution of inter-arrival times and arrival times of events in a Poisson process), one may develop concepts to capture the spatial-temporal growth of random graph processes.

10.2.6.3 Infectious Disease Transmission as a Subgraph in a Random Graph Process

Individuals are represented by vertices. Contacts are represented by edges. (Social) contacts can be regarded as a random graph process by itself. Contacts made over a fixed period of time is a random graph.

An infectious contact is a contact at which a transmission of infection takes place, and hence all infectious contacts during the same period make a subgraph. The geometry of this subgraph is different from the graph that represents social contacts. This subgraph grows along a *tree*, because if three individuals $\{a, b, c\}$ are friends forming a triangle relationship, and if individual a infects both individuals $\{b, c\}$, then b and c can not infect each other. The social contact random graph may contain triangles and loops, as illustrated by broken lines in Fig. 10.6. The subgraph of infectious contacts can not. This tree – like subgraph, illustrated in Fig. 10.6 by edges, may resemble a realization of an embedded Galton–Watson branching process. However, its growth is limited to the social contact network of susceptible individuals. Only when the number of susceptible individuals is very large can the initial growth of the subgraph be approximated by such a branching process.

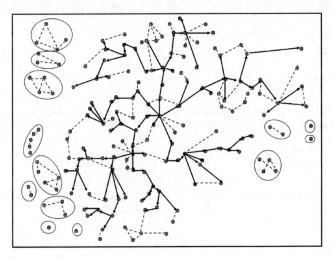

Fig. 10.6 A random graph with several connected components, and a subgraph of infectious contacts

10.3 Formulating the Infectious Contact Process

The infectious contacts arise from a combination of two aspects:

1. *The environment.* The social contact network of individuals in a population as a random graph process, determined by the temporal and spatial network properties, such as whether the network is directed, the

neighbourhood structure of the network, clustering, its growth over time (e.g. stationary increment), etc.

2. *The hosts.* (1) Whether all susceptible individuals are of the same type with equal susceptibility and (2) whether all infectious individuals have equal ability to infect others, so that the probability of transmission per contact, does not vary by contact to contact. If either (1) or (2) is not true, there exists heterogeneity among hosts and this is the intrinsic heterogeneity.

Let x denote the time measured along a typical infectious individual, starting from the beginning of the infectious period at $x = 0$. Following Bartlett [9], and Mode and Sleeman [10], we use a counting process over the time period $\{K(x), x \in [0, \infty)\}$ where

$$K(x) = \text{cumulative number of infectious contacts by time } x.$$

This process is illustrated in Fig. 10.7, which is (c) in Fig. 10.5.

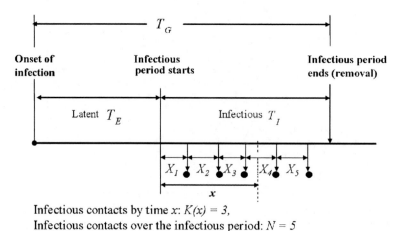

Infectious contacts by time x: $K(x) = 3$,
Infectious contacts over the infectious period: $N = 5$

Fig. 10.7 Event history and the point process for a typical infected individual by time x since the beginning of the infectious period

When an individual is still in the latent period, we write $x < 0$. Given $x \geq 0$, $K(x)$ is non-negative integer random variable with p.g.f. $G_K(s, x) = E[s^{K(x)}] = \sum_{k=0}^{\infty} s^k \Pr\{K(x) = k\}$ and mean value $E[K(x)]$. We use the notation $F_I(x) = \Pr\{T_I < x\}$ for the c.d.f. of the infectious period T_I. The mean infectious period is denoted by $\mu = E[T_I]$. Transmission occurs only when an infected individual is still infectious. The counting process $\{K(x)\}$ stops when the infectious period T_I expires (removal). It is assumed that

$\{K(x)\}$ and T_I are independent. T_I serves as a random stopping time of the process $\{K(x)\}$.

10.3.1 The Expressions for R_0 and the Distribution of N such that $R_0 = E[N]$

Let time τ denote that time since infection of a typical infectious individual. A generic expression for R_0 is given by the equation [5]

$$R_0 = \int_0^\infty \beta(\tau)A(\tau)d\tau \tag{10.21}$$

where $A(\tau)$ is the probability that at time τ since infection the individual is infectious. In other words, $A(\tau) = \Pr\{T_E \leq \tau \cap T_E + T_I > \tau\}$. $\beta(\tau)$ is associated with the property of a counting process that generates new infectious individuals during the infectious period. The product $\beta(\tau)A(\tau)$ is the expected infectivity at time τ after infection.

If T_E and T_I are independent and if at time τ since infection, an individual is infectious, then $\tau \geq T_E$ and $T_I > \tau - T_E$. Therefore,

$$A(\tau) = \int_0^\tau \overline{F}_I(\tau - x)f_E(x)dx = \overline{F}_{E+I}(\tau) - \overline{F}_E(\tau), \tag{10.22}$$

which can be easily proven since $\overline{F}_{E+I}(\tau) = \int_\tau^\infty \int_0^y f_I(y-x)f_E(x)dxdy$ and $\overline{F}_E(\tau) = \int_\tau^\infty f_E(x)dx$. $f_E(x)$ and $f_I(x)$ are the p.d.f.'s of T_E and T_I, respectively. Therefore,

$$\int_0^\infty A(\tau)d\tau = \int_0^\infty \overline{F}_{E+I}(\tau)d\tau - \int_0^\infty \overline{F}_E(\tau)d\tau$$

$$= E[T_E + T_I] - E[T_E] = E[T_I] = \int_0^\infty \overline{F}_I(\tau)dx = \mu.$$

We first show that the formulation of R_0 does not depend on any assumption on the latent period, during which transmission does not occur. For an individual with observed latent period $T_E = t_E$, $\beta(\tau) = 0$, if $\tau \leq t_E$; $\beta(\tau) = \beta(x)$, if $\tau > T_E$, $T_I > \tau - T_E$, where $x = \tau - t_E$. The equation (10.21) becomes $R_0 = \int_{t_E}^\infty \beta(\tau)A(\tau)d\tau = \int_0^\infty \beta(x)A(x + t_E)dx$. One can also show that (exercise), $A(x + t_E) = \overline{F}_I(x)$. Therefore

$$R_0 = \int_0^\infty \beta(x)\overline{F}_I(x)dx. \tag{10.23}$$

Furthermore, if the counting process $\{K(x)\}$ has stationary increment: $\beta(x) \equiv \beta$, then

$$R_0 = \beta\mu. \tag{10.24}$$

$\beta > 0$ is a parameter that captures the transmission rate in the agent–host–environment interface. It can be further sub-modelled to reflect the agent–host–environment interface. A frequently seen expression is $\beta = \lambda p$ so that $R_0 = \lambda p\mu$. p is the probability of transmission per contact between an infectious host and a susceptible host. λ is the average number of contacts among hosts per unit of time in the environment. The control measure for β may be further categorized into interventions designed to alter the social-contact network and designed to alter the transmission probability.

One may also model $\beta = ab$ where a measures the infectivity of an infectious individual who comes in contact with a susceptible individual, for whom, the susceptibility is measured by b. For example, to model the way individuals respond to vaccination, an individual may respond to vaccination with a change in their susceptibility to infection and if ever infected, their infectivity may change from what it is without vaccination. For reference, one may consult Becker and Starczak [11] which leads to references to many more publications in this area.

Formation of R_0 by a multiplicative relationship is useful for statistical analyses and is thus of much public health interest. It leads to a log-linear model to explore the controlled reproduction number R_c:

$$\log R_c = \log R_0 + covariates \tag{10.25}$$

where covariates are associated with intervention measures, such as social distancing, vaccination, treatment of infected individuals with anti-viral drugs, or altering the duration of infectiousness such as rapid isolation of identified infectious individuals. If the probability distribution for N such that $R_0 = E[N]$ is known, the (10.25) leads to the use of generalized linear regression analyses for exploring the effectiveness of these intervention measures aided by statistical packages.

An immediate generalization of (10.24) is when the infectious period can be divided into stages $T_I^{(1)} \to T_I^{(2)} \to \cdots \to T_I^{(l)}$, where the duration $T_I^{(j)}$ is associated with distribution $F_j(x)$ with mean μ_j. Let $\beta_j(x) = \frac{d}{dx}E[K_j(x)]$. It can be shown that $R_0 = \sum_{j=1}^{l} \int_0^\infty \beta_j(x)\overline{F}_j(x)dx$ [10, p. 205]. In this case, the infectious contact process may not have stationary increment over the entire infectious period but has piecewise stationary increment within each stage so that $\beta_j(x) = \beta_j$. (10.24) becomes

$$R_0 = \beta_1\mu_1 + \beta_2\mu_2 + \cdots + \beta_l\mu_l. \tag{10.26}$$

In order to derive the distribution for N, denote

$$N(x) = \begin{cases} K(x), & 0 < x < T_I \\ K(T_I). & x \geq T_I. \end{cases}$$

For fixed x, $N(x)$ is also a random variable with p.g.f.

$$G_N(s, x) = \sum_{j=0}^{\infty} s^j \Pr\{N(x) = j\}.$$

It has been shown that [10, pp. 177–178] the following two recursive formulae hold:

$$E[N(x)] = [1 - F_I(x)] E[K(x)] + \int_0^x E[K(u)]dF_I(u), \qquad (10.27a)$$

$$G_N(s, x) = [1 - F_I(x)] G_K(s, x) + \int_0^x G_K(s, u)dF_I(u). \qquad (10.27b)$$

$N = N(\infty)$ is the cumulative number of infectious contacts generated by an infectious individual throughout its entire infectious period. Let $G_N(s) = \sum_{x=0}^{\infty} s^x \Pr\{N = x\}$ be the p.g.f. for N. $G_N(s)$ uniquely defines the distribution $\Pr\{N = x\}$ following the relationship (10.4). $G_N(s)$ can be derived from (10.27b):

$$G_N(s) = \lim_{x \to \infty} G_N(s, x) = \int_0^{\infty} G_K(s, x)dF_I(x). \qquad (10.28)$$

The mean and variance of N can be evaluated by (10.5), specifically, $R_0 = G'_N(1)$. It can be also shown that (10.23) can be obtained through integration by parts of (10.27a) so that $E[N(x)] = \int_0^x \beta(u)\overline{F}_I(x)du$.

The equation (10.28) implies that the probability distribution of N does not depend on the distribution of the latent period T_E, but depends on:

1. The infectious contact process $\{K(x)\}$, with p.d.f. $G_K(s, x)$ incorporating properties of the contact process and the probability of transmission per contact

2. The infectious period distribution $F_I(x) = \Pr\{T_I \le x\}$

However, differently specified infectious contact processes defined by $G_K(s, x)$ and infectious period distributions $F_I(x)$ may result in the same probability distribution for N and the same R_0. We shall see later in this chapter that for some public health applications, such as determining the risk of whether a large outbreak may occur, or the distribution of the final size η (in $\frac{E[Z]}{n} \to \eta$) should a large outbreak occur, knowing R_0 will be sufficient. For some applications, one requires the knowledge of the distribution for N. Once the distribution for N is determined, one can use theories for the embedded Galton–Watson branching process to calculate quantities such as the probability of a small outbreak π, the final size of the small outbreak $\Pr\{Z = z\}$ and the probability of time to extinction $\Pr\{T_g = g\}$. Yet there are also applications where the detailed distribution for N is not sufficient. One also needs to know the underlying stochastic mechanisms such as the property of the counting process $\{K(x), x \in [0, \infty)\}$, the distribution of the latent period T_E and the distribution of the infectious period T_I.

10.3.2 Competing Risks, Independence and Homogeneity in the Transmission of Infectious Diseases

Competing risk comes from survival analysis on a non-negative random variable $X > 0$ associated with a "lifetime". In a continuous framework, X is associated with a hazard function $h(x)$ for "failure". In the competing risk model, when failure occurs, it may be one of m distinct causes or types, denoted by $J \in \{1, 2, ..., m\}$. The overall hazard function is the sum of type-specific hazard functions $h(x) = \sum_{j=1}^{m} h_j(x)$ where

$$h_j(x) = \lim_{\Delta \to 0} \frac{\Pr\{x < X \leq \Delta; J = j | X > x\}}{\Delta}.$$

Example 3. In Fig. 10.7, if we define $Y_k = X_1 + \cdots + X_k$, $Y_k > 0$ is a random variable. Let $T = \min(Y_k, W)$. $T > 0$ is a random variable representing either the infectious individual is removed or the individual produces its kth infectious contact. The probability distribution for T arises from a competing risk framework.

Example 4. In a contact network, let us consider a vertex v is a susceptible individual attached to m edges, representing m neighbouring vertices with on-going social contacts with vertex v. All the m neighbours are infectious. Let T be the time until vertex v become infected. The type-specific hazard function $h_j(t)$ represents the instantaneous risk of infection being transmitted by the jth neighbouring vertex.

The word *independence* in the study of infectious diseases has several different aspects.

1. The counting process $\{K(x)\}$ and the infectious period T_I are independent. This also implies that in the competing risk framework in Example 3, Y_k and T_I are independent for all k.
2. Viewing the transmission as random graphs, if one randomly chooses a vertex v to be the initial infective, this vertex makes *independent* infectious contacts (i.e. adding edges) with other susceptible vertices throughout its infectious period.
3. If one randomly chooses a vertex v to be a susceptible individual, transmission of infection to this individual by any of its neighbouring infectious individuals is independent from potential transmission from other neighbouring infectious individuals.

If both 2 and 3 hold true, then we say that infectious contacts between different pairs are mutually independent.

The word *homogeneity* also consists of three aspects.

1. *The population is homogeneously mixed.* An individual makes contacts with other individuals with the same probability. As the population size $\to \infty$, the contact process itself is a stationary Poisson process with constant intensity λ.
2. *The hosts are homogeneous.* All individuals are of the same type the probability of transmission per contact between each pair of an infectious individual and a susceptible individual is a constant p.
3. The infectious period from all infected individuals are equal to some constant μ with $F_I(x)$ given by (10.29)

$$F_I(x) = \begin{cases} 0, \, x \le \mu \\ 1, \, x > \mu \end{cases} \tag{10.29}$$

such that $E[T_I] = \mu$ and $var[T_I] = 0$.

Violations to one or a combination of the above assumptions result in heterogeneous transmission.

10.4 Some Models Under Stationary Increment Infectious Contact Process $\{K(x)\}$

If $\{K(x)\}$ satisfies *stationary increment* $E[K(x)] = \beta x$ so that $\beta(x) = \frac{d}{dx} E[K(x)] = \beta$ (constant), (10.23) has a simpler expression $R_0 = \beta \mu$ where $\mu = E[T_I] = \int_0^\infty \overline{F}_I(x) dx$ is the mean infectious period.

10.4.1 Classification of some Epidemics Where N Arises from the Mixed Poisson Processes

Since the distribution of N does not depend on the distribution of the latent period, we restrict discussion to the case where the latent period is absent. In the expression $G_N(s) = \int_0^\infty G_K(s, x) dF_I(x)$, differently specified infectious contact process $\{K(x)\}$ defined by $G_K(s, x)$ and infectious period distribution $F_I(x)$ may result in the same probability distribution for N at the macro level, with very different characteristics at the micro level. Some models are classified below as mixed Poisson processes.

10.4.1.1 The Poisson Epidemic

The assumptions are: (1) the infectious contact process is a Poisson process with $G(s, x) = e^{\beta x(s-1)}$; (2) contacts between different pairs are mutually independent; (3) the infectious period from all infected individuals are equal to some constant μ with $F_I(x)$ given by (10.29). Under this model,

$$G_N(s) = e^{\beta\mu(s-1)} = e^{R_0(s-1)}. \tag{10.30}$$

Hence N is also Poisson distributed.

10.4.1.2 The Randomized Epidemic

This terminology is used in von Bahr and Martin-Löf [12]. The model assumes that the infectious periods from all infected individuals are equal to some constant μ with $F_I(x)$ given by (10.29). β is further sub-modelled as $\beta = \lambda p$ where p is a *random* probability with itself following some distribution F_p so that $\beta = \lambda E[p]$. In this case, it can be shown (later) that $G_K(s, x) = \int_p e^{\lambda px(s-1)} dF_p = E_p\left[e^{\lambda px(s-1)}\right]$. Under this model, $G_N(s) = E_p\left[e^{\lambda p\mu(s-1)}\right]$.

Without losing generality, we extend the above by assuming that factors resulting in heterogeneous mixing and/or heterogeneous transmission are not at present understood and the product $\xi = \lambda p$ is a random variable following some distribution $U(\xi)$ with $E[\xi] = \beta$. Therefore $G_K(s, x) = E_\xi[e^{\xi x(s-1)}]$ and

$$G_N(s) = E_\xi[e^{\xi\mu(s-1)}]. \tag{10.31}$$

Provided that the mixing distribution $U(\xi)$ is specified, $G_N(s)$ uniquely defines the distribution of N as a mixed Poisson distribution.

10.4.1.3 The Generalized Epidemic

This terminology is used in Lefèvre and Utev [13]. The assumptions are: (1) the infectious contact process is a Poisson process with $G(s, x) = e^{\beta x(s-1)}$; (2) contacts between different pairs are mutually independent; (3) the infectious periods from all infected individuals are i.i.d. following an arbitrary distribution $F_I(x)$. Using (10.28),

$$G_N(s) = \int_0^\infty e^{\beta x(s-1)} dF_I(x) = E_{T_I}\left[e^{\beta T_I(s-1)}\right]. \tag{10.32}$$

One notices the duality between (10.31) and (10.32). As far as the distribution for N is concerned, the generalized epidemic and the randomized epidemic models can be unified as one mixed Poisson model with

$$G_N(s) = \int_0^\infty e^{\rho(s-1)} dU(\rho) = E_\rho\left[e^{\rho(s-1)}\right] \tag{10.33}$$

with a mixing distribution $U(\rho)$. In (10.32), ρ is re-scaled as $\rho = \beta T_I$ where β is a constant and T_I is random such that $E[\rho] = \beta\mu$. In (10.31), ρ is re-scaled as $\rho = \xi\mu$ where μ is constant, but ξ is random. It implies that, if one starts with a simple Poisson distribution with p.g.f. $e^{\rho(s-1)}$ and chooses a mixing distribution $U(\rho)$, the resulting distribution obtained by (10.33) can be interpreted as arising either from a generalized epidemic with an appropriately chosen infectious period distribution, or from a randomized epidemic with an appropriately chosen distribution for ξ.

The General, or the Kermack–McKendrick, Epidemic

This is a special case of the generalized epidemic, by assuming $F_I(x) = 1 - e^{-\frac{x}{\mu}}$ in (10.32). The p.d.f. is $F_I(x) = \frac{1}{\mu}e^{-\frac{x}{\mu}}$ with $E[T_I] = \mu$. Under this model, (10.32) becomes

$$G_N(s) = \frac{1}{\mu}\int_0^\infty e^{\beta x(s-1)}e^{-\frac{x}{\mu}}dx = \frac{1}{1 + \beta\mu(1-s)}.$$

This p.g.f. gives the geometric distribution for N with $R_0 = \beta\mu$. This corresponds to the underlying assumption for both the deterministic SIR model and its stochastic counterpart, the bivariate Markov chain SIR model. Bailey [8] gave it the name *the general epidemic*. A better name for this may be the Kermack–McKendrick epidemic to reflect its early origins [14].

10.4.1.4 The Randomized and Generalized Epidemics

The infectious contact process $\{K(x)\}$ is a mixed Poisson process under the same assumptions as that in the randomized epidemic model, with $G_K(s,x) = E_\xi[e^{\xi x(s-1)}]$. The infectious periods from all infected individuals are i.i.d. following an arbitrary distribution $F_I(x)$, as that assumed in the generalized epidemic model, then

$$G_N(s) = \int_0^\infty E_\xi[e^{\xi x(s-1)}]dF_I(x) = E_{T_I}\left[E_\xi[e^{\xi T_I(s-1)}]\right]. \tag{10.34}$$

With respect to (10.34), let $\rho = \xi T_I$. Conditioning on $T_I = x$, ξ is random and hence the conditional distribution for ρ, $U_1(\rho|x)$, depends on parameter x. The integrand $E_\xi[e^{\xi x(s-1)}]$ can be regarded as a mixed Poisson distribution with p.g.f. $\int_\rho e^{\rho(s-1)}dU_1(\rho|x)$. Provided no dependencies between the parameters of the distributions considered, one can re-write (10.34) as

$$\int_\rho e^{\rho(s-1)}\left(\int_x dU_1(\rho|x)dF_I(x)\right) = \int_\rho e^{\rho(s-1)}dU(\rho)$$

so that it can be also unified as a mixed Poisson distribution with the mixing distribution $U(\rho)$ as a random mixture of two mixing distributions $U_1(\rho|x)$ and $F_I(x)$.

The unification among (10.32), (10.31) and (10.34) is helpful to reduce redundancy of proof for some theories. This unification can be only applied to address certain aspects of an outbreak, such as the assessment of the risk of a large outbreak and the final size, of which, the knowledge of the distribution of N, uniquely defined by the p.g.f. $G_N(s)$, is sufficient.

10.4.2 Tail Properties for N

Karlis and Xekalaki [15] pointed out that the shape of the p.m.f. of a mixed Poisson distribution exhibits a resemblance to that of the p.d.f. of the mixing distribution $u(\lambda)$. Lynch [16] proved that mixing carries the form of the mixing distribution over to the resulting mixed distribution in general. Tail properties of some commonly used probability models for continuous non-negative random variables have been summarized in Table 10.3. We now compare the shape of the distributions for N generated by p.g.f. $\int_\rho e^{\rho(s-1)}dU(\rho)$ with that of their mixing distributions $U(\rho)$.

10.4.2.1 N Distributed According to Geometric and Negative-Binomial Distributions

Figures 10.3 and 10.2 illustrate that the shape of the geometric distribution resembles that of the exponential distribution and the shape of the negative binomial distribution resembles that of the gamma distribution. We may re-parametrize the negative-binomial distribution (10.11) as $\Pr\{N = x\} = \frac{\Gamma(x+\frac{1}{\phi})}{\Gamma(x+1)\Gamma(\frac{1}{\phi})}\varsigma^{\frac{1}{\phi}}(1-\varsigma)^x$ by letting $\zeta = \frac{1}{1+\phi\mu}$. The geometric distribution is a special case $\varsigma(1-\varsigma)^x$.

1. The geometric distribution has exponential tail because for any x,
$\overline{F}_N(x) = \sum_{l=x}^{\infty}\varsigma(1-\varsigma)^l = (1-\varsigma)^x = e^{-\frac{x}{\tau}}$ where $\frac{1}{\tau} = -\log(1-\varsigma)$.
Hence $\frac{\overline{F}_N(s+x)}{\overline{F}_N(s)} = e^{-\frac{x}{\tau}}$.

2. The negative binomial distribution has exponential tail when $\phi > 1$. In fact, from page 210 of Johnson, Kotz and Kemp [17]

$$\Pr\{N \geq x\} = \sum_{l=x}^{\infty}\frac{\Gamma(l+\frac{1}{\phi})}{\Gamma(l+1)\Gamma(\frac{1}{\phi})}\varsigma^{\frac{1}{\phi}}(1-\varsigma)^l = f_N(x)\,_2F_1[1, x+\frac{1}{\phi}; x+1; 1-\varsigma]$$

where $f_N(x) = \Pr\{N = x\}$ and $_2F_1[a, b; c; x]$ is the Gaussian hypergeometric function. If $c > b > 0$, $_2F_1[a, b; c; x] = \frac{\Gamma(c)}{\Gamma(b)\Gamma(c-b)}\int_0^1 \frac{v^{b-1}(1-v)^{c-b-1}}{(1-vx)^a}dv$.

When $\phi > 1$,

$$\frac{\overline{F}_N(s+x)}{\overline{F}_N(s)} = \frac{f_N(s+x)}{f_N(s)} \frac{{}_2F_1[1, s+x+\frac{1}{\phi}; s+x+1; 1-\varsigma]}{{}_2F_1[1, x+\frac{1}{\phi}; x+1; 1-\varsigma]}$$

$$= \frac{\int_0^1 \frac{v^{x+s+\frac{1}{\phi}-1}(1-v)^{-\frac{1}{\phi}}}{1-v(1-\varsigma)} dv}{\int_0^1 \frac{v^{x+\frac{1}{\phi}-1}(1-v)^{-\frac{1}{\phi}}}{1-v(1-\varsigma)} dv} (1-\varsigma)^x \to e^{-\frac{x}{\tau}}, \text{ as } s \to \infty.$$

10.4.2.2 A Distribution Commonly Used in Social Networks Modelling for Infectious Diseases

A distribution with supports $1, 2, \cdots$ given by

$$\Pr\{N = x\} = Cx^{-\theta}e^{-\frac{x}{\tau}}, \quad x = 1, 2, 3, \cdots \tag{10.35}$$

where $C = Li_\theta^{-1}\left(e^{-\frac{1}{\tau}}\right)$ is the normalizing factor and $Li_\theta(z) = \sum_{n=1}^\infty \frac{z^n}{n^\theta}$, was mentioned in Brauer [3]. It has been used in modelling the spread of infectious diseases on scale-free networks by Newman [18]; Meyers, Pourbohloul et al. [19] and many others.

- If $\theta \to 0$, (10.35) returns to the (zero-truncated) geometric distribution $\frac{x^{-\theta}e^{-\frac{x}{\tau}}}{Li_\theta\left(e^{-\frac{1}{\tau}}\right)} \to (1 - e^{-\frac{1}{\tau}})e^{-\frac{x-1}{\tau}}, \quad x = 1, 2, 3, \cdots$.
- If $\tau \to \infty$, (10.35) has the limiting distribution as the Zipf distribution $\frac{x^{-\theta}e^{-\frac{x}{\tau}}}{Li_\theta\left(e^{-\frac{1}{\tau}}\right)} \to \frac{1}{\zeta(\theta)}x^{-\theta}$ where $\zeta(\theta) = \sum_{n=1}^\infty \frac{1}{n^\theta}$.

When $\tau < \infty$, the tail of this distribution is exponential. As $s \to \infty$,

$$\frac{\Pr\{N \geq s+x\}}{\Pr\{N \geq s\}} = \frac{\sum_{j=s+x}^\infty j^{-\theta}e^{-\frac{j}{\tau}}}{\sum_{j=s}^\infty j^{-\theta}e^{-\frac{j}{\tau}}} = e^{-\frac{x}{\tau}} \frac{\sum_{j=s}^\infty (j+x)^{-\theta}e^{-\frac{j}{\tau}}}{\sum_{j=s}^\infty j^{-\theta}e^{-\frac{j}{\tau}}} \to e^{-\frac{x}{\tau}}.$$

There are many possible stochastic mechanisms for which a distribution like (10.35) may arise. One of them is a generalized epidemic model arising from mixed Poisson distributions p.g.f. $\int_0^\infty e^{\beta x(s-1)}dF_I(x)$, where the infectious period W arises from a competing process. For some individuals, the infectious period arises from an exponential distribution with constant hazard $h_e(x) = \gamma$ as the removal rate. For other individuals, the infectious period follows a generally very short but highly skewed distribution, which may be fitted well with a Pareto form, with the hazard function expressed by $h_p(x) = \frac{\alpha\theta}{1+\alpha x}$. This scenario may arise if identified infectious individuals are aggressively isolated during an outbreak. The survivor function of the infectious period has a shape of that of a Pareto distribution with exponential cut-off:

$$\overline{F}_I(x) = \exp\left(-\int_0^x \left(\gamma + \frac{\alpha\theta}{1+\alpha t}\right) dt\right) = \frac{\alpha^\theta}{(1+\alpha x)^\theta} e^{-\gamma x}, \quad \alpha, \gamma, \theta > 0.$$

It has an exponential tail:

$$\frac{\overline{F}_I(s+x)}{\overline{F}_I(x)} = \frac{(1+\alpha s)^\theta}{(1+\alpha s + \alpha x)^\theta} e^{-\gamma x} \to e^{-\gamma x}, \quad \text{as } s \to \infty.$$

Although the exact distribution for N,

$$\Pr\{N = j\} = \frac{\alpha^\theta}{j!} \int_0^\infty \frac{(\beta x)^j \left(\theta\alpha + \gamma\left(1+\alpha x\right)\right)}{(1+\alpha x)^{\theta+1}} e^{-(\beta+\gamma)x} dx$$

may not be easy to calculate, we may use the results in Karlis and Xekalaki [15] to argue that it carries the shape of the mixing distribution $\overline{F}_I(x)$ which is a continuous analogue of (10.35).

10.4.2.3 Discrete Power-Law Distributions as Mixed Distributions

The term "scale-free" [20] refers to networks with degrees following a power-law distribution with $\Pr\{N = x\} \propto \frac{1}{x^{\theta+1}}$. Scale-free models have gained much attention in infectious disease literature. Liljeros et al. [21] used the power-law distribution to model the web of sexual contacts with implication of studying sexually transmitted infections. A discrete power-law distribution is the Zipf distribution $\frac{1}{\varsigma(\theta+1)} x^{-(\theta+1)}$, a discrete analogue to the continuous Pareto distribution. In a general sense, any distribution such that $\lim_{x\to\infty} \frac{F(x)}{x^{\theta+1}} = A$, for some $\theta > 0, A > 0$ is a power-law distribution. Romillard and Theodorescu [22] showed that certain Poisson mixtures are power-law.

A class of power-law distributions have the form with p.m.f.

$$\Pr\{N = x\} = \begin{cases} \theta \frac{\Gamma(x+\alpha)\Gamma(\alpha+\theta)}{\Gamma(\alpha)\Gamma(x+\alpha+\theta+1)}, & \theta > 0, \ \alpha > 0 \\ \theta \frac{\Gamma(x+1)\Gamma(\theta+1)}{\Gamma(x+\theta+2)}, & \text{special case } \alpha = 1 \end{cases} \quad (10.36)$$

which is the Waring distribution given by (6.149) and the special case Yule distribution corresponding to (6.139) when $\alpha = 1$ of Johnson, Kotz and Kemp [17]. The mean and variance may not exist. The first moment exists: $E[N] < \infty$ only if $\theta > 1$. Using the Barnes expansion [17, p. 6],

$$\frac{\Gamma(x+\alpha)}{\Gamma(x+\alpha+\theta+1)} \approx \frac{1}{x^{\theta+1}} \left(1 - \frac{(\theta+1)(\theta+2\alpha)}{2x} + \cdots\right)$$

it shows that when x is sufficiently large, the Waring and the Yule distribution follow the power-law property: $\propto \frac{1}{x^{\theta+1}}$ where the parameter α plays little role.

The Waring (Yule) distribution can be generated as a mixture of the negative binomial distribution of the general form

$$\frac{\Gamma(x+\frac{1}{\phi})}{\Gamma(x+1)\Gamma(\frac{1}{\phi})}\varsigma^{\frac{1}{\phi}}(1-\varsigma)^x, \ 0 < \varsigma < 1. \tag{10.37}$$

If one takes a one-parameter Beta distribution $u(\varsigma) = \theta\varsigma^{\theta-1}$ as the mixing distribution onto a negative binomial, then

$$\Pr\{N=x\} = \frac{\Gamma(x+\frac{1}{\phi})}{\Gamma(x+1)\Gamma(\frac{1}{\phi})}\int_0^1 \varsigma^{\frac{1}{\phi}}(1-\varsigma)^x\,\theta\varsigma^{\theta-1}d\varsigma \tag{10.38}$$

$$= \theta\frac{\Gamma(x+\alpha)\Gamma(\alpha+\theta)}{\Gamma(\alpha)\Gamma(x+\alpha+\theta+1)}, \ \theta > 0, \ \alpha = \frac{1}{\phi} > 0.$$

which returns to the Waring distribution. The Waring distribution can be justified as arising from a mixed Poisson [23, 24].

It can be also formulated as arising from a infectious contact process that itself arises from a mixed negative binomial (10.37) with Pareto distribution as its mixing distribution. Then one can use the results in Lynch [16] to justify why (10.38) carries the Pareto shape. In fact, if we write $\varsigma = \frac{1}{1+\phi\rho}$, then $d\varsigma = -\frac{\phi}{(1+\rho\phi)^2}d\rho$, and $u(\varsigma) = \theta\varsigma^{\theta-1} = \theta\left(\frac{1}{1+\rho\phi}\right)^{\theta-1}$. Hence (10.38) becomes

$$\frac{\Gamma(x+\frac{1}{\phi})}{\Gamma(x+1)\Gamma(\frac{1}{\phi})}\int_0^\infty \left(\frac{1}{1+\phi\rho}\right)^{\frac{1}{\phi}}\left(1-\frac{1}{1+\phi\rho}\right)^k\frac{\theta\phi}{(1+\rho\phi)^{\theta+1}}d\rho$$

where $u_1(\rho) = \frac{\theta\phi}{(1+\rho\phi)^{\theta+1}}$ is the Pareto mixing distribution.

10.5 The Invasion and Growth During the Initial Phase of an Outbreak

$S(t)$ is the random number of susceptible individuals at time t with expected value $E[S(t)]$. S_0 is a fixed number of initial susceptible individuals at $t = 0$. n is the total number of individuals in the population. The *initial phase* of an outbreak is the period of time when depletion of susceptible individuals can be ignored, which is $\left\{t : \frac{E[S(t)]}{n} \approx 1\right\}$. During this phase, there are two approximations to describe the disease spread:

1. The branching process approximation: A typical infectious individual is
   associated with a generation time $T_G = T_E + T_I$ and by its end, it
   produces a random number of N new infections. The mean number of
   these new infections is R_0. If $R_0 \leq 1$, with certainty (i.e. with probability
   one), this branching process will become extinct, resulting in a handful
   of total infections. The extinction is intrinsic.
2. The exponential growth approximation: If $R_0 > 1$, the expected number
   of infectious individuals at time t follows the Malthus' Law $E[I(t)] \propto$
   e^{rt} [25,26], characterized by a parameter r. The growth is intrinsic and
   r is known as the *intrinsic growth rate*, also known as the Malthusian
   parameter.

There is an Euler–Lotka equation [5]

$$\int_0^\infty e^{-r\tau} \beta(\tau) A(\tau) d\tau = 1, \qquad (10.39)$$

where $\beta(\tau)$ and $A(\tau)$ are defined the same way as in (10.21).

10.5.1 Invasion and the Epidemic Threshold

Kendall [27] considered a continuous random variable $Y = \frac{Z}{n}$ defined on
$(0, 1]$, where $Z = n - S_\infty$ is the final size of an outbreak. The shape of this
distribution may have one of the two shapes: the J-shape and the U-shape.
The term J-shape refers to a distribution that is monotonically decreasing so
that it has a mode at zero. The distribution is said to have U-shape if it is
bimodal. As shown in Fig. 10.8, the U-shaped distribution can be thought of
as a weighted average between a J-shaped distribution with weight $0 < \pi < 1$
and a uni-modal distribution of a bell-shape with weight $1 - \pi$.

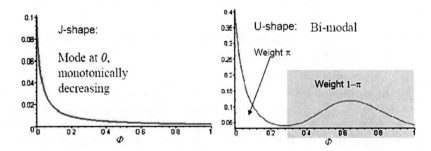

Fig. 10.8 J-shaped and U-shaped distributions for Y

Näsell [28] describes the threshold as a value at which the distribution of Y makes a transition from J-shape to U-shape. In a population with finite size n, the transition occurs in the vicinity of 1 at the value $1 + \frac{K}{n^{\frac{1}{3}}}$, where K is a constant independent of n, but depends on the number of initially infected individuals I_0.

When $n \to \infty$, a *sufficient condition* for a J-shaped distribution of Y is $R_0 < 1$, under which, $\pi = 1$ (see Fig. 10.8). When $R_0 > 1$, either an outbreak dies out with a handful of cases with probability $\pi < 1$, or starts an exponential growth into a large outbreak with probability $1 - \pi$. Metz [29] showed that the outbreak can be either small or large with no middle road in between.

A small outbreak corresponds to a J-shaped distribution. The opposite is not true. When n is finite and R_0 lies within the vicinity of 1 of the order of $O(n^{-\frac{1}{3}})$, then with probability $\pi = 1$, $Y = \frac{Z}{n}$ will have a J-shaped distribution with $\frac{E[Z]}{n} \to 0$ meanwhile $E[Z] \to \infty$. In this case, the outbreak size is neither small, nor large. Martin-Löf [30] shows that $Y^* = \frac{Z}{n^{\frac{2}{3}}}$ has a limit distribution as $n \to \infty$. It may have different shapes, from J-shape to a bimodal U-shape, to uni-modal with mode not at zero, or with a shape that is rather flat. This makes the final size unpredictable.

10.5.2 The Risk of a Large Outbreak and Quantities Associated with a Small Outbreak

10.5.2.1 The Risk of a Large Outbreak $1 - \pi$

When $n \to \infty$, a necessary condition for a non-zero probability $1 - \pi > 0$ is that R_0 exists and $R_0 > 1$. It is possible that for some distributions for N, $R_0 = E[N]$ does not exist. Public health control measures against an outbreak are often centred around the reduction of R_0 to some controlled reproductive number R_c by a reduction factor c so that

$$R_c = (1 - c) R_0 < R_0. \tag{10.40}$$

Ideally, one wants $R_c < 1$ so $1 - \pi = 0$. If this is not achievable, it is important to investigate what other aspects of the distribution for N such as $var[N]$ as well as aspects of the underlying stochastic mechanisms that manifest the p.g.f. $G_N(s)$ contribute to the reduction of the risk of a large outbreak $1 - \pi$.

Using standard results from the (embedded) Galton–Watson branching process, for any $R_0 > 0$ (if it exists), there is always a probability π so that the branching process will become extinct and produce a small outbreak. π is the smallest root of the Fixed-Point Equation

$$s = G_N(s). \tag{10.41}$$

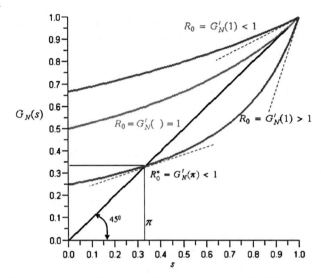

Fig. 10.9 The smallest root of the Fixed-Point equation $s = G_N(s)$

From the properties of $G_N(s)$ given in (10.3) and the fact that $R_0 = G'_N(1)$, the smallest root of (10.41) is illustrated in Fig. 10.9.

When $R_0 = G'_N(1) \leq 1$, the smallest root is $\pi = 1$. In this case, a large outbreak will not occur. When $R_0 = G'_N(1) > 1$, there is an unique solution π in the open interval $(0, 1)$ and the risk of a large outbreak is $1 - \pi$. If the outcome happens to be a small outbreak, the observed branching process will be indistinguishable from that as if arising from a different "reproduction number" $R_0^* = G'_N(\pi) < 1$ with its p.g.f. obtained by taking the graph Fig. 10.9 over $[0, \pi]$ and re-scaling to make the domain and range $[0, 1]$.

Application to Some Specific Distributions of N

Poisson   In the Poisson epidemic, $G_N(s) = e^{R_0(s-1)}$. π is the smallest root of the Fixed-Point Equation (10.41). If $R_0 > 1$, the risk of a large outbreak is calculated by

$$\pi = e^{-R_0(1-\pi)} \tag{10.42}$$

If the outcome happens to be a small outbreak, the observed branching process will be indistinguishable from that as if arising from a different "reproduction number"

$$R_0^* = G'_N(\pi) = R_0 e^{R_0(\pi-1)} = \pi R_0 < 1.$$

Geometric   If the p.g.f. for N is $G_N(s) = \frac{1}{1+R_0(1-s)}$, then N has the geometric distribution. One of the stochastic mechanisms for producing the geometric distribution is the Kermack–McKendrick epidemic, with all the underlying assumptions governing the deterministic or stochastic bivariate Markov chain SIR models. Another stochastic mechanism for producing the geometric distribution is the randomized distribution with the infectious contact process arising from a mixed Poisson process with exponential mixing distribution. There are other stochastic mechanisms that also produce N with geometric distribution. However, for studying the initial behaviour using the embedded Galton–Watson branching process, it is only the p.g.f. $G_N(s)$ that matters. If $R_0 > 1$, the risk of a large outbreak is calculated by $\pi = \frac{1}{1+R_0(1-\pi)}$ with exact solution $\pi = \frac{1}{R_0}$. The risk of a large outbreak is $1 - \frac{1}{R_0}$. It can be also shown that $R_0^* = G'_N(\pi) = \frac{1}{R_0}$.

Negative binomial If the p.g.f. for N has the form $G_N(s) = \dfrac{1}{[1+R_0\phi(1-s)]^{\frac{1}{\phi}}}$ for some $\phi > 0$, then N has the negative binomial distribution. It may arise as a generalized epidemic with gamma distributed infectious period, or as a randomized epidemic with the infectious contact process arising from a mixed Poisson process with gamma mixing distribution. There are also other stochastic mechanisms that can produce N with the negative binomial distribution. In addition, $\dfrac{1}{[1+R_0\phi(1-s)]^{\frac{1}{\phi}}} \to e^{R_0(s-1)}$, as $\phi \to 0$. If $R_0 > 1$, the risk of a large outbreak is calculated by

$$\pi = \frac{1}{[1 + R_0\phi(1 - \pi)]^{\frac{1}{\phi}}}. \tag{10.43}$$

For later use, we write down the derivatives

$$G'_N(s) = \frac{R_0}{(1+\phi R_0(1-s))^{\frac{1}{\phi}+1}}, \quad G''_N(s) = \frac{(1+\phi)R_0^2}{(1+\phi R_0(1-s))^{\frac{1}{\phi}+2}}. \tag{10.44}$$

Proposition 2. *Within the negative binomial family of distributions,* $var[N] = R_0 + \phi R_0^2$, $G_N(s)$ *is ordered by the parameter* ϕ. *That is, for each fixed* $s \in (0,1)$, $\phi_1 < \phi_2$ *implies* $\dfrac{1}{[1+R_0\phi_1(1-s)]^{\frac{1}{\phi_1}}} < \dfrac{1}{[1+R_0\phi_2(1-s)]^{\frac{1}{\phi_2}}}$.

Corollary 1. *If* $R_0 > 1$ *and* $\pi \in (0,1)$ *is the Fixed-Point as the smallest root of (10.43), then* π *is an increasing function of* ϕ. *The larger the value of* ϕ, *the larger the probability that an initial outbreak will die out without evolving into a large outbreak and the smaller the risk.*

An alternative way to show that π is an increasing function of ϕ is to use implicit differentiation for (10.43) to directly show that $\frac{d\pi}{d\phi} > 0$. In fact

$$\frac{d\pi}{d\phi} = \frac{1}{1 - \dfrac{R_0}{(\phi R_0(1-\pi)+1)^{\frac{1}{\phi}+1}}} \cdot \frac{(1+R_0\phi(1-\pi))\log(1+R_0\phi(1-\pi)) - R_0\phi(1-\pi)}{\phi^2(\phi R_0(1-\pi)+1)^{\frac{1}{\phi}+1}}.$$

From (10.44), $\dfrac{R_0}{(1+\phi R_0(1-\pi))^{\frac{1}{\phi}+1}} = G'_N(\pi) < 1$. One only needs to show

$$(1 + R_0\phi(1-\pi))\log(1 + R_0\phi(1-\pi)) - R_0\phi(1-\pi) > 0$$

which is true because $\log(1+x) > \frac{x}{1+x}$ for any $x > 0$.

Corollary 2. *If $R_0 > 1$, the Poisson epidemic, as $\phi \to 0$, gives the smallest probability π and hence the highest risk of a large outbreak, compared with any other model within the negative binomial family of distributions with the same R_0.*

This statement can be extended beyond the negative binomial family. For all randomized and generalized epidemics that can be formulated as mixed Poisson distributions, the Poisson epidemic always produces the smallest probability π.

Proposition 3. *Let π_0 be the smallest root of $s = e^{R_0(s-1)}$ corresponding to the Poisson epidemic. Let π_ρ be the smallest root of $s = \int e^{\rho(s-1)} dU(\rho)$ corresponding to an epidemic arising from some stochastic mechanisms such that $G_N(s) = E_\rho\left[e^{\rho(s-1)}\right]$ can be interpreted as the p.g.f. of a mixed Poisson distribution. If the latter epidemic has finite mean R_0 which equals that of the Poisson epidemic, then $\pi_0 < \pi_\rho$.*

Since from any distribution with p.g.f. given by $G_N(s)$, $\Pr\{N = 0\} = G_N(0)$ which is the probability that, after introducing the initially infected individuals into a susceptible population, transmission never occurs. This leads to the next Corollary, which is a well-known result that was originally proven by Feller [31] in 1943.

Corollary 3. $e^{R_0(s-1)} < E_\rho\left[e^{\rho(s-1)}\right]$ *implies* $e^{-R_0} < G_N(0) = E_\rho\left[e^{-\rho}\right]$. *Therefore, the probability of $N = 0$ is always higher under a mixed Poisson distribution than under a simple Poisson distribution with the same mean.*

It implies that variances imposed by infectious period, heterogeneous contact environment, heterogeneous transmission among individuals or combinations of them, all reduce the risk of an large outbreak.

10.5.2.2 Boom or Bust: Generations to Extinction in Small Outbreaks

Using the branching process approximation in the early stage of the epidemic and from the theory of branching process (Chap. 8 of [32]), $\Pr\{T_g \leq g\}$ is uniquely determined by the p.g.f. $G_N(s)$ and is calculated by

$$\Pr\{T_g \leq g\} = G_N^g(0) \stackrel{\text{def.}}{=} \underbrace{G_N(G_N(\cdots G_N}_{g \text{ times}}(0)\underbrace{\cdots)}_{g \text{ times}}. \tag{10.45}$$

If there is no secondary transmission, the initial infective individuals make the first generation, hence $\Pr\{T_g = 1\} = \Pr\{N = 0\} = G_N(0)$. $\Pr\{T_g \leq g\}$

is a non-decreasing function of g. Starting from $\Pr\{T_g = 1\} = G_N(0)$, with the limit

$$\lim_{g \to \infty} \Pr\{T_g \leq g\} = \pi \begin{cases} = 1, \text{ if } R_0 < 1, \\ < 1, \text{ if } R_0 > 1. \end{cases}$$

For any $R_0 > 0$, the probability of a small outbreak that becomes extinct in a few generations is π, the smallest root of $s = G_N(s)$, $s \in (0, 1]$. The recursive procedure and convergence of (10.45) are illustrated in Fig. 10.10.

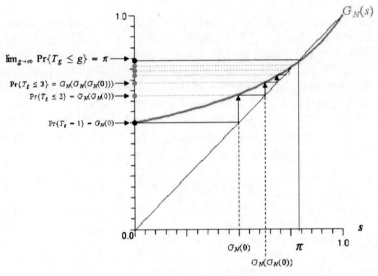

Fig. 10.10 Graphic presentation of $\Pr\{T_g \leq g\} = G_N^g(0) \to \pi$, as $g \to \infty$

$\frac{1}{\pi}(1 - \Pr\{T_g \leq g\}) = \frac{\pi - G_N^g(0)}{\pi}$ is the conditional probability that, given the outbreak being small, it has become extinct after g generations. It can be shown that for a suitable positive constant A,

$$\pi - G_N^g(0) = \pi - \Pr\{T_g \leq g\} \sim A \left[G_N'(\pi)\right]^g$$

and $0 < G_N'(\pi) < 1$. It implies that if extinction is going to occur, the smaller the value of $G_N'(\pi)$, the more likely it will happen quickly with very few generations.

When N Follows the Negative Binomial Distribution

When N arises from the negative-binomial distribution, (10.45) becomes

$$\Pr\{T_g = 1\} = \Pr\{N = 0\} = (1 + \phi R_0)^{-\frac{1}{\phi}}$$

$$\Pr\{T_g \leq 2\} = \left(1 + \phi R_0 - \phi R_0 \left(1 + \phi R_0\right)^{-\frac{1}{\phi}}\right)^{-\frac{1}{\phi}}$$

$$\vdots$$

$$\Pr\{T_g \leq g\} = \frac{1}{[1 + \phi R_0 \left(1 - \Pr\{T_g \leq g - 1\}\right)]^{\frac{1}{\phi}}},$$

with the limit $\lim_{g \to \infty} \Pr\{T_g \leq g\} = \pi$.

It can be shown that, T_g is ranked by stochastic order in the sense of the survivor function $\overline{F}_{Tg}(g) = \Pr\{T_g \geq g\} = 1 - \Pr\{T_g \leq g - 1\}$, according to the parameter ϕ. The larger the ϕ, the shorter the distribution $\Pr\{T_g \geq g\}$. For N with negative-binomial distribution with mean value R_0, not only large variance $var[N] = \phi R_0^2 + R_0$ lowers the risk of large outbreaks $1 - \pi$, but also that if extinction is going to occur, it is likely to happen quickly with very few generations. An extreme case is when $\phi \to \infty$. It can be shown that $\lim_{\phi \to \infty} (1 + \phi R_0)^{-\frac{1}{\phi}} = 1$ and hence $\lim_{\phi \to \infty} (1 + \phi R_0 - \phi R_0 \Pr\{T_g \leq g - 1\})^{-\frac{1}{\phi}} = 1$, thus

$$\lim_{\phi \to \infty} \Pr\{T_g \leq g\} = 1, \text{ any } g \geq 2.$$

Special Cases as $\phi \to 0$ and $\phi = 1$

The special cases are summarized in Table 10.4.

Table 10.4 The distributions for generation time to extinction

	Poisson distribution $\phi \to 0$	Geometric distribution $\phi = 1$
$G_N(s) =$	$\exp(-R_0(1 - s))$	$\frac{1}{1 + R_0(1-s)}$
$\Pr\{T_g = 1\} =$	e^{-R_0}	
$\Pr\{T_g \leq 2\} =$	$e^{R_0(e^{-R_0} - 1)}$	$\Pr\{T_g \leq g\} = \frac{R_0^g - 1}{R_0^{g+1} - 1}$
$\Pr\{T_g \leq 3\} =$	$e^{R_0\left(e^{R_0(e^{-R_0} - 1)} - 1\right)}$	
$\vdots$	$\vdots$	

At $\phi = 1$,

$$\Pr\{T_g \leq g\} = \frac{R_0^g - 1}{R_0^{g+1} - 1} \to \begin{cases} 1, & R_0 < 1 \\ \frac{1}{R_0}, & R_0 > 1 \end{cases} \text{ as } g \to \infty.$$

To demonstrate that T_g is stochastically longer when $\phi \to 0$ compared to when $\phi = 1$, Fig. 10.11 illustrates the survivor function $\Pr\{T_g \geq g\}$

at $R_0 = 0.667$. Under the same R_0, the probability of lasting more than 5 generations before extinction if N is Poisson distributed, is approximately 1.4 times of that if N has a geometric distribution.

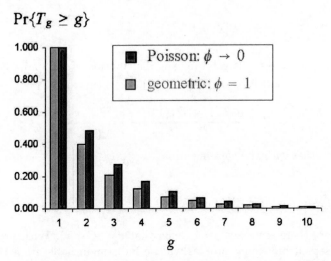

Fig. 10.11 Under the same R_0, Poisson distributed N provides a stochastically longer generation time to extinction T_g than that based on a geometric distribution for N

10.5.2.3 Distribution of the Final Size of Small Outbreaks

For the integer-valued random variable Z corresponding to the final size of a small minor outbreak, the probability

$$\sum_{z=1}^{\infty} \Pr\{Z = z\} = \pi \begin{cases} = 1, \text{ if } R_0 \leq 1 \\ < 1, \text{ if } R_0 > 1 \end{cases}$$

where π is the probability of a minor outbreak as a solution of (10.41). Let $G_Z(s) = \sum_{z=1}^{\infty} s^z \Pr\{Z = z\}$. It was shown [10, p. 193] that $G_Z(s) = s \cdot G_N(G_Z(s))$. In general, $G_Z(s)$ is not a p.g.f. since $G_Z(1) = \pi \leq 1$. Taking derivatives with respect to s,

$$G'_Z(s) = G_N(G_Z(s)) + sG'_N(G_Z(s))G'_Z(s), \tag{10.46a}$$

$$G''_Z(s) = 2G'_N(G_Z(s))G'_Z(s) + sG''_N(G_Z(s))\left(G'_Z(s)\right)^2 \tag{10.46b}$$
$$+ sG'_N(G_Z(s))G''_Z(s).$$

Mean and Variances for Z When $R_0 < 1$

In this case, $\pi = 1$ and $G_N(1) = G_Z(1) = 1$. Since $G'_N(1) = R_0 < 1$, letting $s = 1$ in (10.46a), one gets $G'_Z(1) = 1 + R_0 G'_Z(1)$. Hence

$$E[Z] = G'_Z(1) = \frac{1}{1 - R_0}. \tag{10.47}$$

Therefore, $E[Z] = \frac{1}{1-R_0}$ is valid for any distribution of N. Similarly, letting $s = 1$ in (10.46b) and using $var[Z] = G''_Z(1) + G'_Z(1) - (G'_Z(1))^2$, one gets

$$var[Z] = \frac{G''_N(1) + R_0(1 - R_0)}{(1 - R_0)^3} = \frac{var[N]}{(1 - R_0)^3}. \tag{10.48}$$

$var[Z]$ increases with the variance of N.

Mean and Variances for Z When $R_0 > 1$

In this case, there exits $\pi \in (0,1)$ such that $\pi = G_N(\pi)$. With probability π, the observed branching process will be indistinguishable from that as if arising from a different "reproduction number" $R_0^* = G'_N(\pi) < 1$. (10.47) and (10.48) can be extended as the following conditional expectation and conditional variance:

$$E[Z|\text{small outbreak}] = \frac{1}{1 - R_0^*} \tag{10.49a}$$

$$var[Z|\text{small outbreak}] = \frac{\pi G''_N(\pi) + R_0^*[1 - R_0^*]}{(1 - R_0^*)^3} \tag{10.49b}$$

where (10.49a) and (10.49b) are extensions of (10.47) and (10.48) in the sense that they are valid for both $R_0 > 1$ and $R_0 < 1$. For the latter, $\pi = 1$ and $G'_N(\pi) = R_0$.

The Distribution of Z

We treat $G_Z(s) = \sum_{z=1}^{\infty} s^z \Pr\{Z = z\}$ as if a p.g.f. for both $R_0 > 1$ and $R_0 < 1$, such that $\Pr\{Z = z\} = \frac{1}{z!} G_Z^{(z)}(0)$ and $G_Z^{(z)}(0) = \frac{d^z}{ds^z} G_Z(s)\big|_{s=0}$. If $R_0 > 1$,

$$\Pr\{Z = z|\text{small outbreak}\} = \frac{1}{\pi} \Pr\{Z = z\}$$

is the conditional distribution of the final size starting from one initial infective, conditioning on the outcome being a small outbreak.

For any given p.g.f. $G_N(\cdot)$, a closed analytic form of $G_Z(s)$ may not always exist if one wants to solve $G_Z(s) = sG_N(G_Z(s))$. However, $G_Z^{(z)}(0)$ can be sometimes solved recursively starting from $\Pr\{Z = 1\} = G'_Z(0)$. We use the convention that

$$\Pr\{Z = 1\} = G'_Z(0) = G_N(0) = \Pr\{N = 0\}$$

because the event $\{Z = 1\}$ implies that there is no secondary transmission in the population. There is also a convention that $G_Z(0) = \Pr\{Z = 0\} = 0$ as there must be at least one infective individuals to start an outbreak. The recursive procedure can be demonstrated for $\Pr\{Z = 2\}$ and $\Pr\{Z = 3\}$. From (10.46b), $G''_Z(0) = 2G'_N(G_Z(0))G'_Z(0) = 2G'_N(0)G_N(0)$, which gives

$$\Pr\{Z = 2\} = \frac{1}{2}G''_Z(0) = \Pr\{N = 1\}\Pr\{N = 0\}.$$

It is the probability that the index case gives transmission to one individual with probability $\Pr\{N = 1\}$ and the second individual does not transmit with probability $\Pr\{N = 0\}$. With a bit more calculus, $G_Z^{(3)}(0) = 3G''_N(0)[G_N(0)]^2 + 6[G'_N(0)]^2 G_N(0)$ so that $\Pr\{Z = 3\} = \frac{1}{6}G_Z^{(3)}(0)$ can be expressed as

$$\Pr\{Z = 3\} = \frac{1}{2}G''_N(0)[G_N(0)]^2 + [G'_N(0)]^2 G_N(0)$$
$$= \Pr\{N = 2\}(\Pr\{N = 0\})^2 + (\Pr\{N = 1\})^2 \Pr\{N = 0\}.$$

It implies that either the index case produces two secondary cases with probability $\Pr\{N = 2\}$ and neither of the secondary cases produces further transmission with probability $(\Pr\{N = 0\})^2$; or the index case produces one transmission and the secondary case produce one transmission with joint probability $(\Pr\{N = 1\})^2$ and the third case does not transmit with probability $\Pr\{N = 0\}$.

The Case of Negative-Binomial Distribution for N

When $G_N(s)$ has a relatively simple form, one can use $G_N(s)$ and the recursive procedure to generate the entire distribution $\frac{1}{\pi}\Pr\{Z = z\}$. Let $G_N(s) = \frac{1}{[1+R_0\phi(1-s)]^{\frac{1}{\phi}}}$. $G_Z(s) = sG_N(G_Z(s))$ can be written as:

$$G_Z(s) = \frac{s}{[1 + R_0\phi(1 - G_Z(s))]^{\frac{1}{\phi}}}. \tag{10.50}$$

For $z = 1$, $\Pr\{Z = 1\} = G_Z'(0) = \dfrac{1}{[1+R_0\phi]^{\frac{1}{\phi}}}$. We can calculate recursively

for $z \geq 2$, $G_Z^{(z)}(0) = \displaystyle\prod_{j=0}^{z-2}(j\phi + z)\left(\dfrac{1}{\phi R_0 + 1}\right)^{\frac{z}{\phi}}\left(\dfrac{R_0}{\phi R_0 + 1}\right)^{z-1}$ and hence $\Pr\{Z = z|\text{small outbreak}\}$ is

$$\frac{1}{\pi}\Pr\{Z = z\} = \frac{1}{\pi z!}\prod_{j=0}^{z-2}(j\phi + z)\left(\frac{1}{\phi R_0 + 1}\right)^{\frac{z}{\phi}}\left(\frac{R_0}{\phi R_0 + 1}\right)^{z-1}, \qquad (10.51)$$

where π needs to be numerically calculated from the equation $\pi = \dfrac{1}{[1+R_0\phi(1-\pi)]^{\frac{1}{\phi}}}$ if $R_0 > 1$. When $R_0 < 1$, the mean and variance with respect to (10.51) is

$$E[Z] = \frac{1}{1 - R_0}, \quad var[Z] = \frac{(1 + \phi)R_0^2 + R_0(1 - R_0)}{(1 - R_0)^3}. \qquad (10.52)$$

Example 5. When $\phi \to 0$ and $R_0 < 1$, (10.51) is

$$\Pr\{Z = z\} = \frac{(R_0 z)^{z-1}}{z!}e^{-R_0 z}. \qquad (10.53)$$

This distribution was first discovered by Borel (1942) and called the Borel–Tanner distribution. It is discussed in Durrett [33] as Corollary 2.6.2. Its variance is $var[Z] = \frac{R_0}{(1-R_0)^3}$.

Example 6. When $\phi = 1$ and $R_0 < 1$, (10.51) is

$$\Pr\{Z = z\} = \frac{1}{z!}\prod_{j=0}^{z-2}(j + z)\left(\frac{1}{1 + R_0}\right)^{z}\left(\frac{R_0}{1 + R_0}\right)^{z-1}. \qquad (10.54)$$

This distribution can be found as (6.619) in Mode and Sleeman [10]. Its variance is $var[Z] = \frac{R_0(R_0+1)}{(1-R_0)^3}$.

When $R_0 > 1$, to calculate the condition mean $E[Z|\text{small outbreak}]$ and the conditional variance $var[Z|\text{small outbreak}]$, one uses (10.49a) and (10.49b) for the calculation and substitutes $R_0^* = G_N'(\pi) = \dfrac{R_0}{(1+\phi R_0(1-\pi))^{\frac{1}{\phi}+1}}$ and $G_N''(\pi) = \dfrac{(1+\phi)R_0^2}{(1+\phi R_0(1-\pi))^{\frac{1}{\phi}+2}}$. In the geometric distribution where $\phi = 1$ and $\pi = \frac{1}{R_0}$, one gets simple expressions for $R_0 > 1$ [10, p. 196]

$$E[Z|\text{small outbreak}] = \frac{R_0}{R_0 - 1}, \quad var[Z|\text{small outbreak}] = \frac{R_0(R_0 + 1)}{(R_0 - 1)^3}.$$

10.5.3 Behaviour of a Large Outbreak in its Initial Phase: The Intrinsic Growth

For the assessment of the risk of a large outbreak $1 - \pi$ and quantities associated with a small outbreak, it is sufficient to know the distribution for N, without the necessity of knowing the distributions for the latent and infectious periods. On the other hand, should a large outbreak occur, the intrinsic growth rate r, also known as the Malthusian parameter, for $E[I(t)] \propto e^{rt}$, depends crucially on the latent and infectious periods distributions. In other words, even detailed information of the distribution for N is not sufficient. One needs to know the stochastic mechanisms that manifest the distribution for N.

Yan [34] re-writes the equation (10.39) under the following conditions:

1. The infectious contact process $\{K(x)\}$ has stationary increment $\beta(x) \equiv \beta$ over the infectious period
2. T_E and T_I are independent
3. The Laplace transforms for the latent period and the infectious period: $\mathcal{L}_E(r) = \int_0^\infty e^{-rx} f_E(x) dx$ and $\mathcal{L}_I^*(r) = \int_0^\infty e^{-rx} \overline{F}_I(x) dx$ exist, then

$$\beta \mathcal{L}_E(r) \mathcal{L}_I^*(r) = 1, \qquad (10.55)$$

implying that, the property of the infectious contact process $\{K(x)\}$ as summarized by the parameter β, the distribution of the latent period as summarized by $\mathcal{L}_E(r)$ and the distribution of the infectious period as summarized by $\mathcal{L}_I^*(r)$, separately shape the shape of $E[I(t)]$ during the initial phase of an outbreak and a general relationship between the intrinsic growth rate r and the basic reproduction number R_0.

In fact, $\int_0^\infty e^{-r\tau} \beta(\tau) A(\tau) d\tau = 1$ becomes

$$\beta \int_0^\infty e^{-r\tau} \left[\overline{F}_{E+I}(\tau) - \overline{F}_E(\tau) \right] d\tau = \beta \left[\mathcal{L}_{E+I}^*(r) - \mathcal{L}_E^*(r) \right] = 1,$$

where $\mathcal{L}_{E+I}^*(r) = \int_0^\infty e^{-r\tau} \overline{F}_{E+I}(\tau) d\tau$ and $\mathcal{L}_E^*(r) = \int_0^\infty e^{-r\tau} \overline{F}_E(\tau) d\tau$. Since T_E and T_I are independent, the Laplace transform of the generation time $T_E + T_I$ is $\mathcal{L}_{E+I}(r) = \mathcal{L}_E(r) \mathcal{L}_I(r)$. By writing $\mathcal{L}_{E+I}^*(r) = \frac{1 - \mathcal{L}_E(r) \mathcal{L}_I(r)}{r}$ and $\mathcal{L}_E^*(r) = \frac{1 - \mathcal{L}_E(r)}{r}$, one immediately gets $\beta \mathcal{L}_E(r) \mathcal{L}_I^*(r) = 1$ in (10.55). Since $R_0 = \beta \mu$, one further gets

$$R_0 = \frac{\mu}{\mathcal{L}_E(r) \mathcal{L}_I^*(r)} = \begin{cases} \frac{1}{\mathcal{L}_E(r)}, & \text{if } E[T_I] \to 0, \\ \frac{\mu}{\mathcal{L}_I^*(r)}, & \text{if } E[T_E] = 0. \end{cases} \qquad (10.56)$$

The necessary condition for $r > 0$ is $R_0 > 1$. Together with $R_0 = \beta \mu$, one gets $\mu > \beta^{-1}$. In this case, the relationship (10.55) can be illustrated in

Fig. 10.12, where $\mathcal{L}_E(r)\mathcal{L}_I^*(r) = \int_0^\infty e^{-r\tau}A(\tau)d\tau$ is a decreasing function of r, satisfying $\lim_{r\to 0}\mathcal{L}_E(r)\mathcal{L}_I^*(r) = \mu > \frac{1}{\beta}$ and $\lim_{r\to\infty}\mathcal{L}_E(r)\mathcal{L}_I^*(r) = 0$.

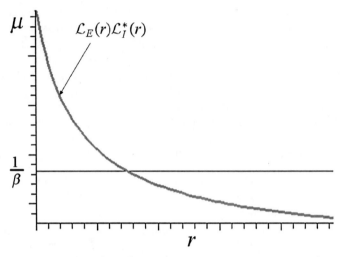

Fig. 10.12 Expression (10.55) as illustrated

Recalling the discussions on stochastic ordering and the relationship in (10.8), one can immediately see that when β is fixed, stochastic ordering of the latent and infectious periods determines the intrinsic growth rate r. Therefore we have the following two propositions.

Proposition 4. *When β and the infectious period distribution $F_I(x)$ are given, the longer the latent period T_E, the smaller the intrinsic growth r. In this statement, the word longer refers to conditions so that T_E is larger in Laplace transform order.*

Figure 10.12 shows that, when $\mathcal{L}_E(r)$ is smaller, the root for r in the equation $\mathcal{L}_E(r)\mathcal{L}_I^*(r) = \beta^{-1}$ is smaller. By consequence, the existence of a latent period produces a smaller initial growth rate than that produced by a model in the absence of a latent period. Since the distribution of T_E does not affect the basic reproduction number R_0, we also have:

Corollary 4. *For fixed infectious period distribution $F_I(x)$ and empirically observed r, if one takes a distribution for the latent period which is shorter in Laplace order than it really is, one underestimates R_0.*

Proposition 5. *When β and the latent period distribution $F_E(x)$ are given, among the distributions for the infectious period with equal mean value $\mu = E[T_I]$, the longer the infectious period T_I, the larger is the intrinsic growth rate r. In this statement, the word longer refers to conditions so that T_I is larger in Laplace transform order.*

Figure 10.12 shows that, when $\mathcal{L}_I^*(r) = \frac{1-\mathcal{L}_I(r)}{r}$ is larger, the root for r in the equation $\mathcal{L}_E(r)\mathcal{L}_I^*(r) = \beta^{-1}$ is larger. In addition, the comparison is restricted to infectious periods with equal mean value $\mu = E[T_I]$ because $R_0 = \beta\mu$ so that comparisons can be made under the same basic reproduction number.

Corollary 5. *Given empirically observed r and mean infectious period μ, if one takes a distribution for the infectious period which is larger in Laplace transform order than what it should be, one underestimates R_0.*

10.5.3.1 The Expressions (10.55) and (10.56) When the Distributions of T_E and T_I Arise from Some Specific Parametric Families

The gamma distribution with p.d.f. given in (10.7), re-parametrized by $\alpha = \frac{\kappa}{\mu}$ gives $f_X(x) = \frac{\alpha(\alpha x)^{\kappa-1}}{\Gamma(\kappa)}e^{-\alpha x} = \frac{\frac{\kappa}{\mu}\left(\frac{\kappa}{\mu}x\right)^{\kappa-1}}{\Gamma(\kappa)}e^{-\frac{\kappa}{\mu}x}$ such that $E[X] = \mu$ and $var[X] = \frac{\mu^2}{\kappa}$. The Laplace transform exists and has a simple expression $\mathcal{L}_X(r) = \left(1 + \frac{r\mu}{\kappa}\right)^{-\kappa}$ and hence $\mathcal{L}_X^*(r) = \frac{1-\left(1+\frac{r\mu}{\kappa}\right)^{-\kappa}}{r}$.

If two random variables T_E and T_I are both gamma distributed, we reserve $\mu = E[T_I]$ for the mean infectious period. We introduce $\nu = E[T_E]$ for the mean latent period. Thus,

$$\mathcal{L}_E(r) = \left(1 + \frac{r\nu}{\kappa_1}\right)^{-\kappa_1}, \quad \mathcal{L}_I^*(r) = \frac{1 - \left(1 + \frac{r\mu}{\kappa_2}\right)^{-\kappa_2}}{r}$$

where κ_1 and κ_2 are the shape parameters for the latent and infectious periods when both are gamma distributed. In this case, The expressions (10.55) and (10.56) become

$$\frac{\beta}{r}\left(1 + \frac{r\nu}{\kappa_1}\right)^{-\kappa_1}\left(1 - \left(1 + \frac{r\mu}{\kappa_2}\right)^{-\kappa_2}\right) = 1 \qquad (10.57a)$$

$$R_0 = \frac{r\mu\left(1 + \frac{r\nu}{\kappa_1}\right)^{\kappa_1}}{1 - \left(1 + \frac{r\mu}{\kappa_2}\right)^{-\kappa_2}}. \qquad (10.57b)$$

The expression (10.57b) was originally given by Anderson and Watson [35] and recently re-visited by Wearing et al. [36] It has many special cases as summarized by Table 10.5.

In particular, Cases C1, C4, C5, C7 and C11 are highlighted below

Table 10.5 Special cases with gamma distributed latent and infectious periods

Cases	ν	μ	κ_1	κ_2	$\mathbf{R_0}$
C1	$\nu \to 0$		$\to \infty$	κ_2	$e^{r\nu} = e^{rT_G}$
C2	$\nu \to 0$		$= 1$	κ_2	$1 + r\nu = 1 + rE[T_G]$
C3	$\nu \to 0$		κ_1	κ_2	$\left(1 + \frac{r\nu}{\kappa_1}\right)^{\kappa_1}$
C4	0	μ	κ_1	$\to \infty$	$\frac{r\mu}{1-e^{-r\mu}} = \frac{rT_G}{1-e^{-rT_G}}$
C5	0	μ	κ_1	$= 1$	$1 + r\mu = 1 + rE[T_G]$
C6	0	μ	κ_1	κ_2	$\frac{r\mu}{1-\left(1+\frac{r\mu}{\kappa_2}\right)^{-\kappa_2}}$
C7	ν	μ	$\to \infty$	$\to \infty$	$e^{r\nu}\frac{r\mu}{(1-e^{-r\mu})}$
C8	ν	μ	$= 1$	$\to \infty$	$(1+r\nu)\frac{r\mu}{1-e^{-r\mu}}$
C9	ν	μ	κ_1	$\to \infty$	$\left(1+\frac{r\nu}{\kappa_1}\right)^{\kappa_1}\frac{r\mu}{1-e^{-r\mu}}$
C10	ν	μ	$\to \infty$	$= 1$	$e^{r\nu}(1+r\mu)$
C11	ν	μ	$= 1$	$= 1$	$(1+r\nu)(1+r\mu)$
C12	ν	μ	κ_1	$= 1$	$\left(1+\frac{r\nu}{\kappa_1}\right)^{\kappa_1}(1+r\mu)$
C13	ν	μ	$\to \infty$	κ_2	$e^{r\nu}\frac{r\mu}{1-\left(1+\frac{r\mu}{\kappa_2}\right)^{-\kappa_2}}$
C14	ν	μ	$= 1$	κ_2	$(1+r\nu)\frac{r\mu}{1-\left(1+\frac{r\mu}{\kappa_2}\right)^{-\kappa_2}}$
General	ν	μ	κ_1	κ_2	$\left(1+\frac{r\nu}{\kappa_1}\right)^{\kappa_1}\frac{r\mu}{1-\left(1+\frac{r\mu}{\kappa_2}\right)^{-\kappa_2}}$

$$R_0 = e^{rE[T_G]} \tag{10.58a}$$

$$R_0 = \frac{r\mu}{1 - e^{-r\mu}} \tag{10.58b}$$

$$R_0 = 1 + r\mu \tag{10.58c}$$

$$R_0 = e^{r\nu}\frac{r\mu}{1 - e^{-r\mu}} \tag{10.58d}$$

$$R_0 = (1 + r\nu)(1 + r\mu)$$
$$= 1 + rE[T_G] + f(1-f)(rE[T_G])^2, \tag{10.58e}$$

where $f = \frac{E[T_E]}{E[T_G]} = \frac{\nu}{\nu+\mu}$. Of these expressions, (10.58a) is the most intuitive and appears in ecology textbooks (e.g. [37]), assuming that the variation in the generation time T_G is negligible and only if $T_G = T_E$ so that transmission only occurs at an instantaneous moment at the end of the latent period. In the absence of a latent period, a non-random infectious period gives (10.58b) which can be found in Anderson and May [38]. The most commonly encountered relationship is (10.58c) which can be derived from the deterministic or stochastic SIR models assuming an exponentially distributed infectious period without a latent period, as in Chap. 2 of this book. If one assumes that both $T_E = \nu$ and $T_I = \mu$ are not random, one gets (10.58d). In recent years, (10.58e) is frequently seen in the literature, such as Lipsitch et al. [39] in the application to the SARS epidemic in Singapore; Chowell et al. [40] for influenza, etc. The underlying assumption in (10.58e) is that both the latent and infectious periods are exponentially distributed.

For gamma distributed latent period, the Laplace transform is $\mathcal{L}_E(r) = \left(1 + \frac{r\nu}{\kappa_1}\right)^{-\kappa_1}$. Both ν and κ_1 rank T_E according to Laplace transform order. For fixed κ_1

$$\frac{d}{d\nu}\mathcal{L}_E(r) = -\frac{r}{\left(\frac{1}{\kappa_1}(\kappa_1 + r\nu)\right)^{\kappa_1+1}} < 0,$$

implying that the longer the mean latent period, the larger the latent period in Laplace transform order. On the other hand, for the same mean latent period ν,

$$\frac{d}{d\kappa_1}\mathcal{L}_E(r) = -\frac{\kappa_1^{\kappa_1}}{(\kappa_1 + r\nu)^{\kappa+1}}\left[(\kappa_1 + r\nu)\log\frac{\kappa_1 + r\nu}{\kappa_1} - r\nu\right] < 0,$$

implying that the larger the value of the shape parameter κ_1 (hence the smaller the variance $var[T_E] = \frac{\nu^2}{\kappa_1}$), the smaller the value of $\mathcal{L}_E(r)$ and hence the longer the latent period in Laplace transform order. In both cases, if β and $\mathcal{L}_I^*(r)$ are fixed, a longer latent period in Laplace transform order implies a smaller value of r.

Two benchmarks are $\kappa_1 = 1$ and $\kappa_1 \to \infty$. Among a subset of the gamma distribution with $\kappa_1 \geq 1$, for the same R_0, the exponentially distributed latent period gives the largest r. When the latent period distribution degenerates to a fixed point with $T_E = \nu$, one gets the smallest r. Conversely, if r is empirically observed, then the exponentially distributed latent period yields the smallest R_0 whereas the fixed latent period yields the largest R_0.

To examine the effects of the infectious period distribution on r with a given latent period distribution $F_E(x)$, since $R_0 = \beta\mu$, we only consider distributions for the infectious period with equal mean value $\mu = E[T_I]$. Under this condition, the longer the infectious period T_I in Laplace transform order, the larger is the intrinsic growth rate r. For gamma distributed infectious period, $\mathcal{L}_I^*(r) = \frac{1-\left(1+\frac{r\mu}{\kappa}\right)^{-\kappa}}{r}$ and

$$\frac{d}{d\kappa_2}\mathcal{L}_I^*(r) = \frac{\kappa_2^{\kappa_2}}{r(\kappa_2 + r\mu)^{\kappa_2+1}}\left[(\kappa_2 + r\mu)\log\frac{\kappa_2 + r\mu}{\kappa_2} - r\mu\right] > 0.$$

In this case, since we keep the mean $\mu = E[T_I]$ fixed and $var[T_I] = \frac{\mu}{\kappa_2}$, a longer infectious period T_I in Laplace transform order implies that the variance is smaller. In other words, under the same R_0 and fixed latent period distribution, a gamma distributed infection period with smaller variance produces larger initial growth rate. Among a subset of the gamma distribution with $\kappa_2 \geq 1$, for the same R_0, the exponentially distributed infectious period ($\kappa_2 = 1$) gives the smallest r and the fixed infectious period ($\kappa_2 \to \infty$) gives the largest r. Conversely, if r is empirically observed and if the mean infectious period μ is given (along with the given latent period distribution), the larger the value κ_2 (hence the smaller the variance), the smaller the R_0. This can be also seen by taking the derivative of (10.57b) with respect to κ_2:

$$\frac{d}{d\kappa_2}R_0 = -\frac{r\mu\kappa_2^{\kappa_2}(\kappa_1 + r\nu)^{\kappa_1}\left[(\kappa_2 + r\mu)\log\frac{1}{\kappa_2}(\kappa_2 + r\mu) - r\mu\right]}{\kappa_1^{\kappa_1}(\kappa_2 + r\mu)^{1-\kappa_2}\left(\kappa_2^{\kappa_2} - (\kappa_2 + r\mu)^{\kappa_2}\right)^2} < 0.$$

Example 7. Let us consider an SIR model without the latent period. Let us assume that the infectious contacts arise from a Poisson process and that the infectious period has mean $\mu = 4.098$. In this example, let $R_0 = \beta\mu = 1.386$. The infectious period T_I has the distribution $f_I(x) = \frac{\kappa^\kappa x^{\kappa-1}}{\mu^\kappa \Gamma(\kappa)}e^{-\frac{\kappa x}{\mu}}$, $\mu > 0, \kappa = 1, 2, 3, \cdots$, a special case of the gamma distribution (the Erlang distribution) with integer-valued κ. It can be viewed as the distribution of the sum of κ i.i.d. exponential random variables with mean $\frac{\mu}{\kappa}$. A "linear chain trick" is to consider a compartment model as given by Fig. 10.13 and to use deterministic ordinary differential equations to numerically calculate $I(t) \approx E[I(t)]$. Figure 10.14 shows results with $\kappa = 1, 2, 3, 4$. When $\kappa = 1$,

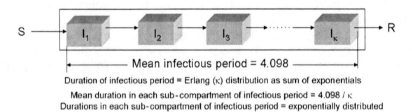

Mean infectious period = 4.098

Duration of infectious period = Erlang (κ) distribution as sum of exponentials

Mean duration in each sub-compartment of infectious period = 4.098 / κ
Durations in each sub-compartment of infectious period = exponentially distributed

Fig. 10.13 Illustration of the linear chain trick

the model reduces to the SIR model with exponentially distributed infectious period with $r = \frac{R_0 - 1}{\mu} = \frac{0.386}{4.098} = 0.094192$ The larger the κ, the steeper the initial increase one expects. If $\kappa = 4$, r can be solved via the equation: $\frac{1.386}{4.098r}\left(1 - \left(1 + \frac{4.098r}{4}\right)^{-4}\right) = 1$ for the non-negative real value. The solution is $r = 0.14122$.

One can replace the gamma distribution for the latent or the infectious period by other distributions, provided that their Laplace transforms exist. For example, one may consider an inverse-Gaussian distribution for a non-negative random variable X with p.d.f.

$$f_X(x) = \sqrt{\frac{\kappa\mu}{2\pi x^3}}\exp\left\{-\frac{\kappa(x-\mu)^2}{2\mu x}\right\}, \quad \mu, \kappa > 0. \tag{10.59}$$

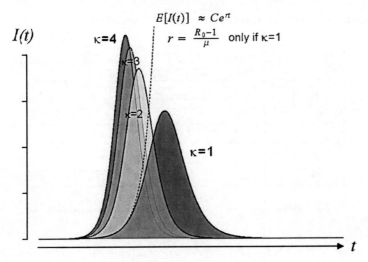

Fig. 10.14 Same R_0 : the larger the κ, the larger the growth rate r

It has been parametrized such that $E[X] = \mu$ and $var[X] = \frac{\mu^2}{\kappa}$, like for the gamma distribution. The Laplace transform is $\mathcal{L}_X(r) = \exp\left\{\kappa\left(1 - \sqrt{\frac{2r\mu+\kappa}{\kappa}}\right)\right\}$.

If two random variables T_E and T_I are both inverse-Gaussian distributed, as before, we reserve $\mu = E[T_I]$ for the mean infectious period and use $\nu = E[T_E]$ for the mean latent period. Then $\mathcal{L}_E(r) = \exp\left\{\kappa_1\left(1 - \sqrt{\frac{2r\nu+\kappa_1}{\kappa_1}}\right)\right\}$ and $\mathcal{L}_I^*(r) = \frac{1-\exp\left\{\kappa_2\left(1-\sqrt{\frac{2r\mu+\kappa_2}{\kappa_2}}\right)\right\}}{r}$. The expressions (10.55) and (10.56) become

$$\frac{\beta}{r}\exp\left\{\kappa_1\left(1 - \sqrt{\frac{2r\nu+\kappa_1}{\kappa_1}}\right)\right\}\left(1 - \exp\left\{\kappa_2\left(1 - \sqrt{\frac{2r\mu+\kappa_2}{\kappa_2}}\right)\right\}\right) = 1,$$

$$R_0 = \frac{1}{\exp\left\{\kappa_1\left(1 - \sqrt{\frac{2r\nu+\kappa_1}{\kappa_1}}\right)\right\}} \times \frac{r\mu}{1 - \exp\left\{\kappa_2\left(1 - \sqrt{\frac{2r\mu+\kappa_2}{\kappa_2}}\right)\right\}}.$$

Similar to Table 10.5, one can make a table of all the special cases; see Table 10.6.

One does not need to assume that the latent and infectious periods arise from the same distribution family. For example, one may take an inverse-Gaussian distributed latent period to combine with a gamma distributed infectious period and derive

$$R_0 = \frac{1}{\exp\left\{\kappa_1\left(1 - \sqrt{\frac{2r\nu+\kappa_1}{\kappa_1}}\right)\right\}} \times \frac{r\mu}{1 - \left(1 + \frac{r\mu}{\kappa_2}\right)^{-\kappa_2}}$$

Table 10.6 Cases with inv.-Gaussian distributed latent and infectious periods

	ν	μ	κ_1	κ_2	$\mathbf{R_0}$	Remarks
C1	$\nu \to 0$	$\to \infty$		κ_2	$e^{r\nu} = e^{rT_G}$	C1 in Table 10.5
C2	$\nu \to 0$	$= 1$		κ_2	$\dfrac{1}{\exp\left(1-\sqrt{1+2r\nu}\right)}$	
C3	$\nu \to 0$	κ_1		κ_2	$\dfrac{1}{\exp\left\{\kappa_1\left(1-\sqrt{\dfrac{2r\nu+\kappa_1}{\kappa_1}}\right)\right\}}$	
C4	0	μ	κ_1	$\to \infty$	$\dfrac{r\mu}{1-e^{-r\mu}}$	C4 in Table 10.5
C5	0	μ	κ_1	$= 1$	$\dfrac{r\mu}{1-\exp\left\{(1-\sqrt{2r\mu+1})\right\}}$	
C6	0	μ	κ_1	κ_2	$\dfrac{r\mu}{1-\exp\left\{\kappa_2\left(1-\sqrt{\dfrac{2r\mu+\kappa_2}{\kappa_2}}\right)\right\}}$	
C7	ν	μ	$\to \infty$	$\to \infty$	$e^{r\nu}\dfrac{r\mu}{1-e^{-r\mu}}$	C7 in Table 10.5
⋮	⋮	⋮	⋮	⋮	⋮	Other cases

or a gamma distributed latent period to combine with an inverse-Gaussian distributed infectious period and derive

$$R_0 = \left(1 + \frac{r\nu}{\kappa_1}\right)^{\kappa_1} \frac{r\mu}{1 - \exp\left\{\kappa_2\left(1 - \sqrt{\frac{2r\mu+\kappa_2}{\kappa_2}}\right)\right\}}.$$

10.5.4 Summary for the Initial Phase of an Outbreak

10.5.4.1 Invasion and Everything about a Small Outbreak

The latent period does not play any role. Knowledge of the distribution for N is sufficient. One uses the p.g.f. $G_N(s)$ to calculate the probability $1-\pi$ for the risk of a large outbreak. If the outbreak is small, $G_N(s)$ further determines the distribution of the generation time to extinction $\Pr\{T_g \leq g\}$ and final size at extinction $\Pr\{Z = z\}$.

If the distribution of N can be derived from observed data, then details with respect to underlying stochastic mechanisms that give rise to $G_N(s)$, such as the property of the infectious contact process $\{K(x)\}$ and infectious period distribution $F_I(x)$, are irrelevant.

On the other hand, with knowledge of the property of the infectious contact process $\{K(x)\}$ and the infectious period distribution $F_I(x)$, one can use the CMJ branching process to derive the p.g.f. $G_N(s)$ for N.

10.5.4.2 When the Outbreak is Large

The initial phase of a large outbreak is characterized by the approximation $E[I(t)] \propto e^{rt}$. Under the condition that $\{K(x)\}$ has stationary increment property, r is determined by (10.55). It shows that β, the distribution of the latent period and the distribution of the infectious period separately shape the shape of $E[I(t)]$ during the initial phase. However, the distribution for N is irrelevant.

10.6 Beyond the Initial Phase: The Final Size of Large Outbreaks

For a large outbreak, the expected final outbreak size number scales linearly with the size of the susceptible population. In other words, $E[Z] \to \infty$ but $\frac{E[Z]}{n} \to \eta$ where η is a positive quantity, $0 < \eta < 1$. Therefore, one considers the continuous random variable $Y = \frac{Z}{n}$, distributed over the range $(0, 1)$, as the final size.

Theorem 1. *When $R_0 > 1$, conditioning on the outcome being a large outbreak and assuming that $\lim_{n \to \infty} \frac{I_0}{n} \to \varepsilon$, let η be the root of the equation*

$$1 - \eta = \exp\left(-R_0\left(\eta + \varepsilon\right)\right). \tag{10.60}$$

Then $\frac{S_\infty - n\eta}{\sqrt{n}}$ has Gaussian limit distribution $N(0, \sigma^2)$ and the asymptotic variance is given by

$$\sigma^2 = \frac{\eta\left(1 - \eta\right)}{\left(1 - R_0\eta\right)^2} + \frac{\eta^2\left(var[N] - R_0\right)\left(1 - \eta + \varepsilon\right)}{\left(1 - R_0\eta\right)^2}. \tag{10.61}$$

This central limit theorem has been developed from different mathematical approaches by von Bahr and Martin-Löf [12], Ludwig [41], Scalia-Tomba [42], Martin-Löf [43] and Lefèvre and Picard [44]. The final size proportion $Y = \frac{Z}{n}$ converges in distribution to a point mass at η, the root of (10.60). The fluctuations around the limit are Gaussian of order $\frac{1}{\sqrt{n}}$, which become large if the variance $var[N]$ is large. Contrary to the study of the intrinsic growth rate r at the initial phase, the final size is more robust with respect to the stochastic mechanisms at the micro level. Even the distribution of N does not play a significant role, except for its first two moments: $R_0 = E[N]$ and $var[N]$.

In von Bahr and Martin-Löf [12], Theorem 1 was first proven under a Reed-Frost epidemic model [45]. For such a model, it is assumed that an individual i is infected at time t. A given individual j is contacted by i and if j is susceptible then j becomes infected at time $t + 1$. Meanwhile, i becomes removed (by immunity or death) and plays no further part in the epidemic process. All contacts are assumed to be independent of each other. At each increment of time from t to $t + 1$, an individual produces a random number of infectious contacts among the available susceptible individuals following a binomial distribution. When n is large, the binomial distribution is approximated by a Poisson distribution. von Bahr and Martin-Löf [12] then extended their proof to the randomized epidemic model where each infectious individual i has its own transmission probability per contact p_i; the p_i are i.i.d. random variables with a given distribution.

Earlier in this chapter we made the statement that as far as the distribution for N is concerned, the generalized epidemic and the randomized epidemic models can be unified as one mixed Poisson model with $G_N(s) = \int_0^\infty e^{\rho(s-1)} dU(\rho) = E_\rho \left[e^{\rho(s-1)} \right]$. This corresponds to Ludwig [41], where it was noted that as far as the distribution of $Y = \frac{Z}{n}$ is concerned, the generalized epidemic model with an arbitrary but specified infectious period distribution can be reformulated as a special case of the randomized epidemic model.

10.6.1 Generality of the Mean Final Size

Subject to the approximation $\frac{S_0}{n} \approx 1$ and if one re-writes $x = 1 - \frac{S_\infty}{S_0}, \varepsilon = \frac{I_0}{S_0}$, the asymptotic mean of a large outbreak (10.60) is in close agreement with that given by Kermack and Mckendrick [14], of which, given (R_0, S_0, I_0), one solves for S_∞ in the equation

$$S_0 - S_\infty + I_0 + \frac{S_0}{R_0} \log \left(\frac{S_\infty}{S_0} \right) = 0. \tag{10.62}$$

From a deterministic perspective, Ma and Earn [46] used integro-differential equations to show that the final size calculated by (10.62) is invariant, including the existence of a latent period, arbitrarily distributed infectious period, any number of distinct infectious stages and/or a stage during which infectious individuals are isolated, as well as the existence of super-spreading events.

From a probabilistic perspective, let us consider a typical susceptible individual ν_s. We define a type-specific hazard function

$$h_j^{(s)}(t - t_j) \begin{cases} > 0, & \text{if } t > t_j \\ = 0, & \text{otherwise,} \end{cases}$$

where t_j is the time of infection of the individual j, representing the instantaneous risk of infection being transmitted by the jth infectious individual. As illustrated by Example 4, the hazard of a susceptible individual ν_s to become infected at time t, denoted by $h^{(s)}(t)$, can be thought as a competing risk problem. Under the independence assumption, $h^{(s)}(t) = \sum_j h_j^{(s)}(t - t_j)$. Then the probability that a susceptible individual ν_s ever gets infected over the course of the epidemic is $\eta_s = 1 - \exp\left\{ -\sum_j \int_0^\infty h_j^{(s)}(t - t_j) dt \right\}$. Hence,

η_s depends on the set of infectious contact processes that correspond to the hazard $h_j^{(s)}$ only through the cumulative hazard $\int_0^\infty h_j^{(s)}(t)dt$.

Then we add two homogeneity assumptions:

1. *Homogeneous mixing.* As population size $\to \infty$, the contact process itself is a stationary Poisson process with constant intensity λ;
2. *Homogeneous hosts.* All individuals are of the same type the probability of transmission per contact between each pair of an infectious individual and a susceptible individual is a constant p.

Under these assumptions, the type-specific hazard is $h_j^{(s)}(t) = \beta^* = \lambda p$. The overall hazard of a susceptible individual ν_s to become infected at time t becomes $h^{(s)}(t) = h(t)$ which does not depend on ν_s (from assumption 2) and $h(t) = \beta^* E[I(t)]$. $E[I(t)]$ is the expected number of infectious individuals at time t. The probability that a susceptible individual ever get infected over the course of the epidemic is

$$\eta = 1 - \exp\left\{-\int_0^\infty h(t)dt\right\} = 1 - \exp\left\{-\beta^* \int_0^\infty E[I(t)]dt\right\}. \quad (10.63)$$

Note that $\int_0^\infty E[I(t)]dt$ is the expected total infectious time per person. $\mu = E[T_I]$ is the average time spent infectious per infection. Thus, $\int_0^\infty E[I(t)]dt = \eta S_0 E[T_I] = \eta S_0 \mu$

$$\Rightarrow \eta = 1 - \exp\left\{-(\beta^* S_0 \mu)\,\eta\right\} = 1 - \exp\left\{-R_0 \eta\right\}$$

which is (10.60) and $R_0 = \beta^* S_0 \mu = \beta \mu$. Ludwig [41] used this analogy in a discrete time setting, and provided a rigorous proof that the final size distribution does not depend on the duration of latency in the individuals, or on the time course of infectivity, but only on the "time-integrated infectivity" in discrete time. The "time-integrated infectivity" is analogous to $\int_0^\infty h_j^{(s)}(t)dt$, for closed populations with homogeneous mixing.

10.6.2 Some Cautionary Remarks

10.6.2.1 The Risk of Large Outbreak = Final Size?

In the literature, one sometimes encounter such a statement that "the risk of large outbreak = the final size of the large outbreak". In general, this statement is not true. On one hand, the final size equations (10.60) and (10.62) are functions of $R_0 = E[N]$, but do not depend on the exact distribution of N. On the other hand, the risk of a large outbreak $1 - \pi$ is determined by the equation (10.41) which is crucially dependent on $G_N(s)$, the probability generating function of N.

To give a counterexample, if N follows a geometric distribution, with $R_0 > 1$, the risk of a large outbreak is calculated with $\pi = \frac{1}{1+R_0(1-\pi)}$ with exact solution $\pi = \frac{1}{R_0}$. The risk of an large outbreak is $1 - \frac{1}{R_0}$. On the other hand, in a Kermack–McKendrick SIR model, N follows a geometric distribution, but the final size equations are (10.60) and (10.62).

The statement holds if the outbreak can be described by a Poisson epidemic. In this case, the probability of a large outbreak is $1 - \pi = e^{-R_0\pi}$ as derived from (10.42) and it is identical to the final size equation (10.60) as $\varepsilon \to 0$. Recall that a Poisson epidemic arises in the following manner.

1. As $n \to \infty$, the social contact network grows in such a way that at any time t, the realization can be regarded as a Bernoulli random graph [47] with Poisson degree distribution with mean λt. If D is the duration of an outbreak, the random graph as observed at D is a large Bernoulli random graph approximated by a Poisson degree distribution with mean $R_0^* = \lambda D$.
2. The infectious contact subgraph also grows according to a Poisson degree distribution with mean $\lambda p t = \beta t$. Every infective has a fixed infectious period μ. By the end of the outbreak, the observed infectious contacts makes a subgraph which is a large Bernoulli random graph with i.i.d. Poisson degree distributions with mean $R_0 = \beta\mu$.

Using Theorem 2.3.2. of Durrett [33] (which can be proven using random walk theory), if $R_0^* > 1$, with probability one, there is only one giant component with size $\sim \eta^* n$ where η^* is root x of the equation $1 - x = \exp(-R_0^* x)$ in $(0, 1)$. All other components are *small* in the sense that there is a constant ω such that the largest of the remaining components can not have more than $\omega \log n$ vertices. The infectious contact subgraph is proportional to the social contact graph with $R_0 \propto R_0^*$ by a factor $\frac{p\mu}{D}$.

There is a difference in concept between a random graph and a realization of a random graph. The random graph belongs to a probability space. A realization of a large outbreak takes place only within the giant component of the social contact graph. Since the relative sizes of all other component in the social contact graph $\to 0$ as $n \to \infty$, the size of a realized large outbreak η is thus proportional to η^* by the same factor $\frac{p\mu}{D}$. Therefore we get the final size equation $1 - \eta = \exp(-R_0\eta)$.

There are two independence properties involved.

1. A transmission from an infectious vertex v_i to a susceptible vertex v_j is independent from infections to other susceptible vertices derived from v_i.
2. All infectious vertices transmit the diseases independently.

A heuristic argument is that, if one randomly chooses vertex v_i to be the initial infective, the probability that it will lead to a large outbreak is the same probability that it belongs to the giant component. And hence, "the risk of a large outbreak = its final size".

However, if there is a random period of time $T_I^{(i)}$ to each vertex so that v_i, where $T_I^{(i)}$'s are i.i.d. with distribution $F_I(x)$, the evolution of the infectious contact subgraph will be very different from the evolution of the subgraph where every vertex has a fixed infectious period, even though the social contact network is the same Bernoulli random graph. If one chooses a vertex v_i to be the initial infective, then each of the other susceptible vertices in the graph has a probability $\int_0^\infty \left(1 - e^{-\lambda px}\right) F_I(x)$ to be ever infected by v_i. It can be shown that the probability of v_i infects v_j is dependent on other infections of other susceptible vertices with v_i as their direct origin of transmission. By the end of the outbreak the degree of the infectious contact subgraph has the mean value $R_0 = \lambda p E[T_I]$. The degree distribution has p.g.f. $G_N(s) = \int_0^\infty e^{\beta x(s-1)} dF_I(x)$. N is associated with a larger variance, with many infectious vertices having few edges due to a shorter infectious period and some infectious vertices having a large number of vertices due to a longer infectious period. Because of the loss of independency and the large variance, the equality between the risk of a large outbreak and the final size of the large outbreak is lost in generalized epidemic models.

10.6.2.2 Situations Where (10.63) may be Violated

Under the independence assumption that transmission of infection to individual ν_s by individual j is independent from potential transmissions from other infectious individuals, then the hazard function for individual ν_s to be infected at time t is $h^{(s)}(t) = \sum_j h_j^{(s)}(t - t_j)$. There are occasions that this *independent competing risk* can be violated. Such an occasion can arise in a highly clustered social contact network where a number of infectious individuals are correlated within a social cluster. Another occasion is that susceptible individuals are removed by quarantine if they have been identified as exposed to known infectives. In these circumstances, final size equations (10.60) and (10.62) may be incorrect.

10.7 When the Infectious Contact Process may not Have Stationary Increment

If the infectious contact process $\{K(x)\}$ has $\beta(x) = \frac{d}{dx} E[K(x)]$ as a function of x, (10.21) and (10.39) do not have simple forms. Most of the results discussed in previous sections will no longer apply.

10.7.1 The Linear Pure Birth Processes and the Yule Process

Let us consider a process without the stationary increment property, with $\beta(x) = \frac{d}{dx}E[K(x)] = \beta e^{\beta\phi x}$ as a function of x. It can be constructed in the following manner. Consider a linear pure birth process, defined by the conditional probability with the Markov property so that $\Pr\{K(x + h) - K(x) = 1|\mathcal{H}^x\}$, $k = 0, 1, 2, \cdots$ is equal to

$$\Pr\{K(x + h) = k + 1|K(x) = k\} = (\beta_1 k + \beta_2)h + o(h). \qquad (10.64)$$

Conditioning on $K(x) = k$, the instantaneous rate of producing the next infectious contact during $[t, t + h)$ can be considered as an independent competing risk hazard: either from a global environment, with constant rate β_2, or from a clustered environment with non-constant rate $\beta_1 k$ and $\beta_1 \neq \beta_2$. The hazard of producing an infectious contact at time x is

$$\lim_{h \to 0} \frac{\Pr\{K(x + h) = k + 1|K(x) = k\}}{h} = \beta_1 k + \beta_2 = \beta\left(k\phi + 1\right).$$

When $\phi = \frac{\beta_1}{\beta_1} \to 0$, the linear pure birth process reduces to a Poisson process with $\beta = \beta_2$

$$\Pr\{K(x + h) = k + 1|K(x) = k\} = \beta h + o(h), \quad k = 0, 1, 2, \cdots. \qquad (10.65)$$

When $\phi = 1$, the linear pure birth process is the Yule process with

$$\Pr\{K(x+h) = k+1|K(x) = k\} = \beta(k+1)h+o(h), \quad k = 0, 1, 2, \cdots. \qquad (10.66)$$

According to (10.66), given that an infectious individual has produced k infectious contacts by time x, the hazard for producing the $(k + 1)^{st}$ contact is a increasing function of k. Starting at $x = 0$, corresponding to the beginning of the infectious period for an infectious individual, the waiting time to producing the first infectious contact is exponentially distributed with mean $E[X_1] = \frac{1}{\beta_1}$; conditioning on the first infectious contact, the waiting time to the second infectious contact is exponentially distributed with mean $E[X_2] = \frac{1}{2\beta_1}; \cdots$; and conditioning on an infectious individual who has produced k infectious contacts, the waiting time to producing the $(k+1)^{st}$ infectious contact is exponentially distributed with mean $E[X_{k+1}] = \frac{1}{(k+1)\beta_1}$. On the surface, it looks as if the more infectious contacts it produces, the more likely it produces more infectious contacts.

10.7.1.1 A Justification of Whether the Linear Pure Birth Processes may be a Sensible Model

This infectious contact process may arise from the following hypothesis. A typical infectious individual resides in a highly clustered environment (household, hospital ward, etc.) but meanwhile has contacts with susceptible individuals from outside this environment. The contact network structures in the clustered environment is different from that in the "global" environment. During its infectious period, this individual may infect:

1. Susceptibles from the highly clustered environment such as household members (if such environment is a household) or nurses and other patients (if such environment is a clustered hospital ward), such that this network manifest data as if arising from a preferential attachment model
2. Susceptibles from a "global" environment that can be approximated by a large Bernoulli random graph

For outbreaks that mainly spread within and among a highly clustered environment, infected individuals directly and indirectly attributed to the index case during the time $x \leq T_I$ (infectious period) are often recorded as exposed and counted as the next generation of the index case. This artificially changes the order of events and creates artificial "generations" of infectives. However, real-life epidemiologic data may arise in this "artificial" order. Let us modify the definition of $\{K(x)\}$ so that $K(x)$ is not only the cumulative number of infectious contacts produced directly by an infectious individual, but also includes those infected cases attributed to itself in later generations, counted up to time x when the original index case is still infectious. By time x, if an infected individual is still infectious, then

$$\lim_{h \to 0} \frac{\Pr\{K(x+h) - K(x) = 1 | K(x) = 0\}}{h} = \beta_1$$

$$\lim_{h \to 0} \frac{\Pr\{K(x+h) - K(x) = 1 | K(x) = 1\}}{h} = 2\beta_1 \qquad (10.67)$$

$$\vdots$$

where (10.67) implies that given the index case has produced one infectious case, and at time x both cases are infectious, the instantaneous rate of a new infectious contact attributed to the index case is due to the contribution of the index case itself and of the other infectious individual. $\{K(x)\}$ is a Yule process which could be manifested by real epidemiologic data, if the index case resides in a highly clustered and connected environment and simultaneously gives exposure (not necessarily transmission) to a large number of people. It is the environment that manifests a large number of infected individuals (i.e. *super-spreading events*) and data "directly link" them to the first case in the environment.

10.7.2 Parallels to the Preferential Attachment Model in Random Graph Theory

In random graph theory, there is a concept of "preferential attachment" [20] of the random graphs over the space of vertices. The linear pure birth process is an analogue to such a concept over time. There is an ongoing debate whether preferential attachment actually happen in growing networks. Liljeros et al. [21] are convinced of preferential attachment as a mechanism for sexual networks, as "people become more attractive the more partners they get." However, Jones and Handcock [48] are skeptical and argue that networks with infinitely large variances but dramatically different structures can manifest the same marginal degree distribution, whereas these different network structures produce different epidemic behaviour. This debate has a longer history. In 1919, Greenwood and Woods [49] put forward three hypotheses into the occurrence of accidents:

1. *Pure chance*, which gives rise to the Poisson process (10.65)
2. *True contagion*, i.e. initially all individuals have the same probability of incurring an accident, but this probability is modified by each accident sustained to give rise to the linear pure birth process
3. *Apparent contagion (proness)*, i.e. individuals have constant but unequal probabilities of having an accident and the resulting process being a mixed Poisson process (10.18)

Xekalaki [50] gives a comprehensive survey of the history of these hypotheses and related models. The arguments on occurrence of accidents can be equally applied to disease transmission: whether observed data can distinguish if the underlying stochastic mechanism is arising from a linear pure birth process or a mixed Poisson process with large variance. The two very different stochastic mechanisms produce quantitatively very different epidemic behaviour.

10.7.3 Distributions for N when $\{K(x)\}$ Arises as a Linear Pure Birth Process

Let us start with an infectious contact process $\{K(x)\}$ with a marginal distribution having the negative binomial form

$$\Pr\{K(x) = k\} = \frac{\Gamma(k + \frac{1}{\phi})}{\Gamma(k+1)\Gamma(\frac{1}{\phi})}\varsigma^{\frac{1}{\phi}}(1 - \varsigma)^k \qquad (10.68)$$

which a parameter $0 < \varsigma < 1$.

If $\{K(x)\}$ arises as a linear pure birth process given by (10.64), from page 274 of Bhattachrya and Waymire [51], the marginal distribution for $K(x)$ is

$$\Pr\{K(x) = k\} = \frac{\Gamma(k + \frac{1}{\phi})}{\Gamma(k + 1)\Gamma(\frac{1}{\phi})} \left(e^{-\beta\phi x}\right)^{\frac{1}{\phi}} \left(1 - e^{-\beta\phi x}\right)^k. \tag{10.69}$$

In this case, $\varsigma = e^{-\beta\phi x}$.

For comparison, a mixed Poisson process in (10.33) has marginal distribution

$$\Pr\{K(x) = k\} = \frac{\Gamma(k + \frac{1}{\phi})}{\Gamma(k+1)\Gamma(\frac{1}{\phi})} \left(\frac{1}{1 + \beta\phi x}\right)^{\frac{1}{\phi}} \left(\frac{\beta\phi x}{1 + \beta\phi x}\right)^k \tag{10.70}$$

where $K(x)$ has p.g.f. $G_K(s, x) = \int_0^\infty e^{\xi x(s-1)} dU(\xi)$ and ξ follows a gamma distribution. In this case, $\varsigma = \frac{1}{1+\beta\phi x}$.

The link functions $\varsigma = \frac{1}{1+\beta\phi x}$ and $\varsigma = e^{-\beta\phi x}$ are two hypotheses for whether $\{K(x)\}$ has stationary increment or not. However, they are often not identifiable through data.

10.7.3.1 Degenerate Distribution of the Infectious Period $T_I = \mu$

An infectious contact process $\{K(x)\}$ with marginal distribution (10.68) can be further stopped by a randomly distributed infectious period to generate distributions for N. If all infectious individuals have constant infectious period μ, then N also has a negative binomial distribution:

$$\Pr\{N = j\} = \frac{\Gamma(j + \frac{1}{\phi})}{\Gamma(j + 1)\Gamma(\frac{1}{\phi})} \varsigma^{\frac{1}{\phi}} (1 - \varsigma)^j.$$

If data are available to suggest a distribution for N which is negative binomial, there may be two further assumptions, either $\varsigma = \frac{1}{1+\beta\phi\mu}$ or $\varsigma = e^{-\beta\phi\mu}$, associated with two completely different stochastic mechanisms for how $\{K(x)\}$ arises. In addition, we have also seen a third way to get the negative distribution for N. It corresponds to $G(s, x) = e^{-\beta x(1-s)}$ with gamma distributed infectious period and $E[T_I] = \mu$. Therefore, there are three stochastic mechanisms that give rise to the identical distribution for N. Observed data can not identify these underlying models.

10.7.3.2 Exponential Distribution of the Infectious Period with $E[T_I] = \mu$

If T_I follows an exponential distribution with p.d.f. $F_I(x) = \frac{1}{\mu}e^{-\frac{x}{\mu}}$ and if the infectious contacts follow a linear pure birth process with $\Pr\{K(x) = k\}$ given by (10.69), the resulting distribution for N is

$$\int_0^\infty \frac{\Gamma(j + \frac{1}{\phi})}{\Gamma(j+1)\Gamma(\frac{1}{\phi})} \left(e^{-\beta\phi x}\right)^{\frac{1}{\phi}} \left(1 - e^{-\beta\phi x}\right)^j \left(\frac{1}{\mu}e^{-\frac{x}{\mu}}\right) dx. \qquad (10.71)$$

Since $\varsigma = e^{-\beta\phi x}$, (10.71) becomes

$$\begin{aligned}
\Pr\{N = j\} &= \frac{\Gamma(j + \frac{1}{\phi})}{\Gamma(j+1)\Gamma(\frac{1}{\phi})} \int_0^1 \varsigma^{\frac{1}{\phi}} (1 - \varsigma)^j \left(\theta\varsigma^{\theta-1}\right) d\varsigma \\
&= \theta \frac{\Gamma(j+\alpha)\Gamma(\alpha+\theta)}{\Gamma(\alpha)\Gamma(j+\alpha+\theta+1)}, \quad \theta > 0, \ \alpha = \frac{1}{\phi} > 0. \qquad (10.72)
\end{aligned}$$

where $\theta = \frac{1}{\beta\phi\mu} > 0$. It can be regarded as a mixing distribution in $\frac{\Gamma(k+\frac{1}{\phi})}{\Gamma(k+1)\Gamma(\frac{1}{\phi})} \int \varsigma^{\frac{1}{\phi}} (1-\varsigma)^k u(\varsigma) d\varsigma$ with $u(\varsigma) = \frac{1}{\beta\phi\mu}\varsigma^{\frac{1}{\phi\beta\mu}-1} = \theta\varsigma^{\theta-1}, 0 < \varsigma < 1$. We have seen it before as (10.36), with the power-law property: $\propto \frac{1}{j^{\theta+1}}$.

Notice that the same Waring distribution can also arise if $\{K(x)\}$ has marginal distribution given by (10.70) combined with a Pareto infections period distribution. Therefore, there are multiple stochastic mechanisms that generate the distribution for N with power-law tail behaviour.

10.7.3.3 A General Observation for Models with Non-stationary Increment Process $\{K(x)\}$ with Random Mixing

If $\{K(x)\}$ arises from a non-stationary increment process described by the linear pure birth process (10.64), the marginal distribution for $K(x)$ is given by a negative-binomial form (10.68) with the link $\varsigma = e^{-\phi\beta x}$. Any further assumption on randomness in $\phi\beta x$, such as a random infectious period, will result in a mixed negative-binomial distribution with some mixing distribution $U(\varsigma)$. It has been shown by Karlis and Xekalaki [52] that $\Pr\{N = j\} \propto j^{-\theta} \sum_{l=0}^j \binom{j}{l} (-1)^{j-l} E\left(\varsigma^{j-\theta}\right)$ which is a finite series of non-integral moments of the mixing distribution $E\left(\varsigma^{j-\theta}\right)$, with a predominant power-law factor. It is heavy-tailed and power-law, if

$$\lim_{j\to\infty} \frac{\sum_{l=0}^{j+1} \binom{j+1}{l} (-1)^{j-l+1} E\left(\varsigma^{j-\theta+1}\right)}{\sum_{l=0}^j \binom{j}{l} (-1)^{j-l} E\left(\varsigma^{j-\theta}\right)} = 1.$$

Acknowledgments

I thank my colleagues at Public Health Agency of Canada who have taught me public health and infectious disease epidemiology. I thank Professors F. Brauer, P. van den Driessche and J. Wu for providing the opportunity to write this chapter. I thank Professor Jianhong Wu from York University and Dr. Fan Zhang at Public Health Agency of Canada for their proof reading previous versions with new suggestions. I thank Dr. John Edmunds from Health Protection Agency, United Kingdom and Professor Zhi-ming Ma from the Institute of Applied Mathematics, Chinese Academy of Science, Beijing, China, for their crucial discussions on some specific aspects of the use of the linear pure birth process (10.69).

References

1. Brauer, F. Compartmental models in epidemiology. Chapter 2 of *Mathematical Epidemiology* (this volume)
2. Allen, L. An introduction to stochastic epidemic models. Chapter 3 of *Mathematical Epidemiology* (this volume)
3. Brauer, F. An introduction to networks in epidemic modeling. Chapter 4 of *Mathematical Epidemiology* (this volume)
4. Dietz, K. Some problems in the theory of infectious diseases transmission and control. In *Epidemic Models: Their Structure and Relation to Data*, ed. Denis Mollison. Cambridge University Press, Cambridge (1995) 3–16
5. Diekmann, O. and Heesterbeek, J.A.P. Mathematical epidemiology of infectious diseases: model building, analysis and interpretation. *Wiley Series in Mathematical and Computational Biology* (2000)
6. Greenwood, M. and Yule, G. An inquiry into the nature of frequency distributions representative of multiple happenings with particular reference to occurence of multiple attacks of diseases or of repeated accidents. *Journal of the Royal Statistical Society* **A83** (1920) 255–279
7. Isham, V. Stochastic models for epidemics. Chapter 1 of *Celebrating Statistics: Papers in Honour of Sir David Cox on his 80th Birthday*, ed. A.C. Davison, Y. Dodge and N. Wermuth. Oxford University Press, Oxford (2005) 27–54
8. Bailey, N.T.J. *The Mathematical Theory of Infectious Diseases and its Applications*. Second Edition. The Griffin & Company Ltd. (1975)
9. Barttlet, M.S. *Stochastic Population Models in Ecology and Epidemiology*. Methuen, London (1961)
10. Mode, C. and Sleeman, C. *Stochastic Processes in Epidemiology, HIV/AIDS, Other Infectious Diseases and Computers*. World Scientific, Singapore (2000)
11. Becker, N.G. and Starczak, D.N. The effect of random vaccine response on the vaccination coverage required to prevent epidemics. *Mathematical Biosciences* **154** (1998) 117–135
12. von Bahr, B. and Martin-Löf, A. Threshold limit theorems for some epidemic processes. *Advances in Applied Probability* **12** (1980) 319–349
13. Lefèvre, C. and Utev, S. Poisson approximation for the final state of a generalized epidemic process. *The Annals of Probability* **23, No. 3** (1995) 1139–1162
14. Kermack, W.O. and McKendrick, A.G. Contributions to the mathematical theory of epidemics, part I. *Proceedings of the Royal Society London* **A115** (1927) 700–721. Reprinted (with parts II and III) as *Bulletin of Mathematical Biology* **53** (1991) 33–55

15. Karlis, D. and Xekalaki, E. Mixed Poisson distributions. *International Statistical Review* **73, No. 1** (2005) 35–58
16. Lynch, J. Mixtures, generalized convexity and balayages. *Scandinavian Journal of Statistics* **15** (1988) 203–210
17. Johnson, N., Kotz, S. and Kemp, A.W. *Univariate Discrete Distributions*. Second Edition. Wiley, New York (1993)
18. Newman, M. Spread of epidemic disease on networks. *Physical Review* **E66** (2002) 016128
19. Meyers, L., Pourbohloul, B. Newman, M. Skowronski, D. and Brunham, R. Network theory and SARS: predicting outbreak diverity. *Journal of Theoretical Biology* **232** (2005) 71–81
20. Barabási, A.L. and Albert R. Emergence of scaling in random networks. *Science* **286** (1999) 509–512
21. Liljeros, F., Edling, C.R., and Amaral, L.A.N. Sexual networks: implications for the transmission of sexually transmitted infections. *Microbes and Infection* **3** (2003) 189–196
22. Romillard, B. and Theodorescu, R. Inference based on empirical probability generating function for mixtures of Poisson distributions. *Statistics and Decisions* **18** (2000) 266–349
23. Irwin, J.O. A unified derivation of some well-known frequency distributions of interest in biometry and statistics. *Journal of Royal Statistical Society (A)* **118** (1955) 389–404
24. Irwin, J.O. The generalized Waring distribution applied to accident theory. *Journal of Royal Statistical Society (A)* **131** (1968) 205–225
25. Malthus, T.R. *An Essay on the Principle of Population as It Affects the Future Improvement of Society, with Remarks on the Speculations of Mr. Godwin, M. Condorcet and Other Writers*. J. Johnson, London (1798)
26. Euler, L. *Introducto in Analysin Infinitorum*. Tomus primus. MarcumMichaelem Bousquet and Socios., *Lausanne* **E101** (1748)
27. Kendall, D. Deterministic and stochastic epidemics in closed populations. *Proceedings of the Fifth Berkeley Symposium on Mathematics and Statistical Probability*, vol. 4. University of California Press, California (1956) 149–165
28. Näsell, I. The threshold concept in stochastic and endemic models. In *Epidemic Models: their Structure and Relation to Data*, ed. Denis Mollison, Cambridge University Press, Cambridge (1995) 71–83
29. Metz, J.A. The epidemic in a closed population with all susceptibles equally vulnerable; some results from large susceptible populations with small initial infections. *Acta Biotheroetica* **27** (1978) 75–123
30. Martin-Löf, A. The final size of a nearly critical epidemic, and the first passage time of a Wienner process to a parabolic barrier. *Journal of Applied Probability* **35** (1988) 671–682
31. Feller, W. *An Introduction to Probability Theory and Its Applications. Volume II.* (1943) Wiley, New York
32. Karlin, S. and Taylor, H.M. *A First Course in Stochastic Processes*, Second Edition. Academic, San Diego (1975)
33. Durrett, R. *Random Graph Dynamics*. Cambridge University Press, Cambridge (2006)
34. Yan, P. Separate roles of the latent and infectious periods in shaping the relation between the basic reproduction number and the intrinsic growth rate of infectious disease outbreaks. *Journal of Theoretical Biology* **251** (2008) 238–252
35. Anderson, D. and Watson, R. On the spread of a disease with gamma distributed latent and infectious periods. *Biometrika* **67 No. 1** (1980) 191–198
36. Wearing, H.J., Rohani, P. and Keeling, M.J. Appropriate models from the management of infectious diseases. *PLoS Medicine* **7** (2005) 621–627
37. Begon, M., Harper, J.L. and Townsend, C.R. *Ecology*. Blackwell, Oxford, UK (1996)
38. Anderson, R. and May, R. *Infectious Diseases of Humans, Dynamics and Control*. Oxford University Press, Oxford (1991)

39. Lipsitch, M., Cohen, T., Cooper, B., et al. Transmission Dynamics and Control of Severe Acute Respiratory Syndrome. *Science* **300** (2003) 1966–1970

40. Chowell, G., Nishiura, H. and Bettencourt, L.M. Comparative estimation of the reproduction number for pandemic influenza from daily case notification data. *Journal of the Royal Society Interface* **4** (2007) 155–166

41. Ludwig, D. Final size distributions for epidemics. *Mathematical Biosciences* **23** (1975) 33–46

42. Scalia-Tomba, G. Asymptotic final size distribution for some chain binomial processes. *Advances in Applied Probability* **17** (1985) 477–495

43. Martin-Löf, A. Symmetric sampling procedures, general epidemic processes and their threshold limit theorems. *Journal of Applied Probability* **23** (1986) 265–282

44. Lefèvre, C. and Picard, P. Collective epidemic processes: a general modelling approach to the final outcome of SIR infectious diseases. In: *Epidemic Models: Their Structure and Relation to Data*, ed. Denis Mollison, Cambridge University Press, Cambridge (1995) 53–70

45. Abby, H. An examination of the Reed Frost theory of epidemics. *Human Biology* **24** (1952) 201–233

46. Ma, J. and Earn, D. Generality of the final size formula for an epidemic of a newly invading infectious disease. *Bulletin of Mathematical Biology* **68** (2006) 679–702

47. Erdös, P. and Rényi, A. On the evolution of random graphs, *Bulletin of Institute of International Statistics* **38** (1961) 343–347

48. Jones, J.H. and Handcock, M.S. An assessment of preferential attachment as a mechanism for human sexual network information. *Proceedings: Biological Sciences, The Royal Society* **270** (2003) 1123–1128

49. Greenwood, M. and Woods, H.M. On the incidence of industrial accidents upon individuals with special reference to multiple accidents. *Report of the Industrial Research Board, No. 4*. His Majesty's Stationery Office, London (1919)

50. Xekalaki, E.A. The univariate generalized Waring distribution in relation to accident theory: proness, spells or contagion? *Biometrics* **39** (1983) 887–895

51. Bhattacharya, R.N. and Waymire, E.C. *Stochastic Processes with Applications*. Wiley, New York (1990)

52. Karlis, D. and Xekalaki, E. The polygonal distribution. *International Conference on Mathematical and Statistical Modelling in Honour of Enrique Castillo* (2006)

Part III
Case Studies

Chapter 11
The Role of Mathematical Models in Explaining Recurrent Outbreaks of Infectious Childhood Diseases

Chris T. Bauch

Abstract Infectious childhood diseases such as measles are characterized by recurrent outbreaks. Mathematicians have long used models in an effort to better understand and predict these recurrent outbreak patterns. This paper summarizes and comments upon those efforts, providing a historical outline of childhood disease models that have been developed since the start of the twentieth century. This paper also discusses the influence of data analysis techniques, such as spectral analysis, on the understanding and modelling of childhood disease dynamics.

11.1 Introduction

Childhood diseases (such as measles, whooping cough, chickenpox, polio, mumps and rubella) have had major and longstanding impacts on public health. Although vaccination programmes have significantly reduced morbidity and mortality from many childhood diseases [1], and are close to eradicating polio altogether [2], to date only smallpox has been completely eradicated through vaccination [3,4]. Moreover, measles continues to kill hundreds of thousands of undernourished and unvaccinated children in lesser-developed countries each year [5]. In the face of persistent vaccine scares that often cause a resurgence in childhood diseases [6–8], an understanding of childhood disease transmission and control remains essential.

The mathematical modelling of childhood diseases continues to be an area of active research. This is due to the significant morbidity and mortality associated with childhood diseases, their relatively simple epidemiology, and the

Department of Mathematics and Statistics, University of Guelph, Guelph, ON, Canada
N1G 2W1
cbauch@uoguelph.ca

widespread availability of time series data describing childhood disease *incidence* (number of new cases per unit time). Time series of childhood disease incidence are the most detailed and complete ecological time series available for any organism, and therefore also provide a way of testing ecological theory [9, 10].

Outbreak patterns for childhood diseases are always recurrent, such that peaks in disease incidence alternate with troughs. However, there is much variation in outbreak patterns across diseases, places and times. Some time series exhibit highly regular annual or biennial outbreaks, some exhibit much longer interepidemic intervals, while others exhibit apparently irregular outbreaks (Fig. 11.1a–d) [11, 12]. Outbreak patterns for a given disease can also change over time, either gradually or through abrupt transitions (Fig. 11.1a–d) [11, 12]. One of the goals of mathematical epidemiology during the twentieth century has been to determine the causes of recurrent outbreaks of childhood diseases and predict the *interepidemic interval*, defined as the time between successive outbreaks [13].

Immunity to childhood diseases is generally lifelong, or long-term (however in the case of whooping cough there may be significant asymptomatic transmission among adults [16]). Susceptible individuals gradually accumulate in the population through births, until there is a sufficient density of susceptibles for an epidemic to start. Epidemiological timescales are fast compared to demographic timescales, hence the epidemic rapidly depletes the susceptible pool and the epidemic eventually dies out. Susceptibles are gradually built up again through new births and the pattern recurs. If this accumulation of susceptibles occurs more slowly, for instance through a reduced birth rate or higher rates of vaccine uptake, then the interepidemic interval increases accordingly.

It is important to emphasize that this 'dissection' of an epidemic cycle is ultimately a description of recurrence, not an explanation (even though it is often presented as such). One could equally well imagine a scenario in which the influx of new susceptibles exactly balances the incidence of disease at all times, so that the susceptible pool is depleted at the same constant rate as it is refilled. This would imply a constant number of cases over time instead of recurrent outbreaks. Without modelling, it is not clear a priori why incidence should exhibit recurrence instead of approaching a stable equilibrium. Modelling is also essential to quantitatively predict the interepidemic interval, and the time of transition between outbreak patterns. Modelling is indispensable not only for understanding recurrence as a natural phenomenon but also for informing public health policy, since it allows one to predict how changes in vaccination coverage and demographic parameters alter the time between outbreaks. Indeed, age-structured measles models show good agreement with pre- and post-vaccination age-stratified case reports and seroprevalence surveys [17], and have been successfully used in this way to predict and plan for upcoming measles epidemics in New Zealand, for example [18].

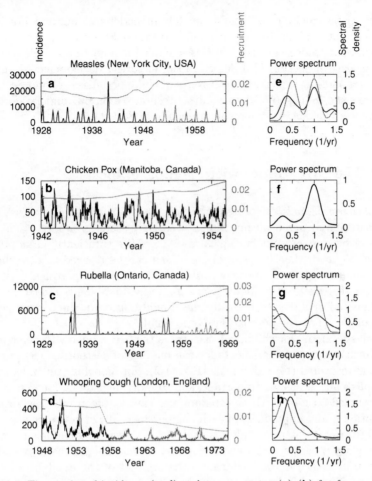

Fig. 11.1 Time series of incidence (**a–d**) and power spectra (**e**)–(**h**) for four common childhood diseases. In each panel, the green line shows annual susceptible recruitment, $\nu(1-p)$, where ν denotes births normalized by 1955 population size and p denotes proportion vaccinated (cf. [11]); recruitment is shown displaced forward in time by the mean age at infection to account for the typical delay between birth and infection (4 years for measles, 7 years for chickenpox, 11 years for rubella and 4 years for whooping cough). Time series are divided into sections based on substantial differences in recruitment rates; the corresponding power spectra are not sensitive to the precise point at which the time series is divided (the chickenpox time series has not been divided because no dramatic change in births occurred during the period covered by the data). The power spectrum is the Fourier transform of the autocovariance function of the time series [14, 15]. Before computation of the power spectrum the data were trend-corrected and tapered with a double cosine bell [14]. The autocovariance function was smoothed with a Tukey window [14] to reduce variance in the power spectrum, which facilitates locating the spectral peaks. The width of the Tukey window was chosen so that the resulting bandwidth was the same (0.35) for both weekly and monthly incidence time series. Figure reproduced from [12]

This paper reviews the role of models in understanding and predicting recurrent outbreak patterns of childhood diseases. Section 11.2 describes the SIR model with demographics (a classic epidemic model applicable to childhood diseases). Section 11.3 describes the historical development of mathematical models used to study recurrence in childhood diseases, and Sect. 11.4 describes the influence of data analysis techniques on our understanding of recurrence. Concluding comments are made in Sect. 11.5.

11.2 The SIR Model with Demographics

Compartmental models allocate individuals into mutually exclusive categories (compartments) based on infection status, age, social group, or other categories of interest or epidemiological relevance. If a sufficiently large population is assumed, then differential equations can be derived to describe the time evolution of the number of individuals in each compartment. Alternatively, a stochastic (e.g. Monte Carlo or individual-based) model can also be simulated or analyzed to study the time evolution of the compartments, particularly when the population size is small.

A classic model for childhood diseases is the SIR model with demographics, which assigns compartments for the number of susceptible (S), infected (I), and recovered (R) individuals (Fig. 11.2). For modelling an endemic disease, birth and death are important, and so individuals in each compartment are assumed to give birth at constant per capita rate ν (all newborns are susceptible), and individuals in each compartment die at constant per capita rate μ (disease-related mortality is neglected). Infected individuals recover at a constant per capita rate γ and retain lifelong immunity. Finally, we assume that the incidence is proportional to the product of the number of susceptible and infected individuals (new cases arise through *mass-action* mixing between susceptible and infected individuals). These assumptions yield the system of ordinary differential equations

$$\dot{S} = \nu N - \beta IS - \mu S, \tag{11.1}$$

$$\dot{I} = \beta IS - \gamma I - \mu I, \tag{11.2}$$

$$\dot{R} = \gamma I - \mu R, \tag{11.3}$$

where $N = S + I + R$ is the population size, β is the mean rate at which an infected individual transmits the infection to a susceptible, and other parameters are as defined above. Note that $1/\gamma$ is the mean duration of infectiousness. It is often assumed that $\nu = \mu$, such that $S + I + R = N$ and the population size is constant. Then, the population size N can be normalized to 1 for simplicity of notation (hence S, I and R become densities). Under these conditions, for $\beta/(\gamma+\mu) \leq 1$, the equations exhibit a disease-free equilibrium $(1, 0, 0)$ that is globally asymptotically stable. Conversely, when $\beta/(\gamma+\mu) > 1$, the disease-free equilibrium is unstable and there is a globally asymptotically stable endemic equilibrium $(\frac{\gamma+\mu}{\beta}, \frac{\mu(\beta-\gamma-\mu)}{\beta(\gamma+\mu)}, \frac{\gamma(\beta-\gamma-\mu)}{\beta(\gamma+\mu)})$. Although this model ignores many aspects of real-world demography and epidemiology, it is pedagogically useful nonetheless, and we will see how it quantitatively captures certain features of real epidemics despite its simplifications.

The SEIR model with demographics is an extension of the SIR model that is also applied to childhood diseases. The 'E' compartment is a latent stage where individuals are infected but not yet infectious. Individuals leave the latent compartment 'E' and enter the infectious compartment 'I' at some constant rate, the inverse of which equals the mean latent period. Latency is common in childhood diseases.

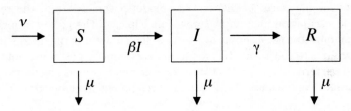

Fig. 11.2 Diagram of SIR model with demographics (11.1)–(11.3). Terms next to *arrows* denote per capita rates of transitions into or out of compartments

The *basic reproductive number*, $\mathcal{R}_0$, is the average number of secondary infections produced by a typical infected individual in a fully susceptible population [19,20]. When $\mathcal{R}_0 > 1$, each infected case more than replaces itself, and incidence grows geometrically, causing an epidemic. However, stochastic effects can cause the disease to fade out in the early stages, even when $\mathcal{R}_0 > 1$ (since the number of infected individuals is initially very small). For childhood diseases, $\mathcal{R}_0$ varies widely depending upon the disease, location, and time. In most Western countries during the twentieth century, $\mathcal{R}_0$ was typically highest for measles and whooping cough ($\mathcal{R}_0 \approx 11 - 18$), in the middle range for chickenpox and mumps ($\mathcal{R}_0 \approx 7 - 14$), and lowest for rubella ($\mathcal{R}_0 \approx 6 - 9$) [19]. For the SIR model with demographics (11.1)–(11.3), it can be shown that $\mathcal{R}_0 = \beta/(\gamma + \mu)$, which is the inverse of the proportion susceptible at the endemic equilibrium.

An accurate model of recurrent outbreaks would be expected to exhibit sustained oscillations (e.g. stable limit cycles). However, stability analysis of the endemic equilibrium of the SIR model shows that it is a stable spiral, therefore solutions converge to it via damped oscillations with period

$$T \approx 2\pi \sqrt{\frac{1}{\gamma\mu(\mathcal{R}_0 - 1)}}. \tag{11.4}$$

Therefore, the simple SIR model alone cannot account for the recurrent outbreaks observed in incidence time series.

However, a number of extensions to the basic SIR model have been found to generate sustained oscillations. These extensions fall into two categories. In first category of *endogenous* mechanisms, the period of oscillation is not explicitly incorporated as a model parameter, but rather it is implicit in how the model is defined and its parameter values set. Endogenous mechanisms usually yield sustained oscillations by destabilizing the endemic equilibrium of the basic model (11.1)–(11.3), giving rise to stable limit cycles with a period of approximately T [19, 21–37]. In contrast, in the second category of *exogenous* mechanisms, oscillations are produced through periodic (seasonal) forcing of model parameters such as the transmission rate [11, 12, 21, 25, 28, 37–40]. The period of forcing is set explicitly as a model parameter, and the resulting periods of oscillation are simply integer multiples of the period of forcing.

Throughout the twentieth century, modellers have experimented with various extensions to the basic SIR model in an attempt to better account for the observed patterns of recurrent outbreaks, using endogenous and/or exogenous mechanisms. We outline this historical development in the next section.

11.3 Historical Development of Compartmental Models

The sources for this historical review include both primary and secondary sources [19, 41–45]. The epidemic modelling literature has become vast, especially in the past thirty years, and modellers have approached the problem of recurrence from many different angles. Indeed, this review itself constitutes just one angle from which the literature can be approached, and is not meant to be comprehensive.

11.3.1 Early Models

Daniel Bernoulli developed what was probably the first compartmental epidemic model in 1760 [46–48]. His model divided the population into susceptible and immune compartments and assumed an age-specific force of infection and case fatality rate, yielding a system of equations with an endemic equilibrium of susceptible and immune individuals. Bernoulli was not specifically interested in the cause of recurrent outbreaks or in predicting the interepidemic interval. Rather, his motivation was to predict the expected gain in life expectancy that would be brought about by applying smallpox control measures. Since variolation was becoming widespread in Europe in the late 1700s, predicting the resulting increase in life expectancy would have been important for pricing annuities [48].

For the next hundred years, apparently little work was done in epidemiological modelling. However in the mid-nineteenth century to the early twentieth century, the subject was broached again by a number of authors. During this time, papers were written on mathematical and statistical models for various types of infectious diseases [21, 49–57] but here we focus specifically on mathematical (including stochastic) models for childhood diseases.

In the nineteenth century, researchers were already seeking the causes of recurrent outbreaks. At this time it was already suspected that the density of susceptibles was an important quantity. As Hirsch claimed in 1883, "the recurrence of the epidemics of measles at one particular place is connected neither with an unknown something (the mystical number of the Pythagoreans), nor with 'general constitutional vicissitudes', as Köstlin thinks; but it depends solely on two factors, the time of importation of the morbid poison, and the number of persons susceptible of it" [58] (quoted from [21]).

However it was probably not until the early 1900s that researchers began using mechanistic models to explain and predict recurrent outbreaks in childhood diseases. At this time, there were at least two competing hypotheses regarding the causes of recurrence. Scientists such as Brownlee hypothesized that seasonal recurrence in diseases such as measles was simply due to seasonal variation in pathogen virulence [51]. By comparison, scientists such as Hamer and Davidson sought an endogenous explanation for recurrence. They suggested that it is unnecessary to invoke seasonal variation in host or pathogen properties, rather, a 'mechanical theory of numbers and density' would suffice [50]. More specifically, Hamer hypothesized in 1906 that incidence is proportional to the product of the densities of susceptible and infected individuals. This is now called the *mass-action mixing* assumption, and is a cornerstone of epidemic modelling.

Hamer geometrically analyzed an average measles epidemic curve to support this hypothesis (Fig. 11.3). He also used this line of analysis to show that an epidemic can die out before all susceptibles have been depleted. Ultimately Hamer's hypothesis was found to be lacking [21], for the same reasons that the SIR model with demographics only exhibits damped oscillations (namely, mass-action mixing alone cannot result in sustained oscillations in models). However, in the context of what was known at the time, and without the benefit of developments in nonlinear mathematics, Hamer's hypothesis is arguably both subtle and original.

Hamer developed a discrete time model, but it was formulated numerically (without the aid of symbolic notation) and hence it was not readily recognizable as such. Epidemic models were established in a more recognizable, modern form through the seminal work of Kermack and McKendrick in the late 1920s [55–57] (reprinted in [59]). They showed that the density of susceptible individuals must exceed a critical threshold in order for an epidemic to

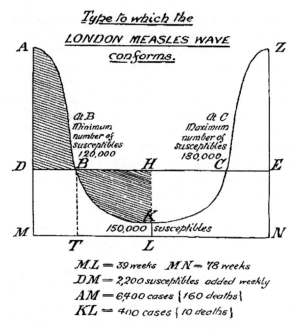

Fig. 11.3 Diagram representing an average measles epidemic curve in London, analyzed in Hamer's 1906 paper [50]

occur. (This result was also discovered by Ross in the early 1900s in the context of discrete-time malaria models [53]. It is also closely related to Hamer's finding that the susceptible pool is not fully depleted by the end of a natural epidemic curve.)

From general assumptions, Kermack and McKendrick derived an integro-differential equation to describe the spread of an epidemic in a closed population,

$$\dot{S}(t) = S(t) \int_0^\infty \overline{A}(\tau) \frac{d}{dt} S(t - \tau) d\tau \,, \tag{11.5}$$

where $\overline{A}(\tau)$ is the expected infectivity of an individual who became infected a time τ ago, and $S(t)$ is the (spatial) density of susceptible individuals [43,55]. For the special case that $\overline{A}(\tau) = \beta \exp(-\gamma\tau)$, and by defining $I(t) \equiv -\frac{1}{\beta} \int_0^\infty \overline{A}(\tau)\dot{S}(t - \tau)d\tau = -\frac{1}{\beta} \int_{-\infty}^t \overline{A}(t - \tau)\dot{S}(\tau)d\tau$ and differentiating, one recovers the SIR model without demographics ($\nu = \mu = 0$ in (11.1)–(11.3) [43].

Soon thereafter (in 1929) Soper, who felt he was 'merely following up the trail blazed by Sir William Hamer more than twenty years ago' [21], formulated and analyzed a discrete-time compartmental model to determine whether mass-action mixing was sufficient to generate sustained oscillations [21]. His model did generate sustained oscillations, however it was necessary to assume that all of an individual's infectiousness is concentrated at a single point in time at the end of the incubation period. In reality, of course, infectivity is distributed over some interval of time. Soper found that for a model with distributed infectivity, oscillations are damped and converged to an endemic equilibrium. As noted in Sect. 11.2, this is also true for the equivalent ordinary differential equation compartmental model (the SIR model with demographics), where the duration of infectiousness is exponentially distributed (Fig. 11.4, top panel).

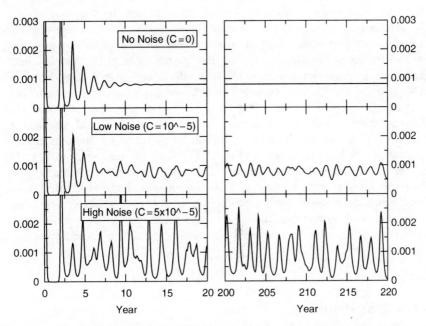

Fig. 11.4 The role of demographic stochasticity in sustaining oscillations. Panels from *top* to *bottom* show the effects of gradually increasing the amount of demographic stochasticity on time series of the density of infecteds, $I(t)$. A Gaussian (white) noise term was added to the birth and death parameters μ and ν in (11.1)–(11.3): the value C in the figure legends is the value of the coefficient of the normalized Gaussian noise term. Other parameters are $\beta = 1{,}240 \text{ year}^{-1}$, $1/\gamma = 15 \text{ days}$, $1/\mu = 1/\nu = 50 \text{ years}$

Although Soper's model failed to produce sustained oscillations except under highly restrictive assumptions, his research stimulated further research on the problem [45]. One of the goals of mathematical epidemiology in the following decades was to correct for this apparent deficiency in Hamer–Soper models by introducing various extensions that would result in sustained oscillations. We review these extensions in the following sections.

11.3.2 Stochasticity

Epidemic modelling started growing substantially in the 1950s [44]. During this time, the challenge of explaining recurrence was taken up by Bartlett, who took an entirely different (stochastic) approach to the problem [22, 60]. Bartlett hypothesized that Hamer–Soper models failed to produce sustained outbreaks because they did not incorporate *demographic stochasticity* (stochastic effects associated with the discrete nature of births, deaths, immigration and emigration processes). These effects are present to some extent in all populations, but are especially significant for smaller population sizes.

Bartlett, using a stochastic model of measles transmission, showed that demographic stochasticity could sustain oscillations that would otherwise be damped. (This can also be seen by adding noise terms to the deterministic SIR model with demographics (Fig. 11.4).) Bartlett also showed that the interepidemic interval of the oscillations in silico agreed approximately with the empirically observed values of the interepidemic interval [22, 61]. It was found more recently (in 1998) that *environmental stochasticity* (stochasticity associated with environmental effects that affect the entire population at once, such as climatic fluctuations) also results in sustained oscillations in a model for pertussis [33].

11.3.3 Seasonality

Seasonality is a prominent feature in many incidence time series, with incidence for some diseases peaking consistently in the winter months (Fig. 11.1 a–d). Measles and chickenpox in particular are strongly seasonal. A number of driving forces have been hypothesized to be behind seasonal recurrence [62]. For instance, the seasonal structure of the school calendar means that transmission rates may be higher in Fall and Winter when school is in session (and children are crowded together) but lower during the summer holiday [21, 63]. Likewise, seasonal variation in host physiology, pathogen virulence, and humidity/rainfall patterns may also play a role [62]. The relative importance of these factors is still unestablished, although evidence indicates that a seasonally-structured school calendar is at least partly responsible [63, 64].

Soper suspected seasonal variation in transmission rates might be important for explaining biennial measles outbreaks, and analyzed a variant of his discrete-time model with a seasonally-varying transmission rate [21].

However, more intensive research on seasonally-forced epidemic models did not begin until the 1970s [25, 38]. Modellers often incorporate seasonality by making the transmission rate β a sinusoidal function of time, e.g. $\beta(t) = \beta_0(1 + \alpha \cos(2\pi t))$ [40]. Some modellers take β to be a step function based explicitly on real school calendars [11, 12, 25]. A seasonally-varying transmission rate yields oscillations at periods that are integer multiples of the period of forcing [11, 12, 40].

11.3.4 Age Structure

Susceptibility, infectivity, morbidity, and mortality rates often vary with age, and available data for these quantities are often broken down into age classes. Age-structured models have been the focus of many studies, and modellers have developed both continuous age-structured (PDE) and discrete age-structured (ODE) models, as well as stochastic and deterministic versions [19,26,29,30,32,44]. Moreover, other features such as seasonality [25,28], stochasticity [65], and strain structure [27] have also been incorporated.

Basic age-structured compartmental models of the type appropriate for endemic childhood diseases, such as the continuous age-structured MSEIR model with demographics (where the 'M' compartment is for maternally-derived immunity) [44], or the continuous age-structured SIR model with demographics [19, 29], either do not exhibit sustained oscillations at all (the endemic equilibrium, if it exists, being stable at all parameter values) [19,29], or exhibit sustained oscillations only under highly restrictive assumptions [30, 44]. Hence, age structure alone is apparently not sufficient to explain recurrence. However, oscillations in age-structured models are typically very weakly damped [19, 27, 29], and so age structure at least facilitates recurrence. For oscillations to be sustained in an age-structured model, additional features must be incorporated, such as seasonality [25, 28], stochasticity [65] or strain structure [27].

11.3.5 Alternative Assumptions About Incidence Terms

Conventionally, it is assumed that new cases are generated through homogeneous mixing, yielding the mass-action incidence term βIS (11.1), (11.2), or the *standard-incidence* term $\beta IS/N$. The assumption of homogeneous mixing may be inaccurate, particularly under certain circumstances. Examples where the incidence does not depend linearly on the number of currently infected individuals include situations where a larger density of infected individuals decreases their per capita infectivity (saturation effects), and situations where multiple exposures to an infected individual are required for a transmission event to occur (threshold effects). Some researchers have shown that abandoning the conventional forms for the incidence term in the model equations can induce sustained oscillations [66–70].

For example, it has been argued that a more realistic term for incidence in many situations would be $\beta I^p S^q / N$ [69, 70]. The case $p > 1$ corresponds to synergistic effects among pathogens, and may occur when viral concentration in the environment must exceed some critical threshold for transmission to occur (if the viral lifespan outside the host is short), or, for vector-borne diseases, when the disease vector needs to attack multiple infected hosts to attain sufficiently high viraemia for transmission to susceptible hosts to occur. When $p > 1$, compartmental models can exhibit sustained oscillations at certain parameter values [69, 70].

11.3.6 Distribution of Latent and Infectious Period

It is conventionally assumed that the latent and infectious periods (i.e. the duration of latency and infectiousness respectively) are distributed exponentially, such that the rate at which a latent individual becomes infectious, or the rate at which an infectious individuals recovers, is constant and independent of how long the individual has been in that compartment. Although this assumption yields relatively tractable ODEs, it is less realistic than other assumptions such as fixed or normally distributed durations.

If a fixed duration of infectiousness (incorporated through a time delay) is used instead, the endemic equilibrium of the basic SIR model can be destabilized through a Hopf bifurcation, yielding stable limit cycles [23, 24, 31, 34]. The resulting interepidemic interval is approximately the same as that predicted by local stability analysis of endemic equilibria of the corresponding models with an exponentially distributed periods (11.4). Results for more general assumptions on the distribution of latent and infectious periods have also been obtained [35, 36].

11.3.7 Seasonality Versus Nonseasonality

The mechanisms described in the previous sections are not mutually exclusive, and it is most likely the case that multiple factors contribute to recurrence. It is difficult to determine the relative importance of these factors, however the fact that the time series exhibit two distinct types of outbreak patterns suggests the existence of at least two distinct mechanisms: some time series are characteristically seasonal (e.g. measles, chickenpox), whereas others are characteristically nonseasonal (e.g. whooping cough).

Models with seasonally-varying transmission rates can predict the interepidemic interval for diseases that are characteristically seasonal [11,12,38,64], but they perform poorly for childhood diseases that are characteristically nonseasonal. For instance, seasonally-forced models predict that whooping cough should have an interepidemic interval of 1 year instead of the observed ≈4 years [12,71,72]. Conversely, the period of damped oscillations T computed from the basic model (11.4) or other models that use only endogenous mechanisms predicts the observed interepidemic interval for diseases that are characteristically nonseasonal [15,19,22,61], but cannot predict seasonal outbreaks. Furthermore, for some diseases, the error of the latter models is enormous: for chickenpox, which is highly seasonal, the observed interepidemic interval in almost all cities is 1 year (e.g. Fig. 11.1b), but models without seasonal forcing predict $T \approx 3$ [12,19].

Using a seasonally-forced SEIR model, it has been demonstrated numerically that solutions for measles remain close to the attractor (a biennial cycle) in the presence of demographic stochasticity, whereas solutions for whooping cough stray far from the attractor (an annual cycle), resulting in a ≈4 year interepidemic interval [12,71,72]. Hence, the relative importance of deterministic (seasonal) versus stochastic (nonseasonal) effects for these two diseases may explain the observed differences between their outbreak patterns. Indeed, a perturbative analysis of the attractors of a seasonally-forced model allows one to predict both seasonal and nonseasonal oscillations in measles, whooping cough, chickenpox and rubella [12]. Additionally, Monte Carlo simulations indicate some of the factors which determine the relative importance of seasonal and nonseasonal effects for a given disease [12]. Seasonality versus nonseasonality in incidence time series will be discussed more fully in the section on spectral analysis (Sect. 11.4).

11.3.8 Chaos

Mathematicians have long sought to determine whether *chaos* (sensitive dependence upon initial conditions) exists in natural and laboratory populations [73]. Because time series of childhood disease incidence are relatively long and accurate (in comparison to many ecological time series), researchers have devoted significant attention to seeking chaos in childhood disease dynamics, particularly for measles [74–82]. The presence or absence of chaos in disease dynamics does not bear directly on the question of what causes recurrent outbreaks, however, the issue has much to do with characterizing outbreak patterns as well as with forecasting future outbreaks [80,82].

Some researchers have attempted to detect the hallmarks of chaos in time series by computing Lyapunov exponents [75,81], deriving return maps from reconstructed phase space trajectories [76], computing correlation dimensions

[75], or applying nonlinear forecasting techniques [80, 82]. Other researchers have constructed epidemic models, pointing out parallels between model dynamics in chaotic parameter regimes and real-world time series. However, existing methods require data sets that are even more extensive than those currently available [75, 81], hence the evidence in support of chaos in measles time series remains equivocal.

11.3.9 Transitions Between Outbreak Patterns

As noted in Sect. 11.1, incidence time series often exhibit sudden or gradual transitions between outbreak patterns for a given disease in a given population. Measles in New York City, USA, for example, exhibited a dramatic shift from an apparently irregular pattern to a regular biennial pattern in the mid 1940s (Fig. 11.1a). Some researchers have suggested that these transitions, particularly for measles, are caused by endogenous mechanisms such as chaos [83] or noise-driven shifts between basins of coexisting attractors [40]. Other researchers have emphasized the impact of exogenous changes in parameters controlling birth rates and vaccine uptake (which move the system to a new set of attractors) and report quantitative agreement between incidence time series and seasonally-forced SEIR models [11, 12].

11.4 Spectral Analysis of Incidence Time Series

The focus of this review so far has been on using models to understand recurrence. In comparison, the present Section focuses on how data analysis can help researchers understand recurrence. Techniques of data analysis are a 'lens' through which a system can be observed. They can sharpen or clarify certain aspects of the data, and thus change our understanding of the system, and how the system is modelled.

An example of how data analysis can clarify understanding comes from considering the interepidemic interval, which is a convenient but potentially misleading concept. Many time series do not have a well-defined time between outbreaks, because the magnitude and timing of outbreaks can vary considerably. For instance, in Liverpool, England, 1950–1967, before the era of mass vaccination (Fig. 11.5), close inspection reveals a complex pattern where small epidemics sometimes follow one another in rapid succession, on the order of months, and significant year-to-year variation in the magnitude of larger epidemics. The 'time between outbreaks' depends upon how large a peak needs to be before it is considered an outbreak. The best that can be said is that the time between outbreaks is, very roughly, 1–2 years. Likewise, for measles in New York City from 1944 to 1962, there exists a regular biennial pattern where small outbreaks alternate with large outbreaks (Fig. 11.1a). For this time series, the interepidemic interval could be equally well argued to be either one year or two years. Similar ambiguities can be seen for other diseases (Fig. 11.1b–d). Hence, the interepidemic interval can be an imprecise and arbitrary concept.

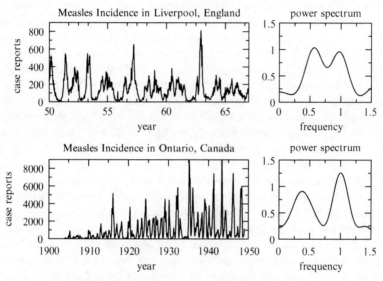

Fig. 11.5 Incidence of measles in Liverpool, England 1950–1967 (*top left panel*) and corresponding power spectrum (*top right panel*), and measles case reports in Ontario, Canada 1904–1948 (*bottom left panel*) and corresponding power spectrum (*bottom right panel*). See Fig. 11.1 caption for details on computation of power spectra

11.4.1 Power Spectra

Some of this ambiguity can be eliminated by applying *spectral analysis* to the time series [14,84]. For instance, starting in the 1980s, epidemic modellers began computing *power spectra* of incidence time series [12, 15, 19, 37, 75]. The power spectrum of a time series describes the relative contributions from each frequency of oscillation to the time series, over some fixed time interval. A single dominant peak at a frequency of $1/Y$ years^{-1} in a power spectrum would therefore correspond to recurrent outbreaks every Y years. Power spectra can be computed, for example, by de-trending the incidence time series, computing the Fourier transform of the autocorrelation function of the detrended time series, and then smoothing with a suitable spectral window such as a Tukey window or a Bartlett window [12, 15, 19]. Figures 11.1, 11.5 and 11.6 show power spectra of childhood disease incidence time series.

Power spectra can reveal much about a time series that is not obvious upon casual inspection. For instance, power spectra of time series of childhood disease incidence reveal that spectral peaks can be either *seasonal* (corresponding to seasonality in the time series, and occurring at frequencies $1/n$ years^{-1}, where n is an integer) or *non-seasonal* (corresponding to nonseasonality in the time series and occurring at any real-valued frequency) [12]. For New York City 1944–1962, spectral peaks appear at frequencies of 1 year^{-1} and $1/2$ years^{-1}, corresponding to the biennium observed in the time series for those years (Fig. 11.1a,e). Power spectra of childhood disease time series generally exhibit both seasonal and nonseasonal peaks, although their relative magnitude may differ significantly (Figs. 11.1, 11.5 and 11.6) [12]. For the apparently irregular time series of measles in Liverpool 1950–1967 (Fig. 11.5) there is both a seasonal peak at a frequency of 1 year^{-1} and a nonseasonal peak occurring at a frequency of $1/1.8$ years^{-1}. For measles in Ontario 1904–1948 (Fig. 11.5) there is both a seasonal peak at 1 year^{-1} and a nonseasonal peak at $1/2.7$ years^{-1}. Smaller populations exhibit larger nonseasonal spectral peaks, perhaps because stochastic effects are more important (see Sect. 11.3.7). This can be seen in chickenpox power spectra for five Canadian provinces with differing population sizes for the years 1942–1955 (Fig. 11.6).

Seasonal peaks are associated with exogenous (seasonal) effects, such as seasonally-varying transmission rates, whereas nonseasonal peaks are associated with endogenous effects, such as stochasticity and other mechanisms discussed in Sect. 11.3. Power spectra provide a way of characterizing outbreak patterns that is less ambiguous than the interepidemic interval, since one can simply identify the frequencies and relative magnitudes of seasonal and nonseasonal peaks in the power spectrum. However, power spectra do not contain phase information, and therefore on their own they are insufficient to completely specify a time series. There is also a danger in choosing a bandwidth for the smoothing window that is too narrow, yielding spurious spectral peaks.

Power spectral analysis illustrates how data analysis can provide new ways of describing and characterizing incidence time series. This, in turn, may lead to a better understanding of disease dynamics and may also influence modelling. For instance, the fact that spectral peaks can be classified into seasonal and non-seasonal types, and the fact that many spectra simultaneously exhibit both types of peaks, suggest that recurrence is not a monolithic phenomenon. Rather, there seem to be two distinct types of recurrence, each with its associated causes, and each of which is present to some extent in all childhood disease time series. Therefore it is important to incorporate multiple mechanisms into epidemic models, in cases where a 'true-to-life'

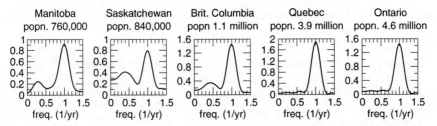

Fig. 11.6 Power spectra of chickenpox incidence time series in five Canadian provinces for the years 1942–1955, showing larger nonseasonal peaks for smaller population sizes. See Fig. 11.1 caption for details of computation of power spectra

representation of recurrence is desirable. Many epidemic models have focused separately on mechanisms such as seasonality [21], age-structure [29], or stochasticity [22]. In the past few decades, modellers have begun studying models with multiple mechanisms, such as seasonality and age structure [25, 28], stochasticity and seasonality [12, 71, 72], or age structure and strain structure [27].

11.4.2 Wavelet Power Spectra

As demographic and epidemiological parameters evolve over time, the frequencies exhibited in the time series can also change. However, the method of computing power spectra as described above assumes that frequencies are stationary over the time interval for which the power spectrum is computed, and thus cannot capture these trends.

A method of spectral analysis that circumvents this problem is *wavelet analysis* [84]. Wavelet analysis also decomposes a time series into component frequencies, however because the method uses wavelets (which are localized in time), it provides a temporally localized description of component frequencies, i.e. it can describe the time evolution of component frequencies.

Wavelet analysis can be applied to childhood disease time series [85], and can reveal interesting temporal trends. For instance, in the wavelet power spectrum for New York City 1928–1966 (Fig. 11.7) we observe a sudden shift from a non-seasonal period of approximately 2.5 years to a seasonal period of 2 years, a transition that was apparently induced by the baby-boom (Fig. 11.1a) [11]. The suddenness of this transition is consistent with bifurcation analysis of a seasonally-forced SEIR model, which for these parameters predicts a shift from an annual attractor to a biennial attractor [11, 12]. Likewise, in London, England 1944–1994, upon the initiation of mass vaccination, we observe a transition from a biennial pattern to a pattern of gradual increase in the non-seasonal period, as vaccine coverage increases over time

(Fig. 11.7). Such obvious patterns do not appear in all wavelet power spectra for childhood diseases, and they can sometimes be difficult to interpret (e.g. for measles in Ontario, Fig. 11.7). Nonetheless, wavelet analysis can refine our understanding of incidence time series and the underlying disease dynamics.

11.5 Conclusions

This paper has reviewed how compartmental epidemic models have been used to understand and predict recurrent outbreaks of childhood diseases. The classic SIR model with demographics was presented, and its shortcomings with respect to explaining recurrence were discussed. The historical development of childhood disease models was then reviewed, with particular emphasis on efforts to correct this apparent deficiency of the classic SIR model, by developing and analyzing models that exhibit sustained oscillations through various mechanisms. The paper concluded with a discussion of the impact of spectral analysis techniques on our understanding of childhood disease dynamics.

Childhood disease modelling has received sustained interest over many decades. Generally speaking, these models, and the methods by which they are analyzed, have become increasingly diverse and sophisticated. Perhaps one reason for this sustained interested and continuing development has been a series of partial successes: models often successfully explain certain features of childhood disease dynamics, but their shortcomings leave still other questions unanswered. In an effort to arrive at a more complete understanding, researchers have developed new models, characterized their properties more completely, and sought closer agreement with data. Today there remain unresolved issues and new questions that continue to attract the attention of mathematicians. One example of an area of continuing research is the influence of spatial structure upon childhood disease dynamics and control [10, 85, 86].

Part of this continuing development is also due to the emergence of new concepts and new techniques of analysis, which often originate in other fields and are then applied to childhood diseases. As new concepts emerge, modellers begin to ask new questions about familiar epidemiological systems, such as: 'Are measles dynamics chaotic?'. Modellers have also applied new methods in time series analysis to childhood disease time series with positive results, as seen in the example of spectral analysis. These new techniques can sharpen understanding of the dynamics behind the time series, and can change the way modellers think about, and discuss, childhood disease dynamics.

These factors, together with the intransigence of childhood diseases, their worldwide health impact, and the relative availability of childhood disease incidence data, suggest that modelling of infectious childhood diseases will

NYC Measles Wavelet Power Spectrum

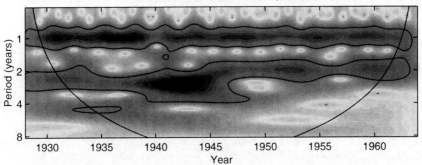

London Measles Wavelet Power Spectrum

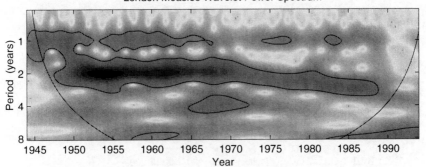

Ontario Measles Wavelet Power Spectrum

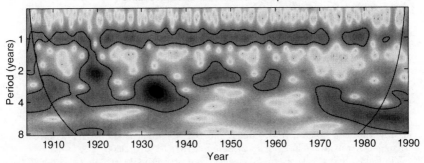

Fig. 11.7 Wavelet power spectra [84, 85] for measles in New York City, USA (*top panel*), London, England (*middle panel*) and Ontario, Canada (*bottom panel*). A complex Morlet wavelet was used. Before computation of the spectra, each time series was log-transformed, its variance normalized to unity, its mean set to zero, and its edges padded with zeros to reduce edge effects [84]. Parabolic lines demarcate the cone of influence (the region of the spectrum in which edge effects are significant). Contours demarcate the 95% confidence level, hence regions within the contours denote significant peak frequencies. As of time of publication, Matlab scripts for wavelet analysis can be downloaded from http://paos.colorado.edu/research/wavelets/

continue to be an area of active research for the foreseeable future, and will continue to present new challenges to biomathematicians.

Acknowledgments

CTB is supported by the Natural Sciences and Engineering Research Council of Canada (NSERC). The spectral analyses that appear in this paper were carried out during the author's tenure as a post-doctoral researcher at McMaster University, 2000–2003, under supervision of Prof. David Earn.

References

1. P. Bonanni. Demographic impact of vaccination: a review. *Vaccine*, 17:S120–S125, 1998
2. CDC. Progress toward Poliomyelitis eradication–Nigeria, January 2003–March 2004. *MMWR Weekly*, 53(16):343–346, 2004
3. F. Fenner, D.A. Henderson, I. Arita, Z. Jezek, and I.D. Ladnyi. *Smallpox and Its Eradication*. World Health Organization, Geneva, 1988
4. F.E. Andre. Vaccinology: past achievements, present roadblocks and future promises. *Vaccine*, 21:593–595, 2003
5. World Health Organization. Fact sheet # 286, March 2005
6. E.J. Gangarosa, A.M. Galazka, C.R. Wolfe, L.M. Phillips, R.E. Gangarosa, E. Miller, and R.T. Chen. Impact of anti-vaccine movements on pertussis control: the untold story. *Lancet*, 351:356–361, 1998
7. J.P. Baker. The pertussis controversy in Great Britain, 1974–1986. *Vaccine*, 21: 4003–4010, 2003
8. V.A. Jansen, N. Stollenwerk, H.J. Jensen, M.E. Ramsay, W.J. Edmunds, and C.J. Rhodes. Measles outbreaks in a population with declining vaccine uptake. *Science*, 301:804, 2003
9. B.T. Grenfell and J. Harwood. (Meta)-Population dynamics of infectious diseases. *Trends Ecol. Evolut.*, 12:395–399, 1997
10. D.J.D. Earn, P. Rohani, and B.T. Grenfell. Persistence, chaos and synchrony in ecology and epidemiology. *Proc. R. Soc. Lond. B*, 265:7–10, 1998
11. D.J.D. Earn, P. Rohani, B.M. Bolker, and B.T. Grenfell. A simple model for complex dynamical transitions in epidemics. *Science*, 287:667–670, 2000
12. C.T. Bauch and D.J.D. Earn. Transients and attractors in epidemics. *Proc. R. Soc. Lond. B*, 270:1573–1578, 2003
13. P.E.M. Fine. The interval between successive cases of an infectious disease. *Am. J. Epidemiol.*, 158:1039–1047, 2003
14. M.B. Priestley. *Spectral Analysis and Time Series*. Academic, London, 1981
15. R.M. Anderson, B.T. Grenfell, and R.M. May. Oscillatory fluctuations in the incidence of infectious disease and the impact of vaccination: Time series analysis. *J. Hyg. Camb.*, 93:587–608, 1984
16. S. Hodder and E. Mortimer. Epidemiology of pertussis and reactions to pertussis vaccine. *Epidem. Rev.*, 14:243–267, 1992
17. H.R. Babad, D.J. Nokes, N.J. Gay, E. Miller, P. Morgan-Capner, and R.M. Anderson. Predicting the impact of measles vaccination in England and Wales: model validation and analysis of policy options. *Epidemiol. Infect.*, 114:319–344, 1995

18. New Zealand Ministry of Health. Modelling measles. predicting and preventing measles epidemics in New Zealand: application of a mathematical model, 1998
19. R.M. Anderson and R.M. May. *Infectious Diseases of Humans*. Oxford University Press, Oxford, 1991
20. O. Diekmann and J.A.P. Heesterbeek. *Mathematical epidemiology of infectious diseases*. Wiley, New York, 2000
21. H.E. Soper. The interpretation of periodicity in disease prevalence. *J. R. Stat. Soc.*, 92:34–73, 1929
22. M.S. Bartlett. Deterministic and stochastic models for recurrent epidemics. *Proceedings of the Third Berkeley Symposium on Mathematical Statistics and Probability*, volume 4, pp. 81–108, 1956
23. D. Green. Self-oscillation for epidemic models. *Math. Biosci.*, 38:91–111, 1978
24. H.L. Smith. Periodic solutions for a class of epidemic equations. *Math. Anal. Appl.*, 64:467–79, 1978
25. D. Schenzle. An age-structure model of pre- and post-vaccination measles transmission. *IMA J. Math. Appl. Med. Biol.*, 1:169–191, 1984
26. H.W. Hethcote. Optimal ages of vaccination for measles. *Math. Biosci.*, 89:29–52, 1988
27. C. Castillo-Chavez, H.W. Hethcote, V. Andreaen, S.A. Levin, and W.M. Liu. Epidemiological models with age structure, proportionate mixing, and cross-immunity. *J. Math. Biol.*, 27:233–258, 1989
28. M.J. Keeling and B.T. Grenfell. Disease extinction and community size: modeling the persistence of measles. *Science*, 275:65–67, 1997
29. D. Greenhalgh. Analytic results on the stability of age-structured epidemic models. *IMA J. Math. Appl. Med. Biol.*, 4:109–144, 1987
30. H.R. Thieme. Persistence under relaxed point-dissipativity (with an application to an endemic model). *SIAM J. Math. Anal.*, 24:407–435, 1993
31. H.W. Hethcote and P. van den Driessche. An SIS epidemic model with variable population size and a delay. *J. Math. Biol.*, 34:177–194, 1996
32. H.W. Hethcote. An age-structured model for pertussis transmission. *Math. Biosci.*, 145:89–136, 1997
33. H.W. Hethcote. Oscillations in an endemic model for pertussis. *Can. Appl. Math. Q.*, 6:61–88, 1998
34. H.W. Hethcote and P. van den Driessche. Two SIS epidemiologic models with delays. *J. Math. Biol.*, 40:3–26, 2000
35. F. Brauer. Models for the spread of universally fatal diseases. *J. Math. Biol.*, 28:451–462, 1990
36. Z. Feng and H.R. Thieme. Endemic models with arbitrarily distributed periods of infection II: fast disease dynamics and permanent recovery. *SIAM J. Appl. Math.*, 61:983–1012, 2000
37. B.T. Grenfell, O.N. Bjornstad, and B.F. Finkenstadt. Dynamics of measles epidemics: scaling noise, determinism, and predictability with the TSIR model. *Ecol. Monogr.*, 72:185–202, 2002
38. W. London and J.A. Yorke. Recurrent outbreaks of measles, chickenpox and mumps. I. Seasonal variation in contact rates. *Am. J. Epidem.*, 98(6):469–482, 1973
39. J.A. Yorke, N. Nathanson, G. Pianigiani, and J. Martin. Seasonality and the requirements for perpetuation and eradication of viruses in populations. *Am. J. Epidem.*, 109:103–123, 1979
40. I.B. Schwartz and H.L. Smith. Infinite subharmonic bifurcation in an SEIR epidemic model. *J. Math. Biol.*, 18:233–253, 1983
41. K. Dietz and D. Schenzle. *Mathematical Models for Infectious Disease Statistics*, pages 167–204. Springer, Berlin Heidelebrg New York, 1985
42. H.W. Hethcote and S.A. Levin. Periodicity in epidemic models. In S.A. Levin, T.G. Hallam, and L.J. Gross, editors, *Biomathematics*, volume 18, pages 193–211. Springer, Berlin Heidelberg New York, 1989

43. O. Diekmann, H. Metz, and H. Heesterbeek. The legacy of Kermack and McKendrick. In D. Mollison, editor, *Epidemic models: their structure and relation to data*, pages 95–118. Cambridge University Press, Cambridge, 1995

44. H.W. Hethcote. The mathematics of infectious diseases. *SIAM Rev.*, 42:599–653, 2000

45. F. Brauer and P. van den Driessche. Some directions for mathematical epidemiology. *Fields Inst. Commun.*, 36:95–112, 2003

46. D. Bernoulli. Essai d'une nouvelle analyse de la mortalite causee par la petite verole. *Mem. Math. Phys. Acad. R. Sci. Paris*, pages 1–45, 1766.

47. K. Dietz and J.A.P. Heesterbeek. Bernoulli was ahead of modern epidemiology. *Nature*, 408:513–514, 2000

48. K. Dietz and J.A.P. Heesterbeek. Daniel Bernoulli's epidemiological model revisited. *Math. Biosci.*, 180:1–21, 2002

49. W. Farr. Progress of epidemics. *Second report of the Registrar General of England*, pages 91–8, 1840

50. W.H. Hamer. Epidemic disease in England. *Lancet*, 1:733–739, 1906

51. J. Brownlee. Statistical studies in immunity: the theory of an epidemic. *Proc. R. Soc. Edn.*, 26:484–521, 1906

52. R. Ross. *Report on the Prevention of Malaria in Mauritius.* London, 1908

53. R. Ross. *The Prevention of Malaria* (2nd ed.). Murray, London, 1911

54. R. Ross and H.P. Hudson. An application of the theories of probably to the study of a priori pathometry, III. *Proc. R. Soc. A*, 93:225–240, 1917

55. W.O. Kermack and A.G. McKendrick. Contributions to the mathematical theory of epidemics I. *Proc. R. Soc. Lond.*, 115:700–721, 1927

56. W.O. Kermack and A.G. McKendrick. Contributions to the mathematical theory of epidemics II. *Proc. R. Soc. Lond.*, 138:55–83, 1932

57. W.O. Kermack and A.G. McKendrick. Contributions to the mathematical theory of epidemics III. *Proc. R. Soc. Lond.*, 141:94–112, 1933

58. A. Hirsch. *Handbook of Geographical and Historical Pathology*, Volume 1, translated by Charles Creighton, New Sydenham Soc., London, 1883

59. W.O. Kermack and A.G. McKendrick. Contributions to the mathematical theory of epidemics I–III. *Bull. Math. Biol.*, 53:33–118, 1991

60. M.S. Bartlett. The critical community size for measles in the United States. *J. R. Statist. Soc.*, 123:37–44, 1960

61. M.S. Bartlett. *Stochastic Population Models in Ecology and Epidemiology.* Methuen, London, 1960

62. S.F. Dowell. Seasonal variation in host susceptibility and cycles of certain infectious diseases. *Emerg. Infect. Dis.*, 7:369–374, 2001

63. P.E.M. Fine and J.A. Clarkson. Measles in England and Wales–I: An analysis of factors underlying seasonal patterns. *Int. J. Epidemiol.*, 11:5–14, 1982

64. B. Finkenstadt and B.T. Grenfell. Time series modelling of childhood infectious diseases: A dynamical systems approach. *J. R. Stat. Soc. C*, 49:187–205, 2000

65. N.M. Ferguson, D.J. Nokes, and R.M. Anderson. Dynamical complexity in age-structured models of the transmission of measles virus. *Math. Biosci.*, 138:101–130, 1996

66. E.B. Wilson and J. Worcester. The law of mass action in epidemiology. *Proc. Natl. Acad. Sci.*, 31:24–34, 1945.

67. E.B. Wilson and J. Worcester. The law of mass action in epidemiology II. *Proc. Natl. Acad. Sci.*, 31:109–116, 1945

68. H.W. Hethcote, H.W. Stech, and P. van den Driessche. Stability analysis for models of diseases without immunity. *J. Math. Biol.*, 13:185–198, 1981

69. W. Liu, S.A. Levin, and Y. Iwasa. Influence of nonlinear incidence rates upon the behavior of SIRS epidemiological models. *J. Math. Biol.*, 23:187–204, 1986

70. W. Liu, H.W. Hethcote, and S.A. Levin. Dynamical behavior of epidemiological models with nonlinear incidence rates. *J. Math. Biol.*, 25:359–380, 1987

71. P. Rohani, M.J. Keeling, and B.T. Grenfell. The interplay between determinism and stochasticity in childhood diseases. *Am. Nat.*, 159:469–481, 2002

72. M.J. Keeling, P. Rohani, and B.T. Grenfell. Seasonally forced disease dynamics explored as switching between attractors. *Physica D*, 148:317–335, 2002

73. R.M. May. Simple mathematical models with very complicated dynamics. *Nature*, 261:459–467, 1976

74. W.M. Schaffer and M. Kot. Nearly one dimensional dynamics in an epidemic. *J. Theor. Biol.*, 112:403–427, 1985

75. L.F. Olsen and W.M. Schaffer. Chaos versus noisy periodicity: Alternative hypothesis for childhood epidemics. *Science*, 249:499–504, 1990

76. M. Kot, D.J. Graser, G.L. Truty, W.M. Schaffer, and L.F. Olsen. Changing criteria for imposing order. *Ecol. Model.*, 43:75–110, 1988

77. W.M. Schaffer. Order and chaos in ecological systems. *Ecology*, 66:93–106, 1985

78. L.F. Olsen, G.L. Truty, and W.M. Schaffer. Oscillations and chaos in epidemics: A nonlinear dyamics study of six childhood diseases in Copenhagen, Denmark. *Theor. Pop. Biol.*, 33:344–370, 1988

79. D.A. Rand and H.B. Wilson. Chaotic stochasticity: a ubiquitous source of unpredictability in epidemics. *Proc. R. Soc. Lond. B*, 246:179–184, 1991

80. G. Sugihara, B.T. Grenfell, and R.M. May. Distinguishing error from chaos in ecological time series. *Philos. Trans. R. Soc. Lond. B*, 330:235–251, 1990

81. S. Ellner, A.R. Gallant, and J. Theiler. Detecting nonlinearity and chaos in epidemic data. In D. Mollison, editor, *Epidemic models: their structure and relation to data*, pages 229–247. Cambridge University Press, Cambridge, 1995

82. B.T. Grenfell, A. Kleczkowski, S.P. Ellner, and B.M. Bolker. Measles as a case study in nonlinear forecasting and chaos. *Philos. Trans. R. Soc. Lond. A*, 348:515–530, 1994.

83. W.M. Schaffer. Can nonlinear dynamics elucidate mechanisms in ecology and epidemiology? *IMA J. Math. Appl. Med. Biol.*, 2:221–252, 1985

84. C. Torrence and G.P. Compo. A practical guide to wavelet analysis. *Bull. Am. Meteor. Soc.*, 79:61–78, 1998

85. B.T. Grenfell, O.N. Bjornstad, and J. Kappey. Travelling waves and spatial hierarchies in measles epidemics. *Nature*, 414:716–723, 2001

86. P. Rohani, D.J.D. Earn, and B.T. Grenfell. Opposite patterns of synchrony in sympatric disease metapopulations. *Science*, 286:968–971, 1999

Chapter 12
Modeling Influenza: Pandemics and Seasonal Epidemics

Fred Brauer

Abstract We describe and analyze compartmental models for influenza, including pre-epidemic vaccination and antiviral treatment. The analysis is based on the final size relation for compartmental epidemic models. We consider models of increasing complexity and compare their predictions using parameter values appropriate to the 1957 pandemic.

12.1 Introduction

Influenza causes more morbidity and more mortality than all other respiratory diseases together. There are annual seasonal epidemics that cause about 500,000 deaths worldwide each year. During the twentieth century there were three influenza pandemics. The World Health Organization estimates that there were 40,000,000–50,000,000 deaths worldwide in the 1918 pandemic, 2,000,000 deaths worldwide in the 1957 pandemic, and 1,000,000 deaths worldwide in the 1968 pandemic. Current fears that the $H5N1$ strain of avian influenza could develop into a strain that can be transmitted readily from human to human and develop into another pandemic, together with a widely held belief that even if this does not occur there is likely to be an influenza pandemic in the near future, have aroused considerable interest in modeling both the spread of influenza and comparison of the results of possible management strategies.

Vaccines are available for annual seasonal epidemics. Influenza strains mutate rapidly, and each year a judgement is made of which strains of influenza are most likely to invade. A vaccine is distributed that protects against the

Department of Mathematics, University of British Columbia, 1984, Mathematics Road, Vancouver BC, Canada V6T 1Z2 brauer@math.ubc.ca

three strains considered most dangerous. However, if a strain radically different from previously known strains arrives, vaccine provides little or no protection and there is danger of a pandemic. As it would take at least 6 months to develop a vaccine to protect against such a new strain, it would not be possible to have a vaccine ready to protect against the initial onslaught of a new pandemic strain. Antiviral drugs are available to treat pandemic influenza, and they may have some preventive benefits as well, but such benefits are present only while antiviral treatment is continued.

Various kinds of models have been used to describe influenza outbreaks. Many public health policy decisions on coping with a possible influenza pandemic are based on construction of a contact network for a population and analysis of disease spread through this network. This analysis consists of multiple stochastic simulations requiring a substantial amount of computer time. We suggest that initially it may be more appropriate to use simpler models until enough data are acquired to facilitate parameter estimation. More complicated models require more parameters, and we believe that the complexity of a model to be used should be influenced by the amount and reliability of data. Our approach is to begin with a simple model and add more structure later.

We begin this chapter by developing a simple compartmental influenza transmission model and then augmenting it to include both pre-epidemic vaccination and treatment during an epidemic. We will then develop a compartmental model with more structure and compare its predictions with those of the simpler model. We will also describe some ways in which the model can be modified to be more realistic, though more complicated. The development follows the treatment in [2, 3]. We will describe the models and the results of their analyses, but omit proofs in order to focus attention on the applications of the models.

12.2 A Basic Influenza Model

Since influenza epidemics usually come and go in a time period of several months, we do not include demographic effects (births and natural deaths) in our model. Our starting point is the SIR epidemic model described in Chap. 2 of Mathematical Epidemiology (this volume). Two aspects of influenza that are easily added are that there is an incubation period between infection and the appearance of symptoms, and that a significant fraction of people who are infected never develop symptoms but go through an asymptomatic period, during which they have some infectivity, and then recover and go to the removed compartment [11]. Thus a model should contain the compartments S (susceptible), L (latent), I (infective), A (asymptomatic), and R (removed). Specifically, we make the following assumptions.

1. There is a small number I_0 of initial infectives in a population of total size K.
2. The number of contacts in unit time per individual is a constant multiple β of total population size N.
3. Latent members (L) are not infective.
4. A fraction p of latent members proceed to the infective compartment at rate κ, while the remainder goes directly to an asymptomatic infective compartment (A), also at rate κ.
5. Infectives (I) leave the infective compartment at rate α, with a fraction f recovering and going to the removed compartment (R) and the remainder dying of infection.
6. Asymptomatics have infectivity reduced by a factor δ, and go to the removed compartment at rate η.

These assumptions lead to the model

$$
\begin{aligned}
S' &= -S\beta(I + \delta A) \\
L' &= S\beta(I + \delta A) - \kappa L \\
I' &= p\kappa L - \alpha I \\
A' &= (1 - p)\kappa L - \eta A \\
R' &= f\alpha I + \eta A \\
N' &= -(1 - f)\alpha I,
\end{aligned}
\tag{12.1}
$$

with initial conditions

$$S(0) = S_0, \ L(0) = 0, \ I(0) = I_0, \ A(0) = 0, \ R(0) = 0, \quad N(0) = S_0 + I_0 = K.$$

In analyzing this model we may remove one variable since $N = S + L + I + A + R$. It is usually convenient to remove the variable R. Our language is ambiguous in that we use S, L, I, A, R, N to denote both the names of the classes and the number of members of the classes, but this should cause no confusion. It is possible to show that the model (12.1) is properly posed in the sense that all variables remain non-negative for $0 \leq t < \infty$. A flow diagram for the model (12.1) is shown in Fig. 12.1. The model (12.1) is the simplest possible description for influenza having the property that there are asymptomatic infections. The question that should be in the back of our minds is whether it is a sufficiently accurate description for its predictions to be useful.

The model (12.1), like the other models that we will introduce later, consists of a system of ordinary differential equations and the number of susceptibles in the population tends to a limit S_∞ as $t \to \infty$. There is a *final size relation* that we may use to find this limit without the need to solve the system of differential equations. If the contact rate β is constant, the final size relation is an equality. It is more realistic to assume saturation of contacts and that β is a function of the total population size N. In general, the final size relation is an inequality. If there are no disease deaths, N is constant and β is constant even with saturation of contacts. If the disease death rate is small, it appears that the final size relation is very close to an equality and it is reasonable to assume that β is constant and use the final size relation as an equation to solve for S_∞.

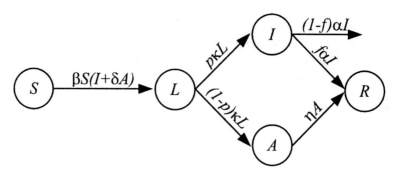

Fig. 12.1 Influenza model flow chart

It is easy to see that there are disease-free equilibria with

$$L = I = A = 0$$

and S arbitrary with $0 \le S \le N(0)$. Since S is a decreasing function, $S(t)$ approaches a limit $S_\infty \ge 0$ as $t \to \infty$. We may use the approach of [16], also described in Chap. 6 of Mathematical Epidemiology (this volume), to calculate the basic reproduction number

$$\mathcal{R}_0 = S_0 \beta \left[\frac{p}{\alpha} + \frac{\delta(1-p)}{\eta} \right]. \tag{12.2}$$

A biological interpretation of this basic reproduction number is that a latent member introduced into a population of S_0 susceptibles becomes infective with probability p, in which case he or she causes $\beta S_0/\alpha$ infections during an infective period of length $1/\alpha$, or becomes asymptomatic with probability $1 - p$, in which case he or she causes $\delta \beta S_0/\eta$ infections during an asymptomatic period of length η.

If $\mathcal{R}_0 > 1$, then the number of infectives first increases before decreasing to zero while if $\mathcal{R}_0 < 1$, then the number of infectives decreases monotonically to zero.

The *final size relation* is given by

$$\ln S_0 - \ln S_\infty = \mathcal{R}_0 \left[1 - \frac{S_\infty}{S_0} \right] + \frac{\beta I_0}{\alpha}. \tag{12.3}$$

Its derivation, similar to the derivation for a simpler case in Chap. 2 of Mathematical Epidemiology (this volume), is given in [2,3]. A very general form of the final size relation that is applicable to each of the models in this chapter is derived in [3]. The final size relation shows that $S_\infty > 0$. This means that some members of the population are not infected during the epidemic. The size of the epidemic, the number of (clinical) cases of influenza during the epidemic, is

$$I_0 + (S_0 - S_\infty).$$

The number of symptomatic cases is

$$I_0 + p(S_0 - S_\infty),$$

and the number of disease deaths is

$$(1 - f)[I_0 + p(S_0 - S_\infty)].$$

While mathematicians view the basic reproduction number as central in studying epidemiological models, epidemiologists may be more concerned with the *attack ratio*, as this may be measured directly. For influenza, where there are asymptomatic cases, there are two attack ratios. One is the clinical attack ratio, which is the fraction of the population that becomes infected, defined as

$$1 - \frac{S_\infty}{N(0)}.$$

There is also the symptomatic attack ratio, defined as the fraction of the population that develops disease symptoms, defined as

$$p\left[1 - \frac{S_\infty}{N(0)}\right].$$

The attack ratios and the basic reproduction number are connected through the final size relation (12.3). If we know the parameters of the model we can calculate $\mathcal{R}_0$ from (12.2) and then solve for S_∞ from (12.3).

We apply the model (12.1) using parameters appropriate for the 1957 influenza pandemic as suggested by Longini et al. [11]. The latent period is approximately 1.9 days and the infective period is approximately 4.1 days, so that

$$\kappa = \frac{1}{1.9} = 0.526, \quad \alpha = \eta = \frac{1}{4.1} = 0.244.$$

We also take

$$p = 2/3, \quad \delta = 0.5, \quad f = 0.98.$$

As in [11] we consider a population of 2,000 members, of whom 12 are infective. In [11] a symptomatic attack ratio was assumed for each of four age groups, and the average symptomatic attack ratio for the entire population was 0.326. This implies $S_\infty = 1,022$. Then, neglecting the small number of initial infectives for ease of calculation, we obtain

$$\mathcal{R}_0 = 1.373$$

from (12.3). Now, we use this in (12.2) to calculate

$$S_0\beta = 0.402.$$

We will use these data as baseline values to estimate the effect that control measures might have had. Our reason for neglecting I_0 in (12.3) is that we can then calculate $\mathcal{R}_0$ directly from (12.3) and $S_0\beta$ from (12.2). If we include I_0, the calculation is more complicated as we must first express (12.3) in terms of the model parameters and then solve for β before calculating $\mathcal{R}_0$ from (12.2).

The number of clinical cases is 978 (including the initial 12), the number of symptomatic cases is 656, again including the original 12, and the number of disease deaths is approximately 13.

The model (12.1) can be adapted to describe management strategies for both annual seasonal epidemics and pandemics.

12.3 Vaccination

To cope with annual seasonal influenza epidemics there is a program of vaccination before the "flu" season begins. Each year a vaccine is produced aimed at protecting against the three influenza strains considered most dangerous for the coming season. We formulate a model to add vaccination to the model described by (12.1) under the assumption that vaccination reduces susceptibility (the probability of infection if a contact with an infected member of the population is made). In addition we assume that vaccinated members who develop infection are less likely to transmit infection, more likely not to develop symptoms, and are likely to recover more rapidly than unvaccinated members.

These assumptions require us to introduce additional compartments into the model to follow treated members of the population through the stages of infection. We use the classes S, L, I, A, R as before and introduce S_T, the class of treated susceptibles, L_T, the class of treated latent members, I_T, the class of treated infectives, and A_T, the class of treated asymptomatics. In addition to the assumptions made in formulating the model (12.1) we also assume

1. A fraction γ of the population is vaccinated before a disease outbreak and vaccinated members have susceptibility to infection reduced by a factor σ_S.

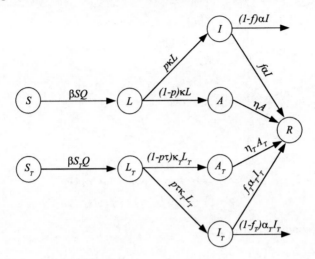

Fig. 12.2 Vaccination model flow chart

2. There are decreases σ_I and σ_A respectively in infectivity in I_T, and A_T; it is reasonable to assume

$$\sigma_I < 1, \quad \sigma_A < 1.$$

3. The rates of departure from L_T, I_T and A_T are κ_T, α_T and η_T respectively. It is reasonable to assume

$$\kappa \le \kappa_T, \quad \alpha \le \alpha_T, \quad \eta \le \eta_T.$$

4. The fractions of members recovering from disease when they leave I and I_T are f and f_T respectively. It is reasonable to assume $f \le f_T$.
5. Vaccination decreases the fraction of latent members who will develop symptoms by a factor τ, with $0 \le \tau \le 1$.

For convenience we introduce the notation

$$Q = I + \delta A + \sigma_I I_T + \delta \sigma_A A_T. \tag{12.4}$$

The resulting model is

$$S' = -\beta SQ$$
$$S'_T = -\beta \sigma_S S_T Q$$
$$L' = \beta SQ - \kappa L$$
$$L'_T = \beta \sigma_S S_T Q - \kappa_T L_T$$
$$I' = p\kappa L - \alpha I \qquad\qquad (12.5)$$
$$I'_T = \tau p \kappa_T L_T - \alpha_T I_T$$
$$A' = (1-p)\kappa L - \eta A$$
$$A'_T = (1 - p\tau)\kappa_T L_T - \eta_T A_T$$
$$R' = f\alpha I + f_T \alpha_T I_T + \eta A + \eta_T A_T$$
$$N' = -(1-f)\alpha I - (1 - f_T)\alpha_T I_T.$$

The initial conditions are

$$S(0) = (1-\gamma)S_0, \quad S_T(0) = \gamma S_0, \quad I(0) = I_0, \quad N(0) = S_0 + I_0$$
$$L(0) = L_T(0) = I_T(0) = A(0) = A_T(0) = 0.$$

corresponding to pre-epidemic treatment of a fraction γ of the population.

We again use N as a variable in the model rather than R for convenience, but $R = N - S - L - I - A - S_T - L_T - I_T - A_T$. A flow diagram for the model (12.5) is shown in Fig. 12.2.

Since the infection now is beginning in a population which is not fully susceptible, we speak of the *control reproduction number* $\mathcal{R}_c$ rather than the basic reproduction number. A computation using the approach of [16] or Chap. 6 of Mathematical Epidemiology (this volume) leads to the control reproduction number

$$\mathcal{R}_c = (1-\gamma)\mathcal{R}_u + \gamma \mathcal{R}_v,$$

with

$$\mathcal{R}_u = S_0 \beta \left[\frac{p}{\alpha} + \frac{\delta(1-p)}{\eta} \right] = \mathcal{R}_0 \qquad\qquad (12.6)$$
$$\mathcal{R}_v = \sigma_S S_0 \beta \left[\frac{p\tau\sigma_I}{\alpha_T} + \frac{\delta(1 - p\tau)\sigma_A}{\eta_T} \right].$$

Then $\mathcal{R}_u$ is a reproduction number for unvaccinated people and $\mathcal{R}_v$ is a reproduction number for vaccinated people. There is a pair of final size relations for the two variables S and S_T, namely

$$S_0[\ln(1-\gamma)S_0 - \ln S_\infty] = \frac{S_0\beta I_0}{\alpha} + \mathcal{R}_u[(1-\gamma)S_0 - S_\infty]$$
$$+ \mathcal{R}_v[\gamma S_0 - S_{T\infty}] \qquad\qquad (12.7)$$
$$S_{T\infty} = \gamma S_0 \left[\frac{S_\infty}{(1-\gamma)S_0} \right]^{\sigma_S}.$$

From (12.7) we calculate that the number of symptomatic disease cases is

$$I_0 + p[(1 - \gamma)S_0 - S_\infty] + p\tau[\gamma S_0 - S_{T\infty}],$$

and the number of disease deaths is

$$(1 - f)[I_0 + p[(1 - \gamma)S_0 - S_\infty] + (1 - f_T)p\tau[\gamma S_0 - S_{T\infty}].$$

By control of the epidemic we mean vaccinating enough people (i. e., taking γ large enough) to make $\mathcal{R}_c < 1$. We use the parameters of Sect. 12.2, with vaccination parameters as suggested in [11],

$$\sigma_S = 0.3, \quad \sigma_I = \sigma_A = 0.2, \quad \kappa_T = 0.526, \quad \alpha_T = \eta_T = 0.323, \quad \tau = 0.4.$$

The estimates of vaccine efficacy are based on those reported in [5]. With these parameter values,

$$\mathcal{R}_u = 1.373, \quad \mathcal{R}_v = 0.047.$$

In order to make $\mathcal{R}_c = 1$, we need to take $\gamma = 0.28$. This is the fraction of the population that needs to be vaccinated to head off an epidemic.

We may solve the pair of final size equations (12.7) with $S(0) = (1 - \gamma)$ $S_0, S_T(0) = \gamma S_0$ for $S_\infty, S_{T\infty}$ for different values of γ. We do this for the parameter values suggested above and we obtain the results shown in Table 12.1, giving the vaccination fraction γ, the number of unvaccinated susceptibles S_∞ at the end of the epidemic, the number of vaccinated susceptibles $S_{T\infty}$ at the end of the epidemic, the number of unvaccinated cases $I_0 + p((1 - \gamma)S_0 - S_\infty)$ and the number of vaccinated cases of influenza $(\gamma S_0 - S_{T\infty})$. The results indicate the benefits of pre-epidemic vaccination of even a small fraction of the population in reducing the number of influenza cases. They also demonstrate the advantage of vaccination to an individual. The attack ratio in the vaccinated portion of the population is much less than the attack ratio in the unvaccinated portion of the population.

Table 12.1 Effect of vaccination

Fraction vaccinated	S_∞	$S_{T\infty}$	Unvaccinated cases	Vaccinated cases
0	1,022	0	656	0
0.05	1,079	84	552	4
0.1	1,149	174	439	7
0.15	1,224	271	323	7
0.2	1,305	375	201	6
0.25	1,395	487	76	3
0.3	1,391	596	13	0

12.4 Antiviral Treatment

If no vaccine is available for a strain of influenza it would be possible to use an antiviral treatment. However, antiviral treatment affords protection only while the treatment is continued. In addition, antivirals are in short supply and expensive, and treatment of enough of the population to control an anticipated epidemic may not be feasible. A policy of treatment aimed particularly at people who have been infected or who have been in contact with infectives once a disease outbreak has begun may be a more appropriate approach. This requires a model with treatment rates for latent, infective, and asymptomatically infected members of the population that we construct building on the structure used for vaccination in (12.5).

Antiviral drugs have effects similar to vaccines in decreasing susceptibility to infection and decreasing infectivity, likelihood of developing symptoms, and length of infective period in case of infection. However, they are likely to be less effective than a well-matched vaccine, especially in the reduction of susceptibility.

Treatment may be given to diagnosed infectives. In addition, one may treat contacts of infectives who are thought to have been infected. This is modelled by treating latent members. In practice, some of those identified by contact tracing and treated would actually be susceptibles, but we neglect this in the model. Although we have allowed treatment of asymptomatics in the model, this is unlikely to be done, and we will describe the results of the model under the assumption $\varphi_A = \theta_A = 0$. However, for generality we retain the possibility of antiviral treatment of asymptomatics in the model. If treatment is given only to infectives, the compartments L_T, A_T are empty and may be omitted from the model.

We add to the model (12.5) antiviral treatment of latent, infective, and asymptomatically infected members of the population, but we do not assume an initial vaccinated class. In addition to the assumptions made earlier we also assume

1. There is a treatment rate φ_L in L and a rate θ_L of relapse from L_T to L, a treatment rate φ_I in I and a rate θ_I of relapse from I_T to I, and a treatment rate φ_A in A and a rate θ_A of relapse from A_T to A.

The resulting model is

$$S' = -\beta SQ$$
$$L' = \beta SQ - \kappa L - \varphi_L L + \theta_L L_T$$
$$L_T' = -\kappa_T L_T + \varphi_L L - \theta_L L_T$$
$$I' = p\kappa L - \alpha I - \varphi_I I + \theta_I I_T \qquad (12.8)$$
$$I_T' = p\tau\kappa_T L_T - \alpha_T I_T + \varphi_I I - \theta_I I_T$$
$$A' = (1-p)\kappa L - \eta A - \varphi_A A + \theta_A A_T$$
$$A_T' = (1-p\tau)\kappa_T L_T - \eta_T A_T + \varphi_A A - \theta_A A_T$$
$$N' = -(1-f)\alpha I - (1-f_T)\alpha_T I_T,$$

with Q as in (12.4). The initial conditions are

$$S(0) = S_0, I(0) = I_0, L(0) = L_T(0) = I_T(0) = A(0) = A_T(0) = 0, N(0) = S_0 + I_0.$$

A flow diagram for the model (12.8) is shown in Fig. 12.3.

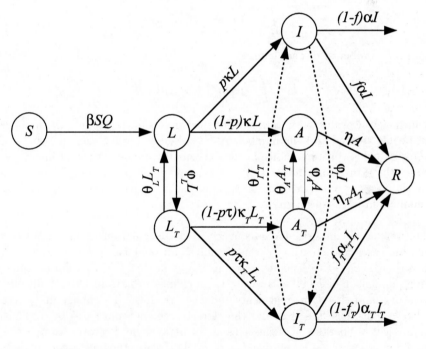

Fig. 12.3 Treatment model flow chart

The calculation of $\mathcal{R}_c$ for the antiviral treatment model (12.8) is more complicated than for models considered previously, but it is possible to show that $\mathcal{R}_c = \mathcal{R}_I + \mathcal{R}_A$ with

$$\mathcal{R}_I = \frac{S_0\beta}{\Delta_I\Delta_L}\left[(\alpha_T + \theta_I + \sigma_I\varphi_I)p\kappa(\kappa_T + \theta_L) + (\theta_I + \sigma_I(\alpha + \varphi_I))p\tau\kappa_T\phi_L\right]$$

$$\mathcal{R}_A = \frac{\delta S_0\beta}{\Delta_L}\left[\frac{(1-p)\kappa(\kappa_T + \theta_L)}{\eta} + \frac{\sigma_A(1-p\tau)\kappa_T\varphi_L}{\eta_T}\right], \tag{12.9}$$

where

$$\Delta_I = (\alpha + \varphi_I)(\alpha_T + \theta_I) - \theta_I\varphi_I$$
$$\Delta_L = (\kappa + \varphi_L)(\kappa_T + \theta_L) - \theta_L\varphi_L.$$

The final size equation is

$$\ln\frac{S_0}{S_\infty} = \mathcal{R}_c\left[1 - \frac{S_\infty}{S_0}\right] + \frac{\beta I_0(\alpha_T + \theta_I + \sigma_I\varphi_I)}{\Delta_I}. \tag{12.10}$$

The number of people treated is

$$\int_0^\infty [\varphi_L L(t) + \varphi_I I(t)]dt$$

and the number of disease cases is

$$I_0 + \int_0^\infty [p\kappa L(t) + p\tau\kappa_T L_T]dt$$

which can be evaluated in terms of the parameters of the model. The number of people treated and the number of disease cases are constant multiples of $S_0 - S_\infty$ plus constant multiples of the (small) number of initial infectives. Since the expressions in terms of $S_0 - S_\infty$ and I_0 are more complicated than in the cases of no treatment or vaccination, we have chosen to give these numbers in terms of integrals.

There is an important consequence of the calculation of the number of disease cases and treatments that is not at all obvious. If $\mathcal{R}_c$ is close to or less than 1, $S_0 - S_\infty$ depends very sensitively on changes in I_0. For example, in a population of 2,000 with $\mathcal{R}_0 = 1.5$, a change in I_0 from 1 to 2 multiplies $S_0 - S_\infty$ and therefore treatments and cases by 1.4 and a change in I_0 from 1 to 5 multiplies $S_0 - S_\infty$ and therefore treatments and cases by 3. Thus, numerical predictions in themselves are of little value. However, comparison of different strategies is valid, and the model indicates the importance of early action while the number of infectives is small.

In the special case that treatment is applied only to infectives, $\mathcal{R}_c$ is given by the simpler expression

$$\mathcal{R}_c = S_0\beta\left[\frac{p(\alpha_T + \theta_I + \sigma_I\varphi_I)}{\Delta_I} + \frac{\delta(1-p)}{\eta}\right].$$

The number of disease cases is $I_0 + p(S_0 - S_\infty)$, and the number of people treated is

$$\frac{\varphi_I(\alpha_T + \theta_I)}{\Delta_I}[I_0 + p(S_0 - S_\infty)].$$

Since the infective period is short, antiviral treatment would normally be applied as long as the patient remains infective. Thus we assume $\theta_I = 0$, and these relations become even simpler. The control reproduction number is

$$\mathcal{R}_c = S_0\beta\left[\frac{p(\alpha_T + \sigma_I\varphi_I)}{\alpha_T(\alpha + \varphi_I)} + \frac{\delta(1-p)}{\eta}\right], \tag{12.11}$$

and the number of people treated is

$$\frac{\varphi_I}{\alpha + \varphi_I}[I_0 + p(S_0 - S_\infty)]. \tag{12.12}$$

Since the basic reproduction number of a future pandemic can not be known in advance, it is necessary to take a range of contact rates in order to make predictions. In particular, we could compare the effectiveness in controlling the number of infections or the number of disease deaths of different strategies such as treating only infectives, treating only latent members, or treating a combination of both infective and latent members. In making such comparisons, it is important to take into account that treatment of latent members must be supplied for a longer period than for infectives. In case of a pandemic, there are also questions of whether the supply of antiviral drugs will be sufficient to carry out a given strategy. For this reason we must calculate the number of treatments corresponding to a given treatment rate. It is possible to do this from the model; the result is a constant multiple of I_0 plus a multiple of $S_0 - S_\infty$.

The results of such calculations appear to indicate that treatment of diagnosed infectives is the most effective strategy [2,3]. However, there are other considerations that would go into any policy decision. For example, a pandemic would threaten to disrupt essential services and it could be decided to use antiviral drugs prophylactically in an attempt to protect health care workers and public safety personnel. A study including this aspect based on the antiviral treatment model given here is reported in [8].

For simulations, we use the initial values

$$S_0 = 1988, \quad I_0 = 12,$$

the parameters of Sect. 12.2, and for antiviral efficacy we assume

$$\sigma_S = 0.7, \quad \sigma_I = \sigma_A = 0.2,$$

based on data reported in [17].

We simulate the model (12.8) with $\theta_I = 0$, assuming that treatment continues for the duration of the infection. We assume that 80% of diagnosed infectives are treated within 1 day. Since the assumption of treatment at a constant rate φ_I implies treatment of a fraction $1 - exp(-\varphi_I t)$ after a time t, we take φ_I to satisfy

$$1 - e^{-\varphi_I} = 0.8,$$

or $\varphi_I = 1.61$. We use different values of $S_0\beta$, corresponding to different values of $\mathcal{R}_0$, and use (12.10) and (12.11), obtaining results shown in Table 12.2.

Table 12.2 Control by antiviral treatment

$S_0\beta$	$\mathcal{R}_0$	$\mathcal{R}_c$	Disease cases	Treatments
0.402	1.37	0.92	64	56
0.435	1.49	1.00	130	113
0.5	1.71	1.15	373	314
0.7	2.39	1.61	865	751

We calculate from (12.11) that $\mathcal{R}_c = 1$ if $S_0\beta = 0.435$ and $\mathcal{R}_0 = 1.49$, corresponding to a symptomatic attack ratio of 39%. This is the critical attack ratio beyond which treatment at the rate specified can not control the pandemic.

In the next section, we will formulate and analyze a compartmental model that describes influenza more precisely. Then we will choose parameters so that the behavior agrees with the behavior of the model (12.1) without treatment as simulated in Sect. 12.2 and compare simulation of a corresponding model with the simulation of Sect. 12.4.

12.5 A More Detailed Model

Recent re-examination of data from past influenza epidemics [1,4,7,8,15] has indicated a more complicated compartmental structure than that of (12.1). It is suggested that susceptibles go first to a non-infectious first latent stage, then either to an asymptomatic stage and removal, or to a second latent stage with some infectivity, followed by two infective stages with different infectivity. Treatment is applied only during the first infective stage. This leads to the following model, taken from [1].

$$S' = -\beta SQ$$
$$L_1' = \beta SQ - \kappa_1 L_1$$
$$L_2' = p\kappa_1 L_1 - \kappa_2 L_2$$
$$I_1' = \kappa_2 L_2 - (\alpha_1 + \varphi)I_1 \qquad (12.13)$$
$$I_2' = \alpha_1 I_1 - \alpha_2 I_2$$
$$I_T' = \varphi I_1 - \alpha_T I_T$$
$$A' = (1-p)\kappa_1 L_1 - \eta A$$
$$N' = -(1-f)\alpha_2 I_2 - (1-f_T)\alpha_T I_T,$$

with

$$Q = \sigma_L L_2 + I_1 + \sigma_I I_2 + \sigma_T I_T + \sigma_A A.$$

The parameters $\sigma_L, \sigma_I, \sigma_T, \sigma_A$ are the relative infectivities in the compartments L_2, I_2, I_T, A respectively. The initial conditions are

$$S(0) = S_0, I_1(0) = I_0, L_1(0) = L_2(0) = I_2(0) = I_T(0) = A(0) = 0, N(0) = S_0 + I_0.$$

A flow diagram for the model (12.13) is shown in Fig. 12.4.

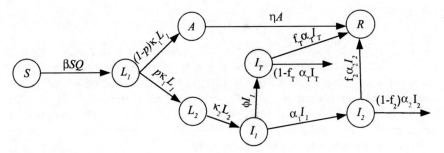

Fig. 12.4 Refined model flow chart

Then we calculate

$$\mathcal{R}_c = S_0\beta\left[\frac{p\sigma_L}{\kappa_2} + \frac{p}{(\alpha_1+\varphi)} + \frac{p\sigma_I\alpha_1}{\alpha_2(\alpha_1+\varphi)} + \frac{p\sigma_T\varphi}{\alpha_T(\alpha_1+\varphi)} + \frac{(1-p)\sigma_A}{\eta}\right].$$
$$(12.14)$$

The basic reproduction number is given by (12.14) with $\varphi = 0$,

$$\mathcal{R}_0 = S_0\beta\left[\frac{p\sigma_L}{\kappa_2} + \frac{p}{\alpha_1} + \frac{p\sigma_I}{\alpha_2} + \frac{(1-p)\sigma_A}{\eta}\right]. \qquad (12.15)$$

The final size relation is

$$\ln \frac{S_0}{S_\infty} = \mathcal{R}_c \left[1 - \frac{S_\infty}{S_0}\right] + \beta I_0 \rho \tag{12.16}$$

with

$$\rho = \frac{1}{\alpha_1 + \varphi} + \frac{\sigma_I \alpha_1}{\alpha_2(\alpha_1 + \varphi)} + \frac{\sigma_T \varphi}{\alpha_T(\alpha_1 + \varphi)}.$$

The number of disease cases is $I_0 + p(S_0 - S_\infty)$, and the number of people treated is

$$\frac{\varphi}{\alpha_1 + \varphi}[I_0 + p(S_0 - S_\infty)].$$

In [8], suggested parameter values are

$$\kappa_1 = 0.8, \quad \kappa_2 = 4.0, \quad \alpha_1 = 1.0, \quad \alpha_2 = \frac{1}{2.85}, \quad \alpha_T = \frac{1}{1.35}$$

$$\eta = \frac{1}{4.1}, \quad \sigma_L = 0.286, \quad \sigma_I = 0.143,$$

and we take

$$\sigma_T = 0.2, \quad \sigma_A = 0.5$$

as in $[2, 3, 11]$. With these parameter values,

$$\mathcal{R}_0 = 1.669 S_0 \beta.$$

In order to obtain $\mathcal{R}_0 = 1.37$ as in the simulation of Sect. 12.2 using (12.15), we take $S_0 \beta = 0.823$ (note that this is not the same value as in Sect. 12.2). Next, we take a treatment rate $\varphi = 1.61$, as in the simulation of Sect. 12.4. Using (12.14) we obtain $\mathcal{R}_c = 0.74$, compared with the value $\mathcal{R}_c = 0.92$ obtained in Sect. 12.4. From (12.16) we obtain $S_\infty = 1,962$, corresponding to 29 cases of disease compared with 64 cases found in Sect. 12.4. The number of people receiving antiviral treatment is 18.

There are significant differences between the predictions of the models (12.8) and (12.13), with the number of disease cases and disease deaths lower for (12.13). Presumably, the model (12.13) is closer to the truth than (12.8). However, use of (12.13) requires more knowledge of the course of the epidemic period and more information about parameter values. Also, it should be noted that the model (12.13), having treatment only during one infective stage, is not suitable for modeling pre-epidemic vaccination.

There are other, probably more serious, omissions in the model. Let us consider another possible direction of generalization.

12.6 A Model with Heterogeneous Mixing

In many past influenza epidemics it has been observed that much of the transmission of infection can be traced to school children, while many of the disease fatalities are in elderly or immune-compromised people who make many fewer contacts. This suggests that in order to give a more accurate description of influenza transmission it is important to separate the population into subgroups with different contact rates and different disease mortality rates. Let us look at the simplest possible version of such a model, an SIR model with treatment of infectives and two subgroups. In the interest of simplicity we do not include latent or asymptomatic compartments but these could easily be added to give a model looking more like a description of influenza. In fact, as it will turn out that the model to be formulated here does exhibit differences in behavior from the analogous one-group model, this should be the next step in constructing a better model for influenza. However, we shall leave this task for another occasion.

A question of considerable importance in the management of influenza is whether one should concentrate control measures, either vaccination or antiviral treatment, on active relatively invulnerable or less active more vulnerable members of the population. A two group model is the simplest possible setting for an analysis of this question. The answer must depend on the parameters of the model, for the following reason: If the disease is universally fatal to members of the second group if infected but has no effect on members of the first group, then it is clearly important to try to protect the second group and the first group can be ignored. On the other hand, if there is no difference in death rates between the two groups, then the goal is to minimize the number of infections and this is achieved by concentrating on the active first group.

Consider two subpopulations of sizes N_1, N_2 respectively, each divided into susceptibles, infectives, and removed members with subscripts to identify the subpopulation. Suppose that group i members make a_i contacts in unit time and that the fraction of contacts made by a member of group i that is with a member of group j is $p_{ij}, i, j = 1, 2$. Then

$$p_{11} + p_{12} = p_{21} + p_{22} = 1.$$

The total number of contacts made in unit time by members of group 1 with members of group 2 is $a_1 p_{12} N_1$ and because this must equal the total number of contacts by members of group 2 with members of group 1, we have a balance relation

$$\frac{p_{12}a_1}{N_2} = \frac{p_{21}a_2}{N_1}.$$

We also make the assumption of proportionate mixing between groups, that is, that the number of contacts between groups is proportional to the relative activity levels. In other words, mixing is random but constrained by the activity levels [13]. We suggest that this is appropriate for diseases in which contacts may be assumed to be essentially random. It would certainly not be appropriate for disease in which contacts include preferences, such as sexually transmitted diseases. Under the assumption of proportionate mixing,

$$p_{ij} = \frac{a_j N_j}{a_1 N_1 + a_2 N_2},$$

and we may write

$$p_{11} = p_{21} = p_1, \quad p_{12} = p_{22} = p_2,$$

with $p_1 + p_2 = 1$. In other words, proportionate mixing means that each group makes a fraction p_j of its contacts with group j for $j = 1, 2$.

Suppose the mean infective period in each group is $1/\alpha$, and the recovery fractions in the two groups are f_1, f_2 respectively. If the death rate in either group is positive, this implies that the total population size in the corresponding group decreases and the mixing proportions may change in time.

The two-group SIR epidemic model is

$$
\begin{aligned}
S_1' &= -\Big[p_1 a_1 \frac{S_1 I_1}{N_1} + p_2 a_1 \frac{S_1 I_2}{N_2}\Big] \\
I_1' &= \Big[p_1 a_1 \frac{S_1 I_1}{N_1} + p_2 a_1 \frac{S_1 I_2}{N_2}\Big] - \alpha I_1 \\
R_1' &= f_1 \alpha I_1 \\
N_1' &= -(1 - f_1)\alpha I_1 \\
S_2' &= -\Big[p_1 a_2 \frac{S_2 I_1}{N_1} + p_2 a_2 \frac{S_2 I_2}{N_2}\Big] \\
I_2' &= \Big[p_1 a_2 \frac{S_2 I_1}{N_1} + p_2 a_2 \frac{S_2 I_2}{N_2}\Big] - \alpha I_2 \\
R_2' &= f_2 \alpha I_2 \\
N_2' &= -(1 - f_2)\alpha I_2.
\end{aligned}
\tag{12.17}
$$

The equations for R_1, R_2 can be discarded from the model since

$$R_i = N_i - S_i - I_i, \quad i = 1, 2.$$

To the model (12.17) we add treatment of infectives at rates φ_1, φ_2 respectively in the two groups that decreases infectivity by a factor σ and changes the rate of departure from the stage α to α_T, presumably with

$$\alpha_T \geq \alpha.$$

In addition, we assume that the recovery fractions for treated members are $f_{T,1} \geq f_1, f_{T,2} \geq f_2$ respectively. This implies the introduction of two treated compartments T_1, T_2 and leads to the model

$$S_1' = -[p_1 a_1 \frac{S_1(I_1 + \sigma T_1)}{N_1} + p_2 a_1 \frac{S_1(I_2 + \sigma T_2)}{N_2}]$$

$$I_1' = [p_1 a_1 \frac{S_1(I_1 + \sigma T_1)}{N_1} + p_2 a_1 \frac{S_1(I_2 + \sigma T_2)}{N_2}] - (\alpha + \varphi_1) I_1$$

$$T_1' = \varphi_1 I_1 - \alpha_T T_1$$

$$R_1' = f_1 \alpha I_1 + f_{T,1} \alpha_T T_1$$

$$N_1' = -(1 - f_1)\alpha I_1 - (1 - f_{T,1})\alpha_T T_1$$

$$S_2' = -[p_1 a_2 \frac{S_2(I_1 + \sigma T_1)}{N_1} + p_2 a_2 \frac{S_2(I_2 + \sigma T_2)}{N_2}] \qquad (12.18)$$

$$I_2' = [p_1 a_2 \frac{S_2(I_1 + \sigma T_1)}{N_1} + p_2 a_2 \frac{S_2(I_2 + \sigma T_2)}{N_2}] - (\alpha + \varphi_2) I_2$$

$$T_2' = \varphi_2 I_2 - \alpha_T T_2$$

$$R_2' = f_2 \alpha I_2 + f_{T,2} \alpha_T T_2$$

$$N_2' = -(1 - f_2)\alpha I_2 - (1 - f_{T,2})\alpha_T T_2.$$

Again, the equations for R_1, R_2 can be discarded from the model.

Using the technique of [16] or Chap. 6 of Mathematical Epidemiology (this volume) we calculate the reproduction number for the model (12.18). Since the model includes treatment, the reproduction number is a control reproduction number, denoted by $\mathcal{R}_c$, and

$$\mathcal{R}_c = a_1 \Gamma_1 + a_2 \Gamma_2, \qquad (12.19)$$

where

$$\Gamma_1 = \frac{p_1(\alpha_T + \sigma \varphi_1)}{\alpha_T(\alpha + \varphi_1)}, \qquad \Gamma_2 = \frac{p_2(\alpha_T + \sigma \varphi_2)}{\alpha_T(\alpha + \varphi_2)},$$

with p_1, p_2 evaluated for $N_1 = N_1(0), N_2 = N_2(0)$. The basic reproduction number $\mathcal{R}_0$ is $\mathcal{R}_c$ with $\varphi_1 = \varphi_2 = 0$,

$$\mathcal{R}_0 = \frac{a_1 p_1 + a_2 p_2}{\alpha} = \frac{a_1^2 N_1 + a_2^2 N_2}{\alpha(a_1 N_1 + a_2 N_2)}.$$

It appears that if the model is simplified by assuming that the population sizes N_1, N_2 are assumed constant, the results are very good approximations to the results given by the full model (12.18) if the disease death rates are small. In fact, use of this simplification and calculation of disease deaths as fractions of the epidemic size gives a very good approximation to the predictions of the full model. The idea is that we use the model (12.20) to estimate the sizes of the epidemic in each group as if there were no disease deaths, and then we estimate the actual number of disease deaths by applying the assumed death rates to these sizes a posteriori. The presumption that the model (12.20) approximates the model (12.18) well amounts to assuming that this logically indefensible procedure gives good results.

Thus, we assume that there are no disease deaths, so that the total population sizes N_1, N_2 of the two groups remain constant. We eliminate N_1, N_2

and consider the model

$$S_1' = -[p_1 a_1 \frac{S_1(I_1 + \sigma T_1)}{N_1} + p_2 a_1 \frac{S_1(I_2 + \sigma T_2)}{N_2}]$$

$$I_1' = [p_1 a_1 \frac{S_1(I_1 + \sigma T_1)}{N_1} + p_2 a_1 \frac{S_1(I_2 + \sigma T_2)}{N_2}] - (\alpha + \varphi_1)I_1$$

$$T_1' = \varphi_1 I_1 - \alpha_T T_1 \tag{12.20}$$

$$S_2' = -[p_1 a_2 \frac{S_2(I_1 + \sigma T_1)}{N_1} + p_2 a_2 \frac{S_2(I_2 + \sigma T_2)}{N_2}]$$

$$I_2' = [p_1 a_2 \frac{S_2(I_1 + \sigma T_1)}{N_1} + p_2 a_2 \frac{S_2(I_2 + \sigma T_2)}{N_2}] - (\alpha + \varphi_2)I_2$$

$$T_2' = \varphi_2 I_2 - \alpha_T T_2.$$

There is a pair of final size equations for (12.20), namely

$$\ln \frac{S_1(0)}{S_1(\infty)} = a_1 [\Gamma_1(1 - \frac{S_1(\infty)}{N_1(0)}) + \Gamma_2(1 - \frac{S_2(\infty)}{N_2(0)})] \tag{12.21}$$

$$\ln \frac{S_2(0)}{S_2(\infty)} = a_2 [\Gamma_1(1 - \frac{S_1(\infty)}{N_1(0)}) + \Gamma_2(1 - \frac{S_2(\infty)}{N_2(0)})].$$

Combining these two equations, we obtain

$$a_2 \ln \frac{S_1(0)}{S_1(\infty)} = a_1 \ln \frac{S_2(0)}{S_2(\infty)}. \tag{12.22}$$

We see from this that if $a_1 > a_2$ then

$$1 - \frac{S_1(\infty)}{S_1(0)} > 1 - \frac{S_2(\infty)}{S_2(0)},$$

that is, the attack ratio of the disease is greater in the more active group. We see also that

$$\frac{S_2(\infty)}{S_2(0)} = [\frac{S_1(\infty)}{S_1(0)}]^{a_2/a_1}.$$

In the general case, with the model (12.18) and disease deaths, there are disease deaths in both untreated and treated classes. The number of disease deaths in the untreated class is

$$(1 - f_1)\alpha \int_0^\infty I_1(t)dt = (1 - f_1)\frac{\alpha}{\alpha + \varphi_1}[N_1(0) - S_1(\infty)] \tag{12.23}$$

and the number of disease deaths in the treated class is

$$(1 - f_{T,1})\alpha_T \int_0^\infty T_1(t)dt = (1 - f_{T,1})\frac{\varphi_1}{\alpha + \varphi_1}[N_1(0) - S_1(\infty)], \tag{12.24}$$

with analogous relations in the second group. Thus the total number of disease deaths is

$$D = \Big[(1 - f_1)\frac{\alpha}{\alpha + \varphi_1} + (1 - f_{T,1})\frac{\varphi_1}{\alpha + \varphi_1}\Big][N_1(0) - S_1(\infty)] \quad (12.25)$$

$$+ \Big[(1 - f_2)\frac{\alpha}{\alpha + \varphi_1} + (1 - f_{T,2})\frac{\varphi_2}{\alpha + \varphi_2}\Big][N_2(0) - S_2(\infty)].$$

The number of members of the population treated is

$$\frac{\varphi_1}{\alpha + \varphi_1}[N_1(0) - S_1(\infty)] + \frac{\varphi_2}{\alpha + \varphi_2}[N_2(0) - S_2(\infty)].$$

If there are disease deaths the final size relation is an inequality

$$\ln\frac{S_1(0)}{S_1(\infty)} \le a_1\Big[\Gamma_1(1 - \frac{S_1(\infty)}{N_1(0)}) + \Gamma_2(1 - \frac{S_2(\infty)}{N_2(0)})\Big].$$

Then $S_1(\infty)$ is smaller than the value obtained in the case of no disease deaths. We conjecture that if the disease death rates are small then $S_1(\infty)$ is close to the value obtained in the case of no disease death rates and thus that the number of deaths can be estimated from (12.25) with $S_1(\infty)$ estimated from (12.21). Although we have not been able to establish a general result of this nature, we give an example to suggest its validity.

12.7 A Numerical Example

In this section we present an example to suggest that results obtained from the model (12.20) assuming constant total population size are good approximations to the results obtained from the full model (12.18). Our example is a compartmental model based on the influenza model of [11], but simplified by omitting the latent and asymptomatic stages and estimating parameters consistent with those of [11]. We divide the population into only two groups, the first consisting of 87.5% of the population aged less than 65 and the second consisting of 12.5% of the population aged 65 or more. The attack ratios assumed in [11] are 0.35 and 0.21 in the two groups, but since 1/3 of the infections are assumed asymptomatic in [11] we assume clinical attack ratios of 0.525 and 0.325 in the two groups. Thus we take $S_1(\infty) = 0.475N_1(0), S_2(\infty) = 0.675N_2(0)$. Then solution of (12.22) gives $a_2 = 0.528a_1$, and the assumption of proportionate mixing gives

$$\frac{p_2}{p_1} = \frac{a_2 N_2}{a_1 N_1} = \frac{12\,a_2}{88\,a_1} = 0.07,$$

so that (approximately)

$$p_1 = 0.93, \quad p_2 = 0.07.$$

We take $\alpha = 1/4$, corresponding to a mean infective period of 4 days and $\alpha_T = 1/3$ corresponding to a decrease to 3 days in the mean infective period due to treatment. In addition, we take

$$\sigma = 0.2, \quad f_1 = 0.99, \quad f_{T,1} = 0.995, \quad f_2 = 0.92, \quad f_{T,2} = 0.96,$$

and initial conditions

$$S_1(0) = 870, \quad I_1(0) = 5, \quad T_1(0) = 0, \quad N_1(0) = 875$$
$$S_2(0) = 125, \quad I_2(0) = 0, \quad T_2(0) = 0, \quad N_2(0) = 125.$$

If there is no treatment, $\varphi_1 = \varphi_2 = 0$, we have

$$\Gamma_1 = \frac{p_1}{\alpha}, \quad \Gamma_2 = \frac{p_2}{\alpha},$$

and since $\alpha = 1/4$ we obtain

$$\Gamma_1 = 3.72, \quad \Gamma_2 = 0.28, \quad \mathcal{R}_0 = 1.41.$$

Then (12.21) and the assumed values

$$S_1(\infty) = 0.475 N_1(0), \quad S_2(\infty) = 0.675 N_2(0)$$

give $a_1 = 0.364, a_2 = 0.192$.

We simulate the model (12.18) with these parameter values, obtaining results shown in Table 12.3. We do not show the results of simulation of the model (12.20) but they are very close to the results shown. Note that in the model (12.20) N_1, N_2 have the constant values $875, 125$ respectively, and the numerical results may be obtained from the final size relations (12.21),(12.22) without solving the system of differential equations.

Although the model (12.20) assumes no disease deaths, we may estimate the number of disease deaths from simulation of (12.18) or, as indicated earlier by using (12.23), (12.24), obtaining the results shown in Table 12.4.

We compare these results with those obtained from a one-group model whose parameters are chosen by averaging those of the two-group model. Thus, we simulate the system

$$\begin{aligned}
S' &= -a\frac{S(I + \sigma T)}{N} \\
I' &= a\frac{S(I + \sigma T)}{N} - (\alpha + \varphi)I \\
T' &= \varphi I - \alpha_T T \\
N' &= -(1 - f)\alpha I - (1 - f_T)\alpha_T T,
\end{aligned} \qquad (12.26)$$

Table 12.3 Final sizes for the model (12.18)

φ_1	φ_2	$S_1(\infty)$	$S_2(\infty)$
0	0	406.4	84.9
0	0.5	429.2	87.3
0	1.0	433.9	87.8
0.25	0	849.1	123.5
0.25	0.5	851.7	123.7
0.25	1.0	852.2	123.7
0.5	0	862.1	124.4
0.5	0.5	862.7	124.5
0.5	1.0	862.8	124.5
1.0	0	865.8	124.7
1.0	0.5	866.1	124.7
1.0	1.0	866.1	124.7

Table 12.4 Diseases and cases for the model (12.18)

φ_1	φ_2	Cases, G1	Deaths, G1	Cases, G2	Deaths, G2
0	0	468.6	4.7	40.1	3.2
0	0.5	445.8	4.5	37.7	2.0
0	1.0	444.1	4.4	37.2	2.0
0.25	0	25.9	1.5	1.5	0.1
0.25	0.5	23.4	0.2	1.3	0.1
0.25	1.0	22.8	0.2	1.3	0.1
0.5	0	12.9	0.1	0.6	0
0.5	0.5	12.3	0.1	0.5	0
0.5	1.0	12.2	0.1	0.5	0
1.0	0	9.2	0.1	0.3	0
1.0	0.5	8.1	0.1	0.3	0
1.0	1.0	8.1	0.1	0.3	0

with parameter values

$$a = 0.342, \quad \sigma = 0.2, \quad \alpha = 1/4, \quad \alpha_T = 1/3, \quad f = 0.9813, \quad f_T = 0.9906,$$

treatment rates $\varphi = 0, 0.5, 1.0$, and initial conditions

$$S(0) = 995, \quad I(0) = 5, \quad T(0) = 0, \quad N(0) = 1,000.$$

The results are shown in Table 12.5.

The numbers of disease cases and disease deaths predicted by the one-group model (12.26) are very close to those predicted by the two-group model (12.18). This raises the question of whether using the two - group model is worth the additional effort. Our simulations suggest that the one - group model gives essentially the same results as the two-group model if treatment rates in the two groups are the same. However, examining the

Table 12.5 Results for the one-group model (12.26)

φ	$N(\infty)$	$S(\infty)$	Cases	Deaths
0	990.7	501.5	498.5	9.3
0.5	999.8	988.9	11.1	0.2
1.0	999.9	991.2	8.8	0.1

effect of using different treatment rates in the two groups demands the use of a two-group model. For the parameters used here, it is clear that treating only group 2 is wasteful of resources. Treating both groups is a more effective use of resources than treating only the first group since we may observe that treating the second group as well as the first reduces the number of disease cases but requires fewer treatments compared to treating only the first group. This is consistent with the simulations reported in [11], where it was concluded that treatment of the most active group in the population was almost as effective as treating the entire population.

The model (12.18) is a generalization of the simple SIR model, not of the models (12.8) or (12.13) in which some aspects of influenza were incorporated. We think of (12.18) as an influenza model that focusses on heterogeneity of mixing. It would not be difficult to formulate a model incorporating the structure of (12.13) and two groups with different activity levels. We suggest that analyzing (12.18) and concentrating on heterogeneity of mixing gives useful information about the importance of heterogeneity in treatment and is a strong argument for carrying out this analysis.

The models (12.18) and (12.20) can be extended easily in several directions. Possible extensions include models with an arbitrary number of groups with different activity levels, models with more stages in the progression through compartments, and models in which there are differences between groups in susceptibility. However, our results are restricted to proportionate mixing. While this appears to be a reasonable assumption for a two-group model, it would be plausible that in a three-group model, two of the groups mix with the third group but not with each other, and this would not be proportionate mixing.

There are still some open questions for models with heterogeneous mixing. For the models studied here, the question of how accurate the approximation of a full model by a model with constant total population sizes when disease death rates are small remains open. Since we believe that the type of model studied here should be considered for modeling future epidemics, such as an anticipated influenza pandemic, these are important questions.

12.8 Extensions and Other Types of Models

We have considered only compartmental models for influenza in this chapter, but there are other useful approaches. Many of the estimates being made for use in making policy decisions for coping with an influenza pandemic are based on network models [6,7,9–12]. These are based on a detailed study of mixing patterns in populations and assume a great deal of knowledge of the contact structure. They divide a population into subgroups having different contact patterns and are able to give very detailed predictions, including the effects of management strategies that treat different segments of the population differently. Their analyses involve large-scale stochastic simulations; while they give a great deal of information it is difficult to estimate their sensitivity to uncertainties in the parameter values. In addition to the use being made now for pandemic influenza preparation, they would also be useful for coping with seasonal influenza epidemics.

In Sect. 12.6 we have introduced heterogeneous mixing as an aspect of a model. Another heterogeneity arises from considering a population as a collection of separate units or patches with travel between them rather than as a single entity.

We have assumed mass action incidence in the models of Sects. 12.2–12.6, but it is probably more realistic to assume some density dependence in the contact rates. In this case, the final size relations are inequalities rather than equations. If there are no disease deaths, the total population remains constant and mass action is equivalent to general incidence. It appears that if the disease death rate is small, mass action incidence is a good approximation, and the use of final size equations in place of actually solving the systems of differential equations is a valid approximation.

We have not built into our models any assumptions of behavioral change by the population during an epidemic. It would be reasonable to expect infectives to withdraw from contact, either because of weakness caused by infection or because of public health encouragement to infectives to reduce contacts in order to avoid spreading infection. This could be modeled by addition of an isolated compartment with reduced contacts. Another response to be expected is that at least part of the population would try to reduce contacts and take hygienic measures to decrease risk of infection. These factors have not been built into models here but are obvious candidates for consideration.

We have assumed rates of transition between compartments to be proportional to compartment sizes, which is equivalent to assuming negative exponential distributions of times in compartments. The assumption of more realistic distributions would lead to more complicated models, formulated as integral or integro-differential equations. An important question is whether such more realistic assumptions, giving models that are much more difficult to analyze, yield a worthwhile return in more accurate information.

A new and important question that is beginning to be studied is the possibility of development of resistance to antiviral drugs [1, 14, 15]. Models currently being developed build on compartmental models of the type studied in this chapter, and there are many questions coming under study.

In studying epidemic models in general there is a trade-off between the ease of analyzing a model and the amount of detail and accuracy provided by the model. The appropriate model for a given situation depends on the nature of the information sought and the amount and reliability of the data.

Acknowledgements

Much of the material in this chapter is based on ongoing work with Julien Arino, P. van den Driessche, James Watmough, and Jianhong Wu, and they are responsible for many of the ideas and much of the wording. In addition, Julien Arino supplied the templates on which the figures are based.

References

1. M.E. Alexander, C.S. Bowman, Z. Feng, M. Gardam, S.M. Moghadas, G. Röst, J. Wu & P. Yan, Emergence of drug resistance: implications for antiviral control of pandemic influenza, Proc. R. Soc. B **274**, 1675–1684 (2007)
2. J. Arino, F. Brauer, P. van den Driessche, J. Watmough & J. Wu, Simple models for containment of a pandemic, J. R. Soc. Interface **3**, 453–457 (2006)
3. J. Arino, F. Brauer, P. van den Driessche, J. Watmough & J. Wu, A model for influenza with vaccination and antiviral treatment, J. Theor. Biol., in press, available online, doi: 10.1016/j.jtbi.2008.02.026
4. S. Cauchemez, F. Carrat, C. Viboud, A.J. Valleron & P.Y. Boëlle, A Bayesian MCMC approach to study transmission of influenza: application to household longitudinal data, Stat. In Med. **232**, 3469–3487 (2004)
5. L.R. Elveback, J.P. Fox, E. Ackerman, A. Langworthy, M. Boyd & L. Gatewood, An influenza simulation model for immunization studies, Am. J. Epidem. **103**, 152–165 (1976)
6. N.M. Ferguson, D.A.T. Cummings, S. Cauchemez, C. Fraser, S. Riley, A. Meeyai, S. Iamsirithaworn & D.S. Burke, Strategies for containing an emerging influenza pandemic in Southeast Asia, Nature **437**, 209–214 (2005)
7. N.M. Ferguson, D.A.T. Cummings, C. Fraser, J.C. Cajka, P.C. Cooley & D.S. Burke, Strategies for mitigating an influenza pandemic, Nature **442**, 448–452 (2006)
8. M. Gardam, D. Liang, S.M. Moghadas, J. Wu, Q. Zeng & H. Zhu, The impact of prophylaxis of healthcare workers on influenza pandemic burden, J. R. Soc. Interface **4**, 727–734 (2007)
9. T.C. Germann, K. Kadau, I.M. Longini & C.A. Macken, Mitigation strategies for pandemic influenza in the United States, Proc. Natl. Acad. Sci. U. S. A. **103**, 5935–5940 (2006)
10. M. Lipsitch, T. Cohen, M. Murray & B.R. Levin, Antiviral resistance and the control of pandemic influenza, PLoS Med. **4**, 111–121 (2007)
11. I.M. Longini, M.E. Halloran, A. Nizam & Y. Yang, Containing pandemic influenza with antiviral agents, Am. J. Epidem. **159**, 623–633 (2004)

12. I.M. Longini, A. Nizam, S. Xu, K. Ungchusak, W. Hanshaoworakul, D.T. Cummings & M.E. Halloran, Containing pandemic influenza at the source, Science **309**, 623–633 (2004)
13. A. Nold, Heterogeneity in disease transmission modeling, Math. Biosc. **52**, 227–240 (1980)
14. R.R. Regoes & S. Bonhoeffer, Emergence of drug-resistant influenza virus: population dynamical considerations, Science **312**, 389–391 (2006)
15. N.I. Stillianakis, A.S. Perelson & F.G. Hayden, Emergence of drug resistance during an influenza epidemic: insights from a mathematical model, J. Infect. Dis. **177**, 863–873 (1998)
16. P. van den Driessche & J. Watmough, Reproduction numbers and subthreshold endemic equilibria for compartmental models of disease transmission, Math. Biosc. **180**, 29–48 (2002)
17. R. Welliver, A.S. Monto, O. Carewicz, E. Schatteman, M. Hassman, M.J. Hedrick, H.C. Jackson, L. Huson, P. Ward & J.S. Oxford, Effectiveness of oseltamivir in preventing influenza in household contacts: a randomized controlled trial, JAMA **285**, 748–754 (2001)

Chapter 13
Mathematical Models of Influenza: The Role of Cross-Immunity, Quarantine and Age-Structure

M. Nuño, C. Castillo-Chavez, Z. Feng and M. Martcheva

Abstract This chapter compiles some of the results on influenza dynamics that involve a single strain, as well as two competing strains. The emphasis is on the role of cross-immunity, quarantine and age-structure as mechanisms capable of supporting recurrent influenza epidemic outbreaks. Quarantine or age-structure alone can support oscillations while cross-immunity enhances the likelihood of strain coexistence and impacts the length of the period. It is the hope that the perspective provided here will instigate others to use mathematical models in the study of disease transmission and its evolution, particularly in a setting that involves highly variable pathogens.

13.1 Introduction

Recent (2003–2004) fatalities attributed to "flu" infections that resulted from the interactions between domestic avian populations and humans in Asia (Cambodia, China, Indonesia, Japan, Laos, South Korea, Thailand and Vietnam) have brought back memories of past "flu" epidemics, particularly those of the 1918 pandemic, which resulted in the deaths of approximately 40 million individuals [8, 27]. The periodic recurrence of epidemic outbreaks has been documented in the context of communicable diseases like measles, rubeola and influenza [1]. The identification of mechanisms capable of generating recurrent, particularly periodic, outbreaks has been investigated using

Department of Biostatistics, UCLA School of Public Health, Los Angeles, CA 90095-1772, USA miriamnuno@ucla.edu

Department of Mathematics and Statistics, Arizona State University, Temple, AZ 85287-1804, USA ccchavez@asu.edu

Department of Mathematics, Purdue University, West Lafayette, IN 47907, USA zfeng@math.purdue.edu

Department of Mathematics, Florida University, Gainesville, FL 32611-8105, USA maia@math.ufl.edu mathematical models [16–18,22]. Here, we look at the role on the qualitative dynamics of communicable diseases played by three mechanisms: quarantine of infectious individuals, strain-specific cross-immunity and age-structure. In addition, we also look at systems facing competing pathogens. The work here is carried on in the context of influenza.

Only three subtypes of influenza type A (H1N1, H2N2 and H3N2) have been found to be responsible for influenza pandemics or epidemics over the past century. Point mutations in specific regions of the HA molecule "continuously" generate new strains *within* a given subtype while major molecular changes have been tied to genetic shifts (new subtypes). High mortality rates are closely tied to the immune system's ability to effectively control infections. In the case of influenza type A (the only type considered here), cross-immunity, that is, the ability of an individual's immune system to tap on its history of prior infections to either reduce the likelihood of new infections (by related strains) or to accelerate "virus control" (within each host) by bolstering the host's immune response, can play a significant role on the pathogen's transmission dynamics at the population level.

Epidemiological studies in various regions of the US have quantified the *natural* reduced susceptibility, gained by individuals who have recovered from infections to specific flu strains, to invasions by related strains generated from the same flu A subtype via point mutations [14, 23–26]. (There is no documented cross-immunity (reduced susceptibility to invasion) to strains from different flu A subtypes [9, 15]]. Hence, only the competition for susceptibles by flu A strains within the same subtype is assumed to be mediated by cross-immunity. The levels of protection (to new infections) offered by cross-immunity are a function of how "close" each invading strain is to the strain responsible for the immune response. Data from experimental studies on various communities [14, 26] have captured population-levels of acquired cross-immunity [6,7]. Data from these studies were used to introduce a coefficient of cross-immunity, $\sigma \in [0, 1]$ [6,7] at the population level. The scale was set as follows: a value of $\sigma = 1$ corresponds to the case when no protection is generated by prior infections while a value of $\sigma = 0$ implies the acquisition of total cross-immunity. In [6,7] it was shown that cross-immunity mediated competition between relate flu strains (both from the same subtype) in a *homogeneous mixing* population can enhance the likelihood of strain coexistence. Typically, the corresponding disease endemic levels are reached via slowly damped oscillations. Furthermore, the "period", that is, the distance between consecutive "damping" peaks turned out to be significantly larger for values of σ close to zero (total cross-immunity). This qualitative "discrepancy" ($\sigma = 0$ vs $\sigma = 1$) is closely tied to the time required (population characteristic) to replenish strain-specific susceptible populations. The closer σ is to zero, the longer it takes to replenish the appropriate susceptible pool.

The theoretical results on epidemics recurrence of Feng and Thieme [13] as well as those of Hethcote and collaborators [19] arose from their studies of modified SIR single-strain epidemic models. These modifications included the possibility of isolating a fraction of the infectious individuals (quarantine class, Q). The introduction of the Q class alters significantly the rate of secondary infections since it does impact the incidence rate. These researchers [12, 19] did show that the isolation of infectious individuals can be enough to support not only disease persistence but also the generation of periodic solutions via a Hopf bifurcation.

In [22] the consequences of incorporating quarantine classes in *two-strain* systems mediated by cross-immunity are systematically explored. Not surprisingly, regions where oscillatory co-existence is possible are identified. It is shown that cross-immunity does in fact promote co-existence to the point that sub-threshold coexistence is possible. That is, by changing the average immunological response of a population, cross-immunity favors the survival of *related* possibly less fit pathogens.

This chapter is organized as follows: Sect. 13.2 outlines some of the results of Feng [12] and Feng and Thieme [13] on the role of a quarantine class as a mechanism capable of generating periodic solutions in $SIQR$ epidemic models; Sect. 13.3 focuses on a flu-inspired two-strain model mediated by cross-immunity that first includes and later excludes the quarantine classes; Sect. 13.4 outlines some results on the role of age-structure and cross-immunity on "flu" dynamics; Sect. 13.5 collects our final thoughts and discusses future possibilities.

13.2 Basic Model

Feng's Ph.D. thesis [12] demonstrated that the introduction of a quarantine class (Q) could destabilize the disease endemic equilibrium in a Susceptible–Infected–Recovered (SIR) model. Hethcote and collaborators [19] further studied the impact of quarantine classes in alternative $SIQR$ frameworks. Their work [19] confirmed the results of Feng and Thieme [13]. In order to state the kind of results that are possible under these models [12, 13, 18] we introduce a simple version of the $SIQR$ models [12, 13]. We let $S(t)$, $I(t)$, $Q(t)$ and $R(t)$ denote the susceptible, infective (assumed to be infectious), quarantined and recovered classes, respectively and assume that the total birth rate equals μN ($N = S+I+Q+R$) where μ also denotes the natural per-capita death rate (assumed to be the same in the four epidemiological classes). We also let γ denote the per-capita recovery rate, δ the per-capita quarantine rate, c the average per-capita contact rate, qc ($0 < q < 1$) the average effective contact rate, and β the per-infective transmission rate. Following the flow diagram in Fig. 13.1 we conclude that the $SIQR$ model of interest is given by the following set of equations:

$$\frac{dS}{dt} = \mu N - \beta S \frac{I}{N-Q} - \mu S,$$
$$\frac{dI}{dt} = \beta S \frac{I}{N-Q} - (\mu + \gamma)I, \tag{13.1}$$
$$\frac{dQ}{dt} = \gamma I - (\mu + \delta)Q,$$

$$\boxed{S \longrightarrow I \longrightarrow Q \longrightarrow R}$$

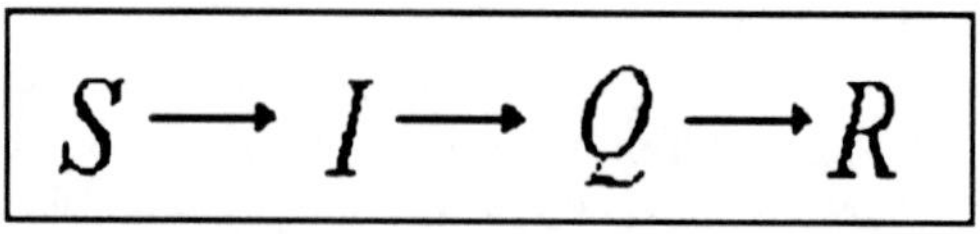

Fig. 13.1 Flow diagram for the classical $SIQR$ model

where $R = N - (S + I + Q)$, $S(0) = N - I_0, I(0) = I_0$, and $Q(0) = 0 = R(0)$. The key difference from standard epidemiological models derives from the fact that the incidence rate has been modified. That is, here it is assumed that the infected proportion "faced" by susceptibles is $\frac{I}{N-Q}$ and *not* $\frac{I}{N}$. Since N is constant, without loss of generality, it is assumed that $N = 1$. The case when $\delta = $ "∞" reduces Model (13.1) to the classical SIR model whose analysis can be found in [16]. The excellent review by Hethcote [18] provides up-to-date references on the theoretical work that has been carried out in SIR models. The basic reproductive number for the classical SIR model is

$$\mathcal{R}_0 = \frac{\beta}{\mu + \gamma} = \beta \times D,$$

where β is the transmission rate and $D = \frac{1}{\mu+\gamma}$ the death-adjusted infectious period [4]. The classical SIR model (two-dimensional system) can support up to two equilibria [16]. The infection-free state is $E_0 = (1, 0)$ and (when $\mathcal{R}_0 > 1$) the endemic state is $E_1 = (\frac{1}{\mathcal{R}_0}, \frac{\mu}{\beta}(\mathcal{R}_0 - 1))$. It can be shown that E_0 is globally asymptotically stable whenever $\mathcal{R}_0 \leq 1$ while E_1 is globally asymptotically stable whenever it exists, that is, when $\mathcal{R}_0 > 1$ (see [18]).

The situation is different when a Q class is introduced ($0 < \delta < \infty$). Model (13.1) supports at most two equilibria. The infection-free state $\mathcal{F}_0 = (1, 0, 0)$ and whenever $\mathcal{R}_0 > 1$ a unique endemic state $\mathcal{F}_1 = (\frac{1}{\mathcal{R}_0}, \frac{\mu}{\beta}[\mathcal{R}_0 - 1]$, $\frac{\delta}{\mu+\gamma}\frac{\mu}{\beta}[\mathcal{R}_0 - 1])$ where $\mathcal{R}_0 = \frac{\beta}{\mu+\gamma+\delta} = \beta D_Q$, β the transmission rate and D_Q the quarantine-death-adjusted infectious period. $\mathcal{F}_1$ is "globally asymptotically" stable but *only* when the quarantine period is either very large or very small. For intermediate values of the quarantine period the endemic state becomes unstable. The existence of periodic solutions can then be established via Hopf-bifurcation (under realistic parameter sets). Reasonable value ranges for the quarantine period ($\frac{1}{\delta}$) capable of generating sustained oscillations were not found in this setting. Significant framework modifications discussed later in this article manage to substantially reduce the $\frac{1}{\delta}$ window where oscillations are possible. A typical mathematical results on the existence of periodic solutions can be summarized [12, 13] as follows:

Theorem 1. *Let θ be the re-scaled per-capita recovery period. Then there is a function: $\zeta_0(\nu)$ defined for small $\nu > 0$*

$$\zeta_0(\nu) = \theta^2(1 - \theta) + O(\nu^{1/2}),$$

with the following properties:
(a) The endemic equilibrium is locally asymptotically stable if $\zeta > \zeta_0(\nu)$ and unstable if $\zeta < \zeta_0(\nu)$, as long as ζ does not become too small.
(b) There is a Hopf bifurcation of periodic solutions at $\zeta = \zeta_0(\nu)$ for small enough $\nu > 0$. The length of periods can be approximated by the formula

$$T = \frac{2\pi}{|\Im w_{\pm}|} \approx \frac{2\pi}{(1 - \theta)^{1/2}\nu^{1/2}} \approx \frac{2\pi}{(\theta y^*)^{1/2}},$$

where y^* is the proportion of infectious individuals at equilibrium.

As it turns out a Hopf-bifurcation occurs in two separate regions determined by the length of the quarantine period (not specifically identified in the above theorem). Figure 13.2 shows a bifurcation occurring for values of the periods of quarantine or isolation (here used interchangeably) in two distinct ranges. Only the region that includes "low" values is of relevance for diseases of interest, like childhood diseases. Feng et al. [12, 13] found data on the length of reported isolation periods during the Scarlet Fever epidemics in England and Wales during 1897–1978 [2] which somewhat supported Model (13.1), that is, they were almost large enough to be within the oscillatory range of this model when relevant epidemiological parameters were fixed. Interesting mathematical problems arise. Numerical simulations of the periodic solutions coming from the left branch exhibited periods with rapidly increasing oscillations. This feature suggested the possibility of a homoclinic bifurcation. Wu and Feng (2000) showed that a relevant perturbed system could indeed support a homoclinic bifurcation (see Fig. 13.3 [29]).

The *practical* relevance of this work [12, 13] follows from the fact that moderate periods quarantine (that is, almost within a realistic range) can destabilize the endemic state. Obviously, "something" was still required to bring $\frac{1}{\delta}$ within an "acceptable" range. In the case of the "flu" cross-immunity (strain competition) or age-structure turn out to be sufficient.

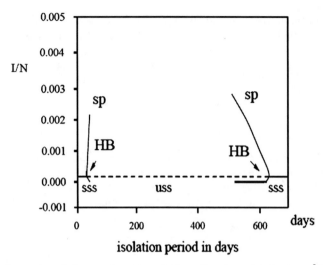

Fig. 13.2 Auto plot of the steady state solutions (fraction of infectives $\frac{I}{N}$) versus the isolation period

13.3 Cross-Immunity and Quarantine

Cross-immunity studies [9, 14, 15, 23, 26, 27] motivated the theoretical studies carried out on immune-mediated strain coexistence which was carried out before [6, 7]. Although simple $SIQR$ models have proved capable of supporting recurrent epidemic single-strain outbreaks, the fact remains that such qualitative behavior seems possible only for unrealistic parameters. Prior work on influenza [6, 7] provided theoretical support to the hypothesis that either competition for hosts by related pathogens or age-structure may be enough to support recurrent, in fact periodic, epidemic outbreaks in regions of parameter space that are relevant to the "flu". Recently, Nuño et al. [22] extended the work in [6, 7, 10] by adding a quarantine class to SIR two-strain models. These SIR two-strain models were capable of supporting coexistence. The system in [22] which includes those in [6, 7, 10] divides the host population into ten epidemiological classes. Susceptible (S) individuals who whenever they become infected by strain i (at a rate β_i) move to the class I_i of infected and infectious individuals. I_i individuals are either isolated, move into the class Q_i at the rate δ_i or move into the recovered (from strain i) class R_i at the rate γ_i. R_i individuals who become infected with strain j do so at the rate $\beta_j \sigma_{ij}$ (a rate reduced by cross-immunity $\sigma_{ij} \in [0, 1]$) move to the class V_j. Finally, individuals who recover from both strains move into the W class at the rates γ_l, l=1 or 2. The model assumes that once an individual is infected with a particular strain, no future infections with the same strain

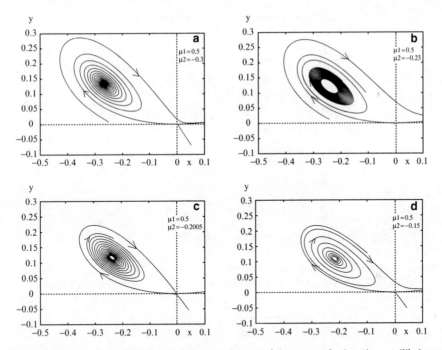

Fig. 13.3 As μ_2 (re-scaled life expectation parameter) increases the interior equilibrium changes from stable to unstable (see (**b**) and (**c**)), and Hopf bifurcation (see (**a**) and (**b**)) and homoclinic bifurcation (see (**b**) and (**c**)) may occur

are possible ($\sigma_{ii} \equiv 0$). It is also assumed that no individuals can carry two infections simultaneously or that the number of such infections is so small that it can be neglected. Furthermore, it is assumed that only I_i-infected individuals are isolated. This last assumption is partially justified from reported studies that show that cross-immunity often reduces "flu" symptoms [14,15]. These assumptions can be easily weakened but such a decision would make our model less tractable. We suspect that the qualitative results would not be too different in the regions of parameter space relevant to the "flu". The system of equations modeling the competition of two strains mediated by competition (see Fig. 13.4) is given by

$$\frac{dS}{dt} = \Lambda - \frac{\beta_1 S(I_1 + V_1)}{A} - \frac{\beta_2 S(I_2 + V_2)}{A} - \mu S,$$

$$\frac{dI_i}{dt} = \beta_i S \frac{(I_i + V_i)}{A} - (\mu + \gamma_i + \delta_i)I_i,$$

$$\frac{dQ_i}{dt} = \delta_i I_i - (\mu + \alpha_i)Q_i,$$

$$\frac{dR_i}{dt} = \gamma_i I_i + \alpha_i Q_i - \beta_j \sigma_{ij} R_i \frac{(I_j + V_j)}{A} - \mu R_i, \qquad j \neq i \qquad (13.2)$$

$$\frac{dV_i}{dt} = \beta_i \sigma_{ij} R_j \frac{(I_i + V_i)}{A} - (\mu + \gamma_i)V_i, \qquad j \neq i$$

$$\frac{dW}{dt} = \gamma_1 V_1 + \gamma_2 V_2 - \mu W,$$

$$A = S + W + I_1 + I_2 + V_1 + V_2 + R_1 + R_1 \quad \text{or} \quad A = 1 - Q,$$

where A denotes the population of non-isolated individuals.

In the study of System (13.2), Nuño et al. [22] show that influenza may

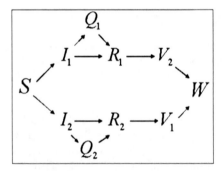

Fig. 13.4 Schematic diagram for the two-strain model with quarantine and cross-immunity

survive in three states or become extinct. It becomes extinct when no strain has the ability of becoming established, that is, when $\mathcal{R}_i < 1$ for $i=1$ and 2 where $\mathcal{R}_i$ denotes the ith strain-specific basic reproductive number. Competitive exclusion is also a possible outcome. It has been shown that the strain with the highest ability to invade (largest $\mathcal{R}_i$, as long as it is greater than one) can persist and eliminate competitors under appropriate conditions. The possibility that boundary endemic (one-strain endemic) equilibria become destabilized giving rise to periodic solutions has been settled. The analysis takes advantage of significant differences in time scales that are particularly relevant in the case of flu (and other "fast" communicable diseases).

The difference in the life-span of the host (several decades) and the lifetime of an individual "flu" infection (few days) allows for the introduction of a small parameter that facilitates the analysis, in particular the establishment of the possibility of periodic solutions via a Hopf bifurcation. Formulae for the approximate periods of these oscillations have also been computed. The dynamics of this model have been explored for various levels of cross-immunity and quarantine period lengths. Numerical simulations show that the use of fixed cross-immunity landscapes with variable (increasing) periods of quarantine that lead to periodic solutions do so with relatively unchanging epidemic periods. In other words, the period of the oscillations does not seem too sensitive to the length of the isolation period. Simulations also show that the amplitude of the outbreaks does increase as $\frac{1}{\delta}$ grows. Various types of additional simulations have been conducted. For example, we have fixed the quarantine period and varied the levels of cross-immunity. Such simulations result in periodic patterns that become irregular as the coefficient of cross-immunity, $\sigma \to 0^+$. Although the analytical findings are carried out only for the case of symmetric cross-immunity($\sigma_{ij} = \sigma_{ji} = \sigma$), the asymmetric case was also investigated, but numerically. The focus of such numerical simulations were for situations where $|\sigma_{ij} - \sigma_{ji}|$ is small. In this slightly asymmetric cases, simulations showed that outbreaks could occur every 10–13 years. The simulated dynamics captured additional differences. For example, alternating strain-specific peaks (highs and lows) for both strains were observed (Fig. 13.5).

The alternating nature of these peaks is the result of competition for a varying pool of susceptibles (their quality is mediated by cross-immunity) "available" for invasion to either strain. Increases and decreases in the effective pool of susceptibles for each strain are closely tied to the differences in levels of cross-immunity and the time of the initial invasion.

Using the stability results for each strain independently (boundary equilibria), Nuño et al. [22] also investigated the possibility that an established strain could be displaced by a non-resident strain. For this question, an appropriate reproductive number $\mathcal{R}_i^j$ had to be identified. $\mathcal{R}_i^j$ describes the number of secondary infections generated by strain i in a population where strain j has become established (resident in an endemic state). In order for strain i to invade (not necessarily replace) resident strain j it is required that $\mathcal{R}_i^j > 1$. The fact that $\mathcal{R}_i^j$ is an increasing function of σ follows directly from (13.3) below.

$$\mathcal{R}_i^j = \frac{\beta_i}{\mu + \gamma_i + \delta_i} \frac{\tilde{S}_j}{\tilde{A}} + \frac{\beta_i \sigma}{\mu + \gamma_i} \frac{\tilde{R}_j}{\tilde{A}}, \tag{13.3}$$

where $\mathcal{R}_i^j$ ($i, j = 1, 2$ and $i \neq j$). Here, $\beta_i/(\mu + \gamma_i + \delta_i)$ gives the number of secondary cases that strain-i infected individual generates in a susceptible population $\tilde{S}_j/\tilde{A}$ (infection prior to cross-immunity) while $\beta_i \sigma/(\mu + \gamma_i)$ gives

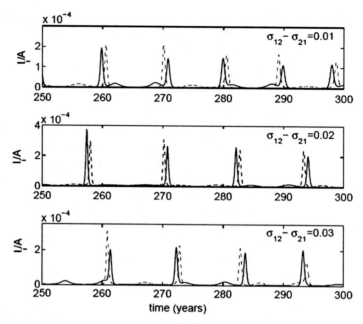

Fig. 13.5 Numerical integration of the model equations. Infective fraction of individuals (non-isolated) with strain 1 (*solid*) and strain 2 (*dashed*). Differences in cross-immunity levels between strains 1 and 2 ($|\sigma_{12} - \sigma_{21}|$) increase (from *top* to *bottom*) 0.01, 0.02 and 0.03. For example cross-immunity for strains 1 and 2 correspondingly are given by $\sigma_{12} = 0.36$ and $\sigma_{21} = 0.33$ (*bottom panel*)

the number of secondary cases that a strain-i infected individual is capable of generating within the "cross-immune" susceptible fraction $\tilde{R}_i/\tilde{A}$. From (13.3), we see that strong cross-immunity ($\sigma \downarrow 0$) reduces (significantly) the likelihood that strain i would successfully invade a population where strain j is endemic. Conversely, the likelihood of coexistence is enhanced when cross-immunity is weak ($\sigma \uparrow 1$), a result that also follows from (13.3). Hence, the dependence of $\mathcal{R}_i^j > 1$ on cross-immunity (σ) enhances the ability of strain i to invade resident strain j.

Numerical simulations were carried out to illustrate oscillatory coexistence for reasonable lengths of the quarantine period. Increasing the length of the periods of quarantine enhanced the amplitude of disease outbreaks. On the other hand, increasing cross-immunity ($\sigma \to 0^+$) for fixed periods of quarantine actually lengthens the distance between epidemic peaks. The dynamics of our two-strain model with quarantine supported reasonable disease patterns with parameter values in acceptable flu regions. It was also shown that cross-immunity facilitates coexistence (strain diversity) since it enhances the survival of less fit strains to survive – a function of the ineffectiveness of a strain to "prevent" future infections by related strains. In other words, some levels of cross-immunity facilitates "group" selection, that is, they enhance the likelihood of survival of related strains which may not survive on their own.

13.4 Age-Structure

Population heterogeneity, particularly age-structure, plays a key role on the dynamics of communicable diseases like influenza, measles, rubella, and others [1]. There has been plenty of theoretical work on *sir* (susceptible–infected–recovered) models with age-structure [18] but relatively few studies in the context of influenza, that have looked at the role of age-structure in maintaining recurrent epidemic outbreaks. The complications arise from the fact that influenza is a highly variable pathogen and, consequently, such questions are only possible in the context of cross-immunity mediated by competition [6,7]. A brief discussion on these topics follows.

The classical *sir* (susceptible–infected–recovered) single-strain age-structured model can be modified to include a quarantine (q) class. Here, we must consider four densities: $s(a,t)$, $i(a,t)$, $q(a,t)$ and $r(a,t)$ where $\int_{a_1}^{a_2} l(a,t)\,da$ denotes the number of l-individuals ($l = s,i,q$, or r) with ages in $[a_1,a_2]$ at time t. The transition rates are assumed to be age-dependent. Hence, $\mu(a)$, $\gamma(a)$, and $\delta(a)$ denote the age-specific mortality, recovery and quarantine rates, respectively. Modeling the incidence rate (new cases of infection per unit time) requires some care as one must model the age-specific interactions between individuals. Following Blythe and Castillo-Chavez [3] and Busenberg and Castillo-Chavez [5], we let $p(a,a',t)$ denote the proportion of contacts of age-a individuals with individuals of age a' at time t given that they had contacts. If we let $n(a,t) = s(a,t) + i(a,t) + q(a,t) + r(a,t)$ and $C(a)$ denote the age-structure per-capita contact rate then we have that $p(a,a',t)$ must satisfy the following properties:

(1) $p(a,a',t) \geq 0$
(2) $\int_0^\infty p(a,a',t)\,da' = 1$
(3) $C(a)n(a,t)p(a,a',t) = C(a')n(a',t)p(a',a,t)$

As was shown in Busenberg and Castillo-Chavez [5] the only separable solution $p(a,a',t) = g(a,t)f(a',t)$ is given by the mixing function commonly referred to as proportionate mixing or $\bar{p}(a')$ where

$$\bar{p}(a',t) = \frac{C(a')n(a',t)}{\int_0^\infty C(l)n(l,t)dl}.$$

Proportionate mixing has often been used as the "prototype" for modeling age-dependent contact rates when it comes to the transmission dynamics of communicable diseases [1,18]. Hence, we shall consider only proportionate mixing, that is, throughout the rest of this discussion we let $p(a,a',t) \equiv \bar{p}(a',t)$. If $m(a)$ denotes the age-specific susceptibility to infection per-contact then the incidence rate under proportionate mixing becomes

$$B(a,t) = \beta(a)s(a,t) \int_0^\infty \bar{p}(a',t)\frac{i(a',t)}{n(a',t) - q(a',t)} \, da'$$

where $\beta(a) = m(a)c(a)$. The remaining equations for our *siqr* model are

$$
\begin{aligned}
\left(\frac{\partial}{\partial a} + \frac{\partial}{\partial t}\right) s(a,t) &= -B(a,t) - \mu(a)s(a,t) \\
\left(\frac{\partial}{\partial a} + \frac{\partial}{\partial t}\right) i(a,t) &= B(a,t) - [\mu(a) + \gamma(a) + \delta(a)]i(a,t) \\
\left(\frac{\partial}{\partial a} + \frac{\partial}{\partial t}\right) q(a,t) &= \delta(a)i(a,t) - [\mu(a) + \gamma(a)]q(a,t) \\
\left(\frac{\partial}{\partial a} + \frac{\partial}{\partial t}\right) r(a,t) &= \gamma(a)[q(a,t) + i(a,t)] - \mu(a)r(a,t)
\end{aligned}
\tag{13.4}
$$

where

$$
\begin{aligned}
s(a,0) &= n_0(a) - i_0(a), \\
i(a,0) &= i_0(a), \\
q(a,0) &= r(a,0) = 0, \\
s(0,t) &= n(0,t) = \int_0^\infty \lambda(a)n(a,t) \, d\bar{a}, \\
i(0,t) &= q(0,t) = r(0,t) = 0.
\end{aligned}
$$

Consequently, $n(a,t)$ satisfies the Kermack and McKendrick [21] initial boundary problem. That is,

$$\left(\frac{\partial}{\partial a} + \frac{\partial}{\partial t}\right) n(a,t) = -\mu(a)n(a,t),$$

$$n(0,t) = \int_0^\infty \lambda(a)n(a,t) \, da,$$

$$n(a,0) = n_0(a),$$

where $\lambda(a)$ is the age-specific fertility rate.

It is known [20] that any solution $n(a,t)$ of the Kermack and McKendrick model approaches (uniformly) as $t \to \infty$ a separable solution $n^*(a)e^{p^*t}$ where p^* is the unique real root of Lotka's characteristic equation [20]. Furthermore, if $z = u + iv$ is also a root of Lotka's characteristic equation then it can be shown that $p^* > u$ (see [20]. Here, we assume that $p^* = 0$ and take $n_0(a) = n^*(a)$ (stable age-distribution). If we ignore the quarantine class (set $\delta(a) = \text{``}\infty\text{''}$ for all a) and use the results of Thieme [28] on asymptotically autonomous systems we can reduce the study of (13.4 to the consideration of the following equivalent system:

$$\left(\frac{\partial}{\partial a} + \frac{\partial}{\partial t}\right) s(a,t) = -\hat{B}(a,t) - \mu(a)s(a,t)$$
$$\left(\frac{\partial}{\partial a} + \frac{\partial}{\partial t}\right) i(a,t) = \hat{B}(a,t) - [\mu(a) + \gamma(a)]i(a,t),$$

(13.5)

and

$$r(a,t) = n^*(a) - s(a,t) - i(a,t),$$

with

$$\hat{B}(a,t) = \beta(a)s(a,t) \int_0^\infty \bar{p}(a')\frac{i(a',t)}{n^*(a')}\,da',$$
$$s(a,0) = n_0^*(a) - i_0(a),$$
$$i(a,0) = i_0(a),$$

and $n(0,t) =$ constant.

Castillo-Chavez et al. [6,7] version of the above model with

$$B(a,t) = \hat{\beta}(a)s(a,t) \int_0^\infty c(a')i(a',t)\,da'$$

is used to support (via some partial analysis and simulation) the hypothesis that single strain models with age-specific activity levels and constant levels of mortality (exponentially distributed survivorship) are incapable of supporting sustained oscillations. However, it was shown that uniform activity levels, $C(a) \equiv$ constant, when combined with age-dependent (non-exponentially distributed survivorship) mortality rates could generate sustained oscillations (Fig. 13.6). The above single-strain model was extended in [6,7] to incorporate the competition of two strains mediated by cross-immunity. Analysis and simulations of such two-strain model in [6,7] supported the possibility of sustained oscillations for some age-specific mortality regimes and wide range of cross-immunity levels. Simulations of a related (discrete in time and age) two-strain age-structure model supported the possibility of periodic solutions for reasonable cross-immunity values [6,7]. In fact, recurrent outbreaks with epidemic periods in the 3–5 year range for intermediate levels of cross-immunity and in the 10–20 year range for strong levels of cross-immunity ($\sigma \downarrow 0$) were observed. These results seem in agreement with those supported by data from limited studies [9,27]. In these numerical studies a significant difference in the amplitudes was observed. Hence, such simulations seem to capture, for intermediate values of cross-immunity, high strain-specific "flu" peaks followed by very low disease levels. The periods were in the 2–3 year range which seems to fit (in a rather crude way) the results reported in [27]. Similarly, the extremely long periods (10–20 years) supported by values of cross-immunity close to zero supported the view that the "same" strain cannot re-appear soon. These results again seemed to be in crude agreement with those reported in [9].

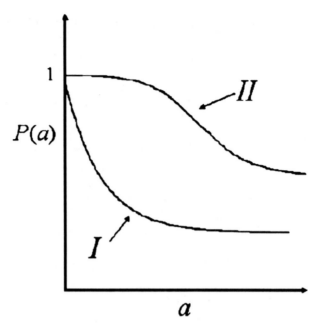

Fig. 13.6 Here $P(a)$ denotes the survivorship function as a function of age. I corresponds to the case when $P(a)$ is a negative exponential while II models the type of survivorship that one could expect in populations with long life-expectancy. The work in [6, 7] supports the hypothesis that as curve I is continuously deformed into curve II the appearance of periodic solutions takes place

13.5 Discussion and Future Work

In this chapter, we provide a personal and limited perspective on the role of quarantine, age structure and cross-immunity on "flu" dynamics, a disease characterized by the presence of two highly distinctive time scales (host's life span and host's infectious period). The set up used here is based on natural extensions of earlier work [6, 7, 12, 18] and references therein. The role of quarantine versus age structure and cross-immunity has been studied independently [6, 7, 12, 13, 19]. It was shown that $SIQR$ models and sir age-structure models were capable of supporting recurrent periodic outbreaks under various scenarios. Evidence to support the view that cross-immunity on its own [6, 7] is not enough to support sustained oscillations in non-structured populations was provided. Our most recent effort [22] investigates the joint impact of cross-immunity and quarantine in a non-structured (homogeneous mixing) population. It is established that strain competition mediated by cross-immunity can support recurrent epidemics with quarantine periods consistent with those implemented when there is a flu epidemic.

Expansions of the work discussed here to models that include a quarantine class and age-structure within single and multiple-strain systems (mediated by cross-immunity) are under way [11]. The analysis of such models is not trivial but we expect again to be able to exploit differences in times scales to gain some understanding of the joint effects of age-structure in models that include strongly or "poorly" isolated (quarantine) classes. We also expect to deepen our understanding of the relationship between cross-immunity, the amplitude of an epidemic outbreak and the length between epidemic peaks from the studies of models where strain-competition is mediated by cross-immunity.

We plan to conduct simulations of the full two-strain model with age-structure under specific scenarios. Seasonality in transmission rates is also an important issue. Its effect is being incorporated following earlier work (see [6, 7]). We hope this discussion will extend the interest in the study of influenza dynamics, a disease with immense pandemic potential.

Acknowledgments

The research of C. Castillo-Chavez and M. Nuño was supported with grants directed toward the Mathematical and Theoretical Biological Institute (MTBI) by the following institutions: National Science Foundation (NSF), National Security Agency (NSA), and Alfred P. Sloan Foundation. This work was also partially supported through the visit of Miriam Nuño and C. Castillo-Chavez to the Statistical and Applied Mathematical Sciences Institute (SAMSI), Research Triangle Park, NC, which is funded by NSF under the grand DMS-0112069. Most recently, UCLA has partially supported M. Nuño under National Institutes of Health grant AI28697. Special regards to go Dr. H. T. Banks for his hospitality, support and learned discussions. The research of the Z. Feng and M. Martcheva was supported by NSF grants DMS-0314575, DMS-0137687, and DMS-0406119.

References

1. R. Anderson and R. May, *Infectious Diseases of Humans: Dynamics and Control*, Oxford University Press, Oxford, 1991
2. G. W. Anderson, R. N. Arnstein and M. R. Lester, *Communicable Disease Control*, Macmillan, New York, 1962
3. S. P. Blythe and C. Castillo-Chavez, *Like with like preference and sexual mixing models*, Math. Biosci., 96, 221–238, 1989
4. F. Brauer and C. Castillo-Chavez, *Mathematical Models in Populations Biology and Epidemiology*, Springer, Berlin Heidelberg New York, 2001

5. S. N. Busenberg and C. Castillo-Chavez, *A general solution of the problem of mixing of subpopulations and its application to risk and age-structure models for the spread of AIDS*, IMA J. Math. Appl. Med. Biol., 8, 1–29, 1991

6. C. Castillo-Chavez, H. W. Hethcote, V. Andreasen, S. A. Levin and W. M. Liu, *Cross-immunity in the dynamics of homogeneous and heterogeneous populations*. In Mathematical Ecology (Trieste, 1986), World Scientific Publishing, NJ, 303–316, 1988

7. C. Castillo-Chavez, H. W. Hethcote, V. Andreasen, S. A. Levin and W. M. Liu, *Epidemiological models with age structure, proportionate mixing, and cross-immunity*, J. Math. Biol., 27, 233–258, 1989

8. CDC (Centers for Disease Control), *Recent Avian Influenza Outbreaks in Asia*, http://www.cdc.gov/flu/avian/outbreaks/asia.htm (April 26, 2005)

9. R. B. Couch and J. A. Kasel, *Immunity to influenza in man*, Annu. Rev. Microbiol., 31, 529–549, 1983

10. K. Dietz, *Epidemiological interference of virus populations*, J. Math. Biol., 8, 291–300, 1979

11. M. Erdem, *Epidemics in structured populations with isolation and cross-immunity*, Ph.D. thesis, Arizona State University-Main, Tempe, August, 2007

12. Z. Feng, *Multi-annual outbreaks of childhood diseases: revisited the impact of isolation*, Thesis, Arizona State University, 1994

13. Z. Feng and H. R. Thieme, *Recurrent outbreaks of childhood diseases revisited: the impact of isolation*, J. Math. Biosci., 128, 93–130, 1995

14. J. P. Fox, C. E. Hall, M. K. Cooney and H. M. Foy, *Influenza virus infections in Seattle families, 1975–1979*, Am. J. Epidemiol., 116, 212–227, 1982

15. W. P. Glezen and R. B. Couch, *Interpandemic influenza in the Houston area, 1974–76*, N. Engl. J. Med., 298, 11, 587–592, 1978

16. H. W. Hethcote, *Qualitative analyses of communicable disease models*, Math. Biosci., 28, 335–356, 1976

17. H. W. Hethcote and S. A. Levin, *Periodicity in epidemiological modeling*. In Applied Mathematical Ecology, Biomathematics 18, Springer, Berlin Heidelberg New York, 193–211, 1989

18. H. W. Hethcote, *The mathematics of infectious diseases*, SIAM Rev., 42(4), 599–653, 2000

19. H. W. Hethcote, M. Zhien and L. Shengbing, *Effects of quarantine in six endemic models for infectious diseases*, Math. Biosci., 180, 141–160, 2002

20. F. Hoppensteadt, *An age dependent epidemic model*, J. Franklin Inst., 297, 325–333, 1974

21. W. O. Kermack and A. G. McKendrick, *Contributions to the mathematical theory of epidemics part I*, Proc. R. Soc. Edinb., A115, 700–721, 1927

22. M. Nuño, Z. Feng, M. Martcheva, C. C. Chavez, *Dynamics of two-strain influenza with isolation and cross-protection*, SIAM J. Appl. Math., 65, 3, 964–982, 2005

23. M. M. Sigel, A. W. Kitts, A. B. Light and W. Henle, *The recurrence of influenza A-prime in a boarding school after two years*, J. Immunol., 64, 33, 1950

24. A. J. Smith and J. R. Davies, *Natural infection with influenza A (H3N2). The development, persistence and effect of antibodies to the surface antigens*, J. Hyg. (Lond.), 77, 271–282, 1976

25. C. H. Stuart-Harris, *Epidemiology of influenza in man*, Br. Med. Bull., 35, 3–8,1979

26. L. H. Taber, A. Paredes, W. P. Glezen and R. B. Couch, *Infection with influenza A/Victoria virus in Houston families, 1976*, J. Hyg. (Lond.), 86, 303–313, 1981

27. S. B. Thacker, *The persistence of influenza in human populations*, Epidemiol. Rev., 8, 129–142, 1986

28. H. R. Thieme, *Asymptotic estimates of the solutions of nonlinear integral equations and asymptotic speeds for the spread of populations*, J. Reine Angew. Math., 306, 94–121, 1979

29. L. Wu and Z. Feng, *Homoclinic bifurcation in an SIQR model for childhood disease*, J. Differ. Equ., 168, 150–167, 2000

Chapter 14
A Comparative Analysis of Models for West Nile Virus

M.J. Wonham and M.A. Lewis

Abstract This chapter describes the steps needed to formulate, analyze and apply epidemiological models to vector-borne diseases. Our models focus on West Nile (WN) virus, an emerging infectious disease in North America, first identified in Africa. We begin by introducing a minimalist model for WN dynamics to illustrate the processes of model formulation, analysis, and application. We then revisit the question of model formulation to examine how two major biological assumptions affect the model structure and therefore its predictions. Next, we briefly compare these different model structures in an introductory exercise of model parameterization, validation, and comparison. Finally, we address model applications in more detail with two examples of how the model output can usefully be connected to public health applications.

14.1 Introduction: Epidemiological Modeling

Investigating and controlling infectious diseases is a complex enterprise that has long been assisted by mathematical modeling (e.g., [2, 23]). In now classic examples, key insights into the dynamics of malaria, influenza, measles, and other infectious diseases have emerged from epidemiological modeling [26, 33, 39]. Today, emerging and re-emerging infectious diseases such as HIV/AIDS, SARS, feline immunodeficiency virus, hoof and mouth, and plant fungi and viruses pose major challenges in public health, wildlife, and agricultural management realms. The increase in outbreak frequency of these diseases demands a rapid and effective management response [9–11, 14, 16, 18].

Centre for Mathematical Biology, Department of Mathematical and Statistical Sciences, University of Alberta, Edmonton, AB, Canada T6G 2G1 mwonham@ualberta.ca
mlewis@math.ualberta.ca

Happily, there are many well-developed mathematical tools for effectively studying disease dynamics. There remain, however, continuing and exciting challenges in both formulating and analyzing biologically relevant disease models.

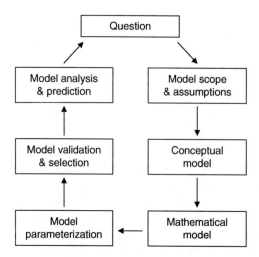

Fig. 14.1 Cartoon of the model development process from an initial question through model formulation and analysis to the generation of further questions

Developing and applying a disease model, as indeed any model, typically follows a series of steps (Fig. 14.1). An initial disease observation prompts questions such as how fast the disease will spread, or how best the outbreak can be controlled. From the initial question, we first define the scope and assumptions of the problem and develop a conceptual hypothesis (1), This is then formulated as a mathematical expression (2), which is parameterized (3), validated and compared (4), analysed (5), and finally applied and used to generate predictions (6), In this view, mathematical modeling follows the familiar scientific method. The model is essentially a formalization of a hypothesis that must be defined (steps 1–2) and tested (steps 3–4) before being used to answer questions or generate predictions (steps 5–6).

If we are lucky, the model's predictions shed light on the original question. They will also likely generate new questions and hypotheses to be addressed by further data collection and a subsequent return to modeling. In this chicken-and-egg fashion, our understanding of disease dynamics develops as empirical study informs modeling which in turn informs further empirical investigation.

The focus of this chapter is primarily on the steps of model formulation (steps 1–2) and model application (step 6) for infectious diseases. Model parameterization, validation, and comparison (steps 3–4) would readily fill another chapter, so we will restrict ourselves to a brief introduction to these important topics and provide references to more detailed resources. The mathematical analysis (step 5) of disease models in general is well treated in this book and elsewhere, so we will keep this aspect to a relative minimum.

We will focus our discussion on one particular type of epidemiological models, the well-studied $S - I$ or Susceptible-Infectious models. These compartmental models, descended from the work of [26], use the dynamics of interactions between S and I individuals to model the rate of emergence of new infectious individuals.

Many excellent texts introduce the philosophy and tools of mathematical modeling in infectious disease systems. For a general presentation of mathematical modeling in biological systems, we find those by [8, 20, 27, 35] particularly helpful. For modeling infectious diseases in particular, [2, 6, 12], provide excellent overviews and detailed examples. For the philosophy and methodology of model selection using maximum likelihood, we refer to [7, 25].

14.2 Case Study: West Nile Virus

Our model formulation and application center around the example of West Nile virus (WN), an emerging infectious disease in North America. WN was first identified in Uganda in 1947, and is widespread in Africa, the Middle East, and Western Asia. Occasional European outbreaks are introduced by migrating birds [21, 38]. In North America, the first recorded epidemic was detected in New York State in 1999 and spread rapidly across the continent. The unprecedented level of bird, horse, and human mortality was attributed to a highly virulent emerging strain of the virus [1, 36].

WN is characterized as an arboviral encephalitis, a designation that refers to its mosquito (arthropod) vector, its viral pathogenic agent, and its encephalitic symptoms. The disease amplifies in a transmission cycle between vector mosquitoes and reservoir-host birds, and is secondarily transmitted to mammals including humans [4, 19, 37]. The North American outbreak has been exceptionally well documented at mosquito, bird, and human levels, making it a prime candidate for mathematical analysis.

We will begin by introducing a minimalist model for WN dynamics to illustrate the processes of model formulation, analysis, and application. We will then revisit the question of model formulation to examine how two major biological assumptions affect the model structure and therefore its predictions.

Next, we will briefly compare these different model structures in an introductory exercise of model parameterization, validation, and comparison. Finally, we will address model applications in more detail with two examples of how the model output can usefully be connected to public health applications.

14.3 Minimalist Model

14.3.1 The Question

It is not often we see a dead bird outdoors in an urban setting. If we had lived in New York City in the summer of 1999, however, we would have been astonished by the unusually high number of dead crows, blue jays, and other birds found in backyards and parks. Later that year, we would have learned that the cause of death was a newly introduced disease, West Nile virus, that was carried by mosquitoes and was killing birds and humans [1, 28]. With those initial reports, we might have begun to ask any number of important questions. How would the disease affect bird populations? How infectious would it be in humans? How fast would it spread from New York to other locations? Would it spread to other animals as well? Was it carried by all mosquito species? Did they transmit it in every bite? How could the disease be controlled? Would mosquito spraying help? Would culling the bird population help?

Some of these questions would best be addressed in field and laboratory studies, others with mathematical modeling, and yet others with both approaches. For now, we will focus on the key question of how best to control a WN outbreak, and take advantage of empirical studies to inform and test our mathematical modeling.

14.3.2 Model Scope and Scale

To formulate a WN disease model , we must make some decisions about its scope and scale. Specifically, we need to think about the model's goals and complexity, and about how to represent time, space, population structure, and natural variation.

In terms first of model goals, are we interested in a more strategic model that simplifies the system to its barest essentials, or a more tactical model with comparatively more detail and complexity [29]? Model choice influences the kind of analysis we can conduct: generally speaking, a simpler model will be more amenable to analytical or more general analysis, whereas a more complex model will be restricted to numerical, or more specific case study analysis. Thus, the strategic approach may provide more qualitative insight into the basic properties of a system, whereas the tactical approach may provide better predictive ability for given species in a given location. Choosing a strategic philosophy is useful at this stage for studying general WN dynamics and control; later we might be interested in a more tactical model of local dynamics in a particular location. Our choice of a strategic or bare bones approach will help inform the remaining decisions about the model scale and scope.

Second, how should we model time and space? Mosquito and bird population dynamics exhibit an annual cycle, so we might consider a discrete-time model with yearly increments. But since we saw a very rapid disease increase in New York City within one summer, it might be interesting to focus on the shorter-term dynamics of a single season. This would allow us to ignore bird vital dynamics, and would require a model mosquito population that reproduces throughout the season. We thus choose a continuous time model that can be formulated as a system of ordinary differential equations (ODEs). For a more detailed discussion of continuous vs. discrete time models of WN virus, see [31]. Since our focal question is not explicitly spatial, we will consider only the changes in populations over time, giving us a nonspatial model. For some spatial approaches to modeling WN, see [30, 32].

Third, how should we represent the mosquito and bird populations? We could treat them in an individual-based framework, in age or stage classes, or as a homogeneous population. In the interests of strategy, we will think of them simply as homogeneous populations of identical individuals that can be represented by a single equation for all ages and stages. (We will revisit this choice in Sect. 14.5.) WN has been reported thus far from ~60 mosquito species and ~280 bird species in North America. Again for strategy, we will model only a single generic mosquito and a single generic bird species, acknowledging that this limits our ability to address broader scale ecological questions in WN dynamics (e.g., [14, 17]). Furthermore, since mammals (including humans) appear not to transmit the disease back to the mosquito population, they are considered secondary hosts [4, 19, 24, 37], so the fundamental disease dynamics do not depend on them. Our model will therefore represent only the mosquito and bird populations.

Finally, are we interested in a stochastic model that can capture the natural variation in model parameters such as birth, death, and infection rates? Or are we interested in a more deterministic model that forsakes the vagaries of realism in favour of clearer, but simplified, analytical results? Given our strategic focus, we will develop a deterministic model. (In the interests of interpreting and applying the model output, however, we will examine the effects of stochastic variation on model predictions in Sect.14.4.3.)

14.3.3 Model Formulation

We have now decided to develop a strategic, continuous-time, single-season, nonspatial, non age-structured deterministic model of WN dynamics. With our adjectives thus in place, we can begin to sketch out the model structure.

We have already chosen to use the well-established $S - I$ epidemiological modeling framework, in which bird and mosquito populations will be divided into classes of susceptible and infectious individuals. We thus have four classes representing the densities of susceptible birds (S_B), infectious birds (I_B), susceptible mosquitoes (S_M), and infectious mosquitoes (I_M) (Fig. 14.2a). Susceptible birds can become infectious when they are bitten by an infectious mosquito; susceptible mosquitoes can become infectious when they bite an infectious bird. For the bird lifecycle, which is one to two orders of magnitude longer than the single season represented by the model, birth and death rates can reasonably be omitted. The mosquito lifecycle, which has length of order one month, is represented by birth and death rates. Since birds die from WN infection, but mosquitoes do not, we include a disease-death rate for birds. This model (Fig. 14.2a) can be expressed as a system of four ordinary differential equations,

$$\underbrace{\frac{dS_B}{dt}}_{\substack{\text{Susceptible} \\ \text{birds}}} = - \underbrace{\alpha_B \beta_B \frac{S_B}{N_B} I_M}_{\text{disease transmission}}$$

$$\underbrace{\frac{dI_B}{dt}}_{\substack{\text{Infectious} \\ \text{birds}}} = \underbrace{\alpha_B \beta_B \frac{S_B}{N_B} I_M}_{\text{disease transmission}} - \underbrace{\delta_B I_B}_{\substack{\text{death from} \\ \text{disease}}}$$

$$\underbrace{\frac{dS_M}{dt}}_{\substack{\text{Susceptible} \\ \text{mosquitoes}}} = \underbrace{b_M N_M}_{\text{birth}} - \underbrace{\alpha_M \beta_B \frac{I_B}{N_B} S_M}_{\text{disease transmission}} - \underbrace{d_M S_M}_{\text{death}}$$

$$\underbrace{\frac{dI_M}{dt}}_{\substack{\text{Infectious} \\ \text{mosquitoes}}} = \underbrace{\alpha_M \beta_B \frac{I_B}{N_B} S_M}_{\text{disease transmission}} - \underbrace{d_M I_M}_{\text{death}}$$

(14.1)

where the total bird population density $N_B = S_B + I_B$ and the total adult female mosquito population density $N_M = S_M + I_M$. At the disease-free equilibrium (DFE), where all individuals are susceptible, the bird and mosquito population densities are denoted N_B^* and N_M^*, respectively. We assume that, at the DFE, the mosquito population is constant so the birth and death

rates are balanced and $b_M = d_M$. The model variables and parameters are
further defined in Tables 14.1 and 14.2. The disease-transmission dynamics
used in this model are known as frequency-dependent. In the Sect. 14.4, we
define this term more fully and compare the effects of modeling different
transmission dynamics.

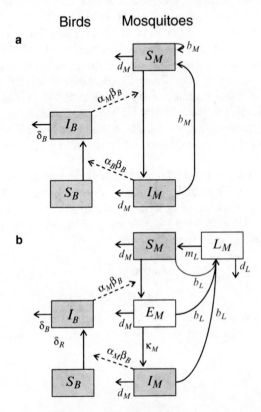

Fig. 14.2 Conceptual model for West Nile disease dynamics in mosquitoes and birds for
(**a**) the minimalist model with only four population classes (14.1) and (**b**) a slightly more
biologically complex and realistic model with two added mosquito compartments (14.9).
Vital and epidemiological dynamics indicated with solid lines and transmission dynamics
with dashed lines. Variables and parameters are defined in Tables 14.1 and 14.2. Adapted
from [43] Fig. 1

14.3.4 Model Analysis

One of the most powerful tools developed for analyzing and interpreting epidemic models is the disease basic reproduction number, $\mathcal{R}_0$ [2, 15, 22, 23]. Conceptually, $\mathcal{R}_0$ is defined as the average number of secondary infections caused by the introduction of a typical infective individual into an otherwise entirely susceptible population [2, 22]. Mathematically, $\mathcal{R}_0$ is defined as the spectral radius of the next generation matrix for new infections [13, 42].

The reproduction number serves as an invasion threshold for predicting disease outbreaks and evaluating control strategies. Quantitatively, it has a threshold value of one. When $\mathcal{R}_0 < 1$, the DFE is locally stable and the introduction of a small number of infectious individuals will not lead to a disease outbreak. When $\mathcal{R}_0 > 1$ the DFE is unstable and an outbreak will occur. The analytical expression for $\mathcal{R}_0$ is also very useful, since it indicates which elements of the disease system can be manipulated to reduce the chance of an outbreak.

To obtain $\mathcal{R}_0$ for model (14.1), we follow [42] in using vector notation to rewrite the equations in which infections appears as the difference between f_j, the rate of appearance of new infectives in class j, and v_j, the rate of transfer of individuals into and out of class j by all other processes. New infectives arise in I_B and I_M only, giving

$$\frac{d}{dt}\begin{bmatrix} I_B \\ I_M \end{bmatrix} = f - v = \begin{bmatrix} \alpha_B \beta_B \frac{S_B}{N_B} I_M \\ \alpha_M \beta_M \frac{I_B}{N_B} S_M \end{bmatrix} - \begin{bmatrix} \delta_B I_B \\ d_M I_M \end{bmatrix}. \tag{14.2}$$

The corresponding Jacobian matrices, F and V, describe the linearization of this reduced system about the DFE (where $S_M = N_M^*$ and $S_B = N_B^*$),

$$F = \begin{bmatrix} 0 & \alpha_B \beta_B \\ \alpha_M \beta_B \frac{N_M^*}{N_B^*} & 0 \end{bmatrix}, V = \begin{bmatrix} \delta_B & 0 \\ 0 & d_M \end{bmatrix}, \tag{14.3}$$

giving the next generation matrix,

$$FV^{-1} = \begin{bmatrix} 0 & \frac{\alpha_B \beta_B}{d_M} \\ \frac{\alpha_M \beta_B N_M^*}{\delta_B N_B^*} & 0 \end{bmatrix}. \tag{14.4}$$

The spectral radius, or spectral bound, of FV^{-1} is the reproduction number,

$$\mathcal{R}_0 = \underbrace{\sqrt{\frac{\alpha_B \beta_B}{d_M}}}_{\substack{\text{mosquito} \\ \text{to bird}}} \underbrace{\sqrt{\frac{\alpha_M \beta_B N_M^*}{\delta_B N_B^*}}}_{\substack{\text{bird to} \\ \text{mosquito}}}. \tag{14.5}$$

The $\mathcal{R}_0$ expression in (14.5) consists of two elements under the square root sign. The first represents the number of secondary bird infections caused by one infected mosquito. The second represents the number of secondary mosquito infections caused by one infected bird. Taking the square root gives the geometric mean of these two terms, which can be interpreted as $\mathcal{R}_0$ for the addition of an average infectious individual, whether mosquito or bird, to an otherwise susceptible system [41].

14.3.5 Model Application

The minimalist WN model (14.1) is a neat, simple, compact model for the disease dynamics in mosquitoes and birds. What we do not know is if this model is any good at capturing empirically observed WN dynamics. This important question will be addressed in the processes of model parameterization, validation, and comparison, to which we will return in Sect. 14.6. For simplicity of presentation, however, we will first consider how this model – assuming it is a good one – might be applied.

Given our model, what is the best strategy for controlling a WN outbreak? In the expression for $\mathcal{R}_0$ (14.5), we can find the answer. The goal for reducing the chance of a WN outbreak is to reduce $\mathcal{R}_0$, which can be accomplished by reducing the mosquito density at the DFE, N_M^*. In contrast, reducing the bird density N_B^* will increase $\mathcal{R}_0$ and therefore increase the chance of outbreak. What is the explanation for this puzzling result? Although it seems counterintuitive at first, we can see that reducing the bird density means the remaining individuals are bitten more often by hungry mosquitoes. In this way, the disease transmission is concentrated through a few highly-bitten birds that are more likely to become infected, and to re-infect the mosquitoes.

By looking more closely at the $\mathcal{R}_0$ expression, we can determine how much control is needed. Here the ratio of mosquitoes to birds at the DFE, $n_m^* = N_M^*/N_B^*$, is the crucial feature. Setting $\mathcal{R}_0$ to its critical value of one and rearranging the expression gives the threshold ratio of mosquito to bird densities,

$$\hat{n}_m^* = d_M \delta_B / \alpha_B \alpha_M \beta_B^2 \tag{14.6}$$

above which an outbreak can occur. Reducing the relative mosquito density below this level will prevent an outbreak. The mechanics of applying this strategy are described in more detail in Sect. 14.7. The structure and control implications of $\mathcal{R}_0$ that we see here follow directly from the assumption in model (14.1) of frequency-dependence disease transmission. In the next section, we will see how the results differ when different transmission dynamics are assumed.

14.4 Biological Assumptions 1: When does the Disease-Transmission Term Matter?

The disease-transmission term in our WN mode represents the contact dynamics between mosquitoes and birds, which depend on the biting rate by a mosquito. The term used to describe the biting rate in an $S - I$ model typically takes one of two forms, frequency dependence or mass action [2,3,34].

14.4.1 Frequency Dependence

The commonly used frequency dependent transmission, shown in (14.1), follows [2] in assuming that the mosquito biting rate is saturated, and not limited by bird density. In other words, the biting rate by an individual mosquito is constant across bird densities (Fig. 14.3a), and the biting rate experienced by an individual bird increases with mosquito density (Fig. 14.3b). Under this assumption, the biting rate by a mosquito is taken to be the maximal rate allowed by the gonotrophic cycle, or, the minimum time required between blood meals for a female to produce and lay eggs, or, the maximum possible number of bites per day made by a single mosquito. This biting rate β_B has unit time^{-1}.

These biological assumptions are captured in the mathematical formulation of frequency dependence, in which the proportional bird densities appear in (14.1). Near the disease-free equilibrium, where both populations are almost entirely susceptible (i.e., $S_M^* = N_M^*$, $S_B^* = N_B^*$, and I_M and I_B are very small) the mosquito-to-bird transmission rate $\beta_B I_M S_B / N_B$ depends only on the biting rate, while the bird-to-mosquito transmission rate $\beta_B S_M I_B / N_B$ depends on the biting rate and also on the ratio of mosquito to bird densities (as well as on the disease transmission probabilities α_M and α_B which for simplicity are not shown here).

14.4.2 Mass Action

Another common disease transmission term is mass action (e.g., [3, 34]). Mass action differs from frequency dependence in assuming that the mosquito

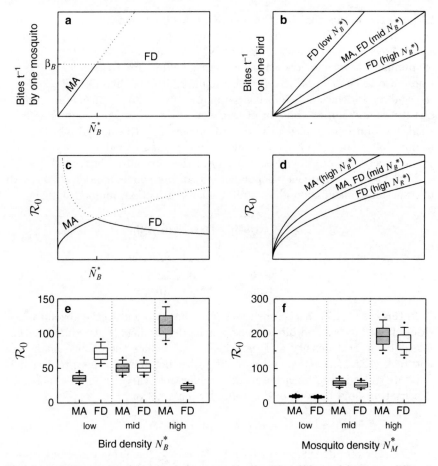

Fig. 14.3 Different disease-transmission terms in the West Nile model assume different biting rates (**a–b**) and lead to qualitatively different reproduction numbers, $\mathcal{R}_0$ (**c–d**) with different numerical values (**e–f**). Biting rates are shown as (**a**) the number of bites per day by a single mosquito as a function of bird density, and (**b**) the number of bites per day on a single bird as a function of mosquito density, for the two disease-transmission terms, frequency dependence FD and mass action MA. The maximum biting rate β_B is reached at the bird density denoted $\tilde{N}_B^*$. The bird densities over which each transmission term applies are indicated by the *solid lines*; using a term at an inappropriate bird density in the *dotted line* regions will give misleadingly high or low $\mathcal{R}_0$ estimates (**a,c**). The reproduction number $\mathcal{R}_0$ is shown as a function of (**c**) bird density and (**d**) mosquito density. At mid bird densities ($N_B^* = \tilde{N}_B^*$), the biting rates (**a,b**) and the $\mathcal{R}_0$ (**c,d**) of MA and FD coincide. At higher bird densities ($N_B^* > \tilde{N}_B^*$), the biting rate (**a,b**) and the $\mathcal{R}_0$ (**c,d**) of FD lie below that of MA, whereas at lower bird densities ($N_B^* < \tilde{N}_B^*$), they lie above ($\mathcal{R}_0$ for this latter scenario not shown in **d**). For numerical $\mathcal{R}_0$ estimates, *vertical dotted lines* separate regions of (**e**) low ($N_B^* < \tilde{N}_B^*$), medium ($N_B^* = \tilde{N}_B^*$), and high ($N_B^* > \tilde{N}_B^*$) bird densities, and (**f**) low, medium, and high mosquito densities. Sample population densities chosen to illustrate these different regions of $\mathcal{R}_0$, expressed as number km^{-2}, are (**a**) $N_M^* = 1,000$, $\tilde{N}_B^* = 100$, and $N_B^* = 50$ (low), 100 (mid), and 500 (high), and (**b**) $\tilde{N}_B^* = 500$, $N_B^* = 550$, and $N_M^* = 100$ (low), 550 (mid), and 5,500 (high). Parameter values as in Table 14.2. *Boxes* show median and 25th and 75th percentiles, *bars* show 10th and 90th percentiles, and *dots* show 5th and 95th percentiles. Adapted from [43] Fig. 2–3

biting rate is limited by the densities of both mosquitoes and birds (Fig. 14.3a,b). Mass action is a biologically sensible assumption only up to some threshold bird density, denoted $\tilde{N}_B^*$. We can understand this limit by examining the disease transmission terms, which are written as $\beta_B' I_M S_B$ (mosquito to bird) and $\beta_B' S_M I_B$ (bird to mosquito), again omitting the terms α_M and α_B for clarity. The biting parameter $\beta'_B = \beta_B/\tilde{N}_B^*$ has units time^{-1} density^{-1}, and can be thought of as the number of bites per day made by a single mosquito, per unit density of birds. Above $\tilde{N}_B^*$, the mosquito biting rate in units bites time^{-1}, $\beta'_B N_B^*$, would exceed β_B, and therefore by definition would exceed the physiological capacity of the mosquito (Fig. 14.3a).

Replacing the frequency-dependence transmission in model (14.1) with mass action transmission gives a different reproduction number, namely

$$\mathcal{R}_0 = \underbrace{\sqrt{\frac{\alpha_B \beta_B'}{d_M}}}_{\substack{\text{mosquito} \\ \text{to bird}}} \underbrace{\sqrt{\frac{\alpha_M \beta_B' N_M^* N_B^*}{\delta_B}}}_{\substack{\text{bird to} \\ \text{mosquito}}}. \tag{14.7}$$

In this case (14.7), $\mathcal{R}_0$ is sensitive not to the ratio, but to the absolute densities of mosquitoes and birds. Thus, the model predicts that reducing either mosquito or bird density will reduce $\mathcal{R}_0$ and reduce the chance of disease outbreak (Fig. 14.3c,d). In terms of the bird population, this prediction is opposite to that of frequency dependence. Setting $\mathcal{R}_0 = 1$ gives the mosquito density threshold for WN outbreak under mass action,

$$\hat{N}_M^* = d_M \delta_B / \alpha_B \alpha_M (\beta_B')^2 N_B^*. \tag{14.8}$$

When the bird density at the DFE is $N_B^* = \tilde{N}_B^*$, the biting rates under mass action and frequency dependence coincide (Fig. 14.3a–b) and the $\mathcal{R}_0$ values are equal (Fig. 14.3c–d). At lower bird densities where $N_B^* < \tilde{N}_B^*$, the biting rate and $\mathcal{R}_0$ of frequency dependence are artificially high, whereas at higher densities where $N_B^* > \tilde{N}_B^*$, it is those of mass action that are artificially high (Fig. 14.3a,c). This is because the frequency-dependent formulation assumes the maximal mosquito biting rate even when the bird density is very low, and the mass action formulation assumes an impossibly high biting rate when the bird density is very high.

We can see the same dynamics when we examine biting rates and $\mathcal{R}_0$ with respect to mosquito density. As expected, the $\mathcal{R}_0$ curves for both transmission terms coincide when $N_B^* = \tilde{N}_B^*$ (Fig. 14.3d, middle curve). At higher bird densities ($N_B^* > \tilde{N}_B^*$) the curve for frequency dependence is lower, but the curve for mass action is higher because the maximal mosquito biting rate is exceeded (Fig. 14.3d). In the opposite case of $N_B^* < \tilde{N}_B^*$, the relative positions of the two curves are reversed (not shown).

14.4.3 Numerical Values of $\mathcal{R}_0$

The analytical results obtained above illustrate how the choice of disease-transmission term can change $\mathcal{R}_0$, thus altering the model's control implications. Do these alterations translate into significant differences in the numerical estimates of $\mathcal{R}_0$? To address this question, we generated quantitative $\mathcal{R}_0$ estimates that incorporated the underlying variation in the constituent model parameters (Table 14.2, Fig. 14.3e–f). For details of this parameter estimation and resampling see [43]; an introduction to these methods is given in [5]. At low bird density where mass action applies, the $\mathcal{R}_0$ of frequency dependence can be significantly too high (Fig. 14.3e). At the threshold bird density $\tilde{N}_B^*$, where both mass action and frequency dependence apply, the $\mathcal{R}_0$ value is the same. At higher bird density where frequency dependence applies, the $\mathcal{R}_0$ of mass action is significantly too high (Fig. 14.3e). Similar comparisons can be made for low, medium, and high mosquito densities (Fig. 14.3f). These numerical results show that, for these parameter values, a transmission term misapplied at an inappropriate host or mosquito population density can significantly over- or underestimate $\mathcal{R}_0$. For the remainder of the chapter, we will use the model formulation with frequency-dependent transmission terms, as in (14.1).

14.5 Biological Assumptions 2: When do Added Model Classes Matter?

The minimalist model (14.1) contains the fewest possible classes for bird and mosquito populations. For mosquitoes in particular, this is a considerable oversimplification of the lifecycle and epidemiology. What is the effect on the model output of incorporating additional biologically realistic classes? We will consider two candidate mosquito classes, and find that one influences $\mathcal{R}_0$ whereas the other does not.

The mosquito lifecycle includes larval and pupal stages, which may represent up to a quarter of the mosquito lifespan. Their inclusion might therefore be expected to slow down the model dynamics. These pre-adult stages can be added to the model as a combined Larval compartment L_M, with associated birth rate b_L and maturation rate m_L (Fig. 14.2b).

Empirical studies of infected mosquitoes show that they undergo a viral incubation period during which they are infected, but not infectious. Only when the virus reaches a sufficiently high concentration, and is disseminated out of the gut and into the salivary glands, is the insect capable of transmitting the disease. This incubation period lasts some 7–12 days and can be modeled as an exposed compartment, E_M, with associated incubation rate

κ_M (Fig. 14.2b). These two compartments can be incorporated into the model's mathematical structure as follows:

$$\underbrace{\frac{dS_B}{dt}}_{\substack{\text{Susceptible} \\ \text{birds}}} = - \underbrace{\alpha_B \beta_B I_M \frac{S_B}{N_B}}_{\text{disease transmission}}$$

$$\underbrace{\frac{dI_B}{dt}}_{\substack{\text{Infectious} \\ \text{birds}}} = \underbrace{\alpha_B \beta_B I_M \frac{S_B}{N_B}}_{\text{disease transmission}} - \underbrace{\delta_B I_B}_{\substack{\text{death from} \\ \text{disease}}}$$

$$\underbrace{\frac{dL_M}{dt}}_{\substack{\text{Larval} \\ \text{mosquitoes}}} = \underbrace{b_L (S_M + E_M + I_M)}_{\text{birth}} - \underbrace{m_L L_M}_{\text{maturation}} - \underbrace{d_L L_M}_{\text{death}}$$

$$\underbrace{\frac{dS_M}{dt}}_{\substack{\text{Susceptible} \\ \text{mosquitoes}}} = - \underbrace{\alpha_M \beta_B S_M \frac{I_B}{N_B}}_{\text{disease transmission}} + \underbrace{m_L L_M}_{\text{maturation}} - \underbrace{d_M S_M}_{\text{death}}$$ (14.9)

$$\underbrace{\frac{dE_M}{dt}}_{\substack{\text{Exposed} \\ \text{mosquitoes}}} = \underbrace{\alpha_M \beta_B S_M \frac{I_B}{N_B}}_{\text{disease transmission}} - \underbrace{\kappa_M E_M}_{\substack{\text{disease} \\ \text{incubation}}} - \underbrace{d_M E_M}_{\text{death}}$$

$$\underbrace{\frac{dI_M}{dt}}_{\substack{\text{Infectious} \\ \text{mosquitoes}}} = \underbrace{\kappa_M E_M}_{\substack{\text{disease} \\ \text{incubation}}} - \underbrace{d_M I_M}_{\text{death}}$$

where the total female mosquito density $N_M = (L_M + S_M + E_M + I_M)$. For this model, the assumption of a constant mosquito population at the DFE is met by the parameter constraint that $b_L = d_M(m_L + d_L)/m_L$.

Following a similar $\mathcal{R}_0$ analysis as that given above (14.2–14.5), we obtain

$$\mathcal{R}_0 = \sqrt{\frac{\alpha_B \beta_B \phi_M}{d_M} \frac{\alpha_M \beta_B \frac{N_M^*}{N_B^*}}{\delta_M}}, \qquad (14.10)$$

where ϕ_M is the proportion of exposed mosquitoes surviving the exposed period to become infectious, $\phi_M = \kappa_M/(\kappa_M + d_M)$. As before, setting $\mathcal{R}_0 = 1$ returns the critical relative mosquito density above which the virus will invade a constant population of susceptible individuals,

$$n_m^* = d_M \delta_B / \alpha_B \alpha_M \beta_B^2 \phi_M. \tag{14.11}$$

By inspecting this $\mathcal{R}_0$ expression (14.10), we can see that the added exposed class alters $\mathcal{R}_0$, reducing it by the fraction $\sqrt{\phi_M}$. In contrast, the added larval class does not influence $\mathcal{R}_0$. We can understand this curious result by recalling the definition of $\mathcal{R}_0$, which applies only to the linearized system around the DFE, and is calculated using only the equations for infected individuals. Recall too that $\mathcal{R}_0$ is simply a ratio and has no time scale, so that although adding a larval compartment may delay the system's dynamics (see Sect. 14.6), it does not affect the average number of secondary infections caused by the introduction of an infectious individual into an otherwise susceptible population – the definition of $\mathcal{R}_0$. The following section shows how both the larval and exposed mosquito compartments can influence the model's numerical outbreak simulations.

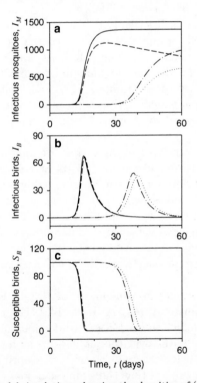

Fig. 14.4 Numerical model simulations showing the densities of (**a**) infectious mosquitoes, (**b**) infectious birds, and (**c**) susceptible birds over time for the minimalist West Nile virus model (*solid line*) and extensions with a larval mosquito class (*dashed line*), an exposed mosquito class (*dash-dot line*), and both exposed and larval classes (*dotted line*). Simulations run for 60 days with initial susceptible densities $N_M^* = 1{,}500$ and $N_B^* = 100$, and a disease inoculum of $I_M(0) = 0.0001$

14.6 Model Parameterization, Validation, and Comparison

In the terms of the scientific method, formulating a model is the formal equivalent of proposing a conceptual hypothesis. Applying a model without testing it (as we did in Sect. 14.3) is like using a hypothesis to make predictions before the hypothesis has been tested. As with hypotheses, evaluating one model is good, but testing and comparing multiple models is even better [7, 25]. The science of model testing – which includes parameterization, validation, and multi-model comparison – is a highly developed statistical enterprise with multiple approaches. One key approach is that of maximum likelihood, which tests the relative abilities of multiple models to fit an independent dataset. This is a widely used and powerful method, but the details are beyond the scope of this chapter and readers are referred to central references such as [7, 25] for further guidance.

Instead, we will take a brief look at how model parameterization can be tackled, and then used for qualitative model comparison. From the two WN models we have formulated, (14.1) and (14.9), we can generate four candidate model structures for this disease: the minimalist model (14.1), two models of intermediate complexity based on model (14.1) with either the larval or the exposed mosquito class added, and the full model with both added classes (14.9). Running numerical solutions can help us compare the predictions of these four models.

Numerical simulation requires first that we obtain parameter values, which can be derived from literature reports of field and laboratory studies (Table 14.2). The more recent studies are readily found through Internet search engines; older studies, which are often gold mines of valuable data, may require a little more legwork and library time. Often, we can find only a mean and range of expected values for our parameters. The mean values give us deterministic model simulations, and the ranges can be used in stochastic simulations to evaluate the effects of natural variation and uncertainty in the estimates (e.g., [43]).

Figure 14.4 shows numerical simulations of all four WN model structures using the mean parameter values in Table 14.2. For a given set of initial conditions, we can predict the densities of infectious mosquitoes, infectious birds, and susceptible birds, over time following the introduction of a small infectious inoculum to an otherwise entirely susceptible population. From these simulations, we can see that the simplest model (14.1) has the fastest dynamics and the earliest outbreak, and the most complex model (14.9) has the slowest and latest (Fig. 14.4). Adding the larval class to model (14.1) makes only the slightest difference in the outbreak timing, but adding the exposed class to model (14.1) makes a substantial difference (Fig. 14.4).

This type of preliminary qualitative model assessment should best be followed by a rigorous quantitative comparison using maximum likelihood or other evaluative techniques to see which model best fits the observed, independent, outbreak data. Some of these methods are discussed for WN models by [43]; more extensive commentary and methodology of model validation and comparison are provided by [7, 25]. For the remaining model analyses in this chapter, we will use the full model with both larval and exposed mosquito classes (14.9).

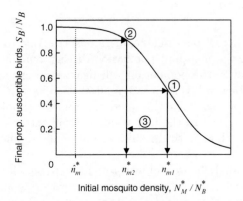

Fig. 14.5 Connecting the WN model output to disease surveillance and control applications. The *curved line* shows the final proportion of surviving birds at the end of the season as a function of the initial mosquito-to-bird density at the beginning of the season. For a season with observed 50% bird mortality, the initial mosquito density can be inferred (1) For a future season with a target bird survival of 90%, the required initial mosquito density can be inferred in the same way (2) The ratio of these two initial mosquito densities (3) gives the relative reduction in the mosquito population required to obtain the target level of bird mortality. Adapted from [41] Fig. 3a

14.7 Model Application #1: WN Control

The model output from (14.9) presents a WN outbreak threshold in terms of the relative densities of mosquitoes and birds at the DFE (14.11). However, this ratio would be extremely difficult to estimate on the ground. Can the model output be better connected to real-life disease management? WN surveillance programs typically track the number of dead birds throughout a season, a datum that can be linked to the initial mosquito-to-bird ratio as follows. By running repeated numerical solutions starting from different initial population densities at the DFE, a relationship can be plotted between the initial ratio of mosquito to bird densities, n_m^*, and the final disease-induced bird mortality at the end of the season (Fig. 14.5). For a given level of mortality observed at the end of the season, the initial value n_m^* can be inferred. For future seasons, a target level of bird loss can be obtained by calculating the required relative reduction in the mosquito population (Fig. 14.5).

14.8 Model Application #2: Seasonal Mosquito Population

Our model is restricted to a single temperate North American season, during which we assume the mosquito population density remains constant. More realistically, however, the population will increase in spring and decline in fall (Fig. 14.6a). How will this variation affect the model predictions? There are a number of ways to tackle this question, from the simple to the complex and from the analytical to the numerical. We will take a simpler approach that will give us both analytical results and a useful graphical interpretation (Fig. 14.6).

To pursue this analysis, we will have to put $\mathcal{R}_0$ to one side and introduce a second major model analysis tool, the disease growth rate λ. Mathematically, λ is the maximum real part of the eigenvalues of the ODE system linearized around the DFE. It has the threshold value $\lambda = 0$, above which a disease outbreak will occur and below which it will not. The parameters λ and $\mathcal{R}_0$ are related, in that the same threshold mosquito density $\hat{n}_m^*$ corresponds to both $\mathcal{R}_0 = 1$ and $\lambda = 0$. An important difference between the two parameters is that $\mathcal{R}_0$ is a dimensionless ratio with no time scale, whereas λ is a rate with unit time^{-1}. This feature is an advantage when we want to consider the effects of different mosquito levels over time.

Note that the disease reproduction number $\mathcal{R}_0$ and the disease growth rate λ are connected by the disease generation time, T_g, the mean time interval between infection of a host individual and the secondary infections it causes, such that $\lambda = \log(\mathcal{R}_0)/T_g$.

Calculation of the disease growth rate is given in the Appendix. Its calculation introduces a slight change of notation that was partially introduced in (14.6) and will prove more convenient for what follows. Specifically, the ratio of the current mosquito density to the initial bird density, N_M/N_{B^*}, becomes n_m, the ratio at the DFE, N_{M^*}/N_{B^*}, becomes n_m^*, and the threshold ratio for disease outbreak with a constant mosquito density is $\hat{n}_m^*$.

Based on empirical observations, we will represent mosquito seasonality as a simple step function (Fig. 14.6a), giving the mean relative mosquito density over the season,

$$\bar{n}_m = \left(t_a n_m^a + t_b n_m^b\right) / \left(t_a + t_b\right), \qquad (14.12)$$

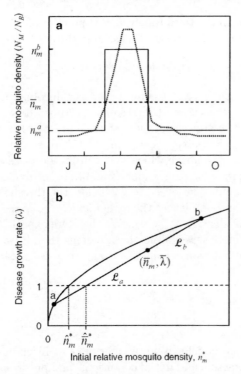

Fig. 14.6 Applying the WN model to a seasonal mosquito population. (**a**) The observed mosquito population levels (from June to October in Boston, USA; data from [40] Fig. 1 replotted here on a linear scale) shown with the dotted line can be represented crudely as a step function from n_m^a to n_m^b to n_m^a (*solid line*). Multiplying by the time spent at levels a and b, periods t_a and t_b respectively, gives the mean mosquito density over the season, $\bar{n}_m$ (*dashed line*). (**b**) For a constant population, the relationship between the disease growth rate λ and the initial mosquito density n_m^* is shown with the curved line. WN outbreak occurs when $\lambda > 0$, i.e., when $n_m^* > \hat{n}_m^*$. For variable mosquito density, the linear relationship between the average growth rate $\bar{\lambda}$ and the average initial disease-free mosquito density $\bar{n}_m^*$ is given by the straight line $L = L_a + L_b$ that connects points a (n_m^a, λ_a) and b (n_m^b, λ_b). A WN outbreak occurs when $\bar{\lambda} > 0$, i.e., when $\bar{n}_m^* > \hat{\bar{n}}_m^*$. For a given season, the point $(\bar{n}_m^*, \bar{\lambda})$ may be calculated from (14.12)–(14.13), or may be obtained graphically as the point along L where the ratio between line segments $L_a:L_b = t_a:t_b$. As long as the lower population density $n_m^a < \hat{n}_m^*$, the threshold mosquito density for disease outbreak will be higher for a seasonal than for a constant mosquito density. Adapted from [41] Fig. 3b–c

where t_a and t_b refer to the total time spent at population levels n_m^a and n_m^b, respectively (Fig. 14.6a). The mean disease growth rate is then given by

$$\bar{\lambda} = (t_a\lambda_a + t_b\lambda_b) / (t_a + t_b), \qquad (14.13)$$

where λ_a and λ_b are the largest eigenvalues of the Jacobian matrix J evaluated at n_m^a and n_m^b, respectively (Fig. 14.6b; see Appendix for details). This gives a mean geometric growth rate for infective mosquitoes over the season of $e^{\bar{\lambda}(t_a+t_b)} = e^{\lambda_a t_a} e^{\lambda_b t_b}$. Setting $\bar{\lambda} = 0$ in (14.11) gives the critical average mosquito level, $\hat{\bar{n}}_m^* > \hat{n}_m^*$, above which WN can invade a seasonally variable population and below which it cannot (Fig. 14.6b). Provided the lower of the two mosquito population levels, n_m^a, is below the threshold $\hat{n}_m^*$, disease-outbreak control requires only that the higher level, n_m^b, be reduced such that the average mosquito density $\bar{n}_m^* < \hat{\bar{n}}_m^*$. We therefore expect WN virus to be easier to control in more seasonal northern regions than in warmer southern regions where the population remains constant above $\hat{n}_m^*$ year-round.

14.9 Summary

In its simplest form, a model can be thought of as a "black box" which takes inputs, such as parameters and initial conditions, and produces outputs, such as disease thresholds, or outbreak levels over time. The conversion from inputs to outputs requires underlying hypotheses about the dynamical relationships between components. These hypotheses are then translated into equations, whose subsequent analysis and simulation yield the model outputs.

As mathematicians we often focus on the details of the black box, fixing the set of model equations, and deriving sophisticated methods for determining the model outputs. This chapter suggests an alternative and complementary activity: analysis of the role of inputs (parameters) and hypotheses (formalized into model structure) in determining the model outputs. Such analyses employ a suite of different models, with uncertain parameters and variable structure. The effects of the parameter uncertainty and model structure on the model outputs (such as predictions of $\mathcal{R}_0$) are then deduced.

We believe that this kind of comparative analysis approach is key for scientists wishing to interface biology with mathematical models, particularly in the area of epidemiology. The goal of this chapter is to demonstrate both the methods and the usefulness of the comparative analysis approach.

Our series of models describes the cross-infection of West Nile virus between birds and mosquitoes. The primary mathematical tool is the basic reproduction number $\mathcal{R}_0$, which is derived from mathematical epidemiology as the spectral radius of the next generation operator [42]. We calculate how $\mathcal{R}_0$ changes with differences in the disease transmission term (frequency dependence versus mass action, Sect. 14.4), with additional model classes (larval and exposed mosquito classes, Sect. 14.5) and with uncertain and variable parameters (Sect. 14.6). Finally, we make two applications of the model, one to WN virus control (Sect. 14.7) and one to outbreaks in seasonal environments (Sect. 14.8).

The calculation of $\mathcal{R}_0$ for the frequency-dependent and mass-action transmission term models shows a striking dependence of the model predictions on model structure. Although both transmission term models have a sound theoretical basis, they yield starkly contrasting predictions as to the effect of bird density on WN virus. When bird densities are low, the frequency-dependent model predicts remaining birds receive more bites and become local hot spots for disease transmission, with each bird having a high probability of becoming infected and passing on the virus. By way of contrast, the mass-action model predicts the disease will die out in regions of low bird density. Thus, while the mass-action model predicts that bird control would be effective in controlling WN, the frequency-dependent model predicts that it would be counterproductive (see [31] for further discussion).

The calculation of $\mathcal{R}_0$ for models with larval and exposed mosquito classes shows how the added complexity of a more realistic model does not always translate into refined model predictions. Here an additional larval class has no effect on the basic reproduction number, and hence on whether an outbreak will occur. Interestingly, the additional larval class does actually change the time-dependent dynamics if an outbreak actually occurs (Fig. 14.4). By way of contrast, the additional exposed class means that some infected mosquitoes may be removed before ever making it to the infective state. As our intuition would lead us to believe, this yields a reduced $\mathcal{R}_0$.

Our experience shows that parameterization of epidemiological models is a substantial task, requiring great familiarity with the biological literature. For example, the parameters shown in Fig. 14.2 (taken from [43]) originally came from 25 different sources, each of which had to be carefully read before the parameter could be extracted. However, as shown in Fig. 14.3, careful model parameterization allows us to incorporate the uncertainty of parameter values into ranges of predictions for $\mathcal{R}_0$, following the methods of [5]. In the case of WN virus models, variation in $\mathcal{R}_0$ arising from model structure was larger than variation arising from parameter uncertainty.

Finally, once a model is tested, via parameterization, validation and multimodel comparison, it is possible to make applications to different epidemiological scenarios. How can a disease be controlled? How is control managed in a seasonal mosquito population? These applications sometime require model extensions (Fig. 14.6) and unique perspective on model outputs (Fig. 14.5). However, it is the applications that allow us to move the science forward and, more often than not, the applications also lead to a new generation of models that promise to keep mathematicians employed for considerable time to come.

References

1. Anderson J., Andreadis T.G., Vossbrink C., Tirrell S., Wakem E., French R., Germendia A. & Van Kruiningen H. (1999) Isolation of West Nile virus from mosquitoes, crows, and a Cooper's hawk in Connecticut. *Science*, 286, 2331–2333.
2. Anderson R.M. & May (1991) *Infectious Diseases of Humans.* Oxford University Press, Oxford.
3. Begon M., Bennett M., Bowers R.G., French N.P., Hazel S.M. & Turner J. (2002) A clarification of transmission terms in host-microparasite models: numbers, densities and areas. *Epidemiology and Infection*, 129, 147–153.
4. Bernard K.A., Maffei J.G., Jones S.A., Kauffman E.B., Ebel G.D., Dupuis A.P., Jr, Ngo K.A., Nicholas D.C., Young D.M., Shi P.-Y., Kulasekera V.L., Eidson M., White D.J., Stone W.B., Team N.S.W.N.V.S. & Kramer L.D. (2001) West Nile virus infection in birds and mosquitoes, New York State, 2000. *Emerging Infectious Diseases*, 7, 679–685.
5. Blower S.M. & Dowlatabadi H. (1994) Sensitivity and uncertainty analysis of complex models of disease transmission: an HIV model, as an example. *International Statistical Review*, 62, 229–243.
6. Brauer F. & Castillo-Chavez C. (2001) *Mathematical Models in Population Biology and Epidemiology.* Springer, Berlin Heidelberg New York.
7. Burnham K.P. & Anderson D. (2002) *Model Selection and Multi-model Inference.* Springer, Berlin Heidelberg New York.
8. Case T.J. (1999) *An Illustrated Guide to Theoretical Ecology.* Oxford University Press, Oxford.
9. Castillo-Chavez C., with Blower S., van den Driessche P., Kirschner D. & Yakubu A.-A. (2002) *Mathematical Approaches for Emerging and Reemerging Infectious Diseases: An Introduction.* Springer, Berlin Heidelberg New York.
10. Chomel B.B. (2003) Control and prevention of emerging zoonoses. *Journal of Veterinary Medical Education*, 30, 145–147.
11. Daszak P. (2000) Emerging infectious diseases of wildlife - threats to biodiversity and human health. *Science*, 287, 1756–1756.
12. Diekmann O. & Heesterbeek H. (2000) *Mathematical Epidemiology of Infectious Diseases.* Wiley, New York.
13. Diekmann O., Heesterbeek J.A.P. & Metz J.A.J. (1990) On the Definition and the Computation of the Basic Reproduction Ratio $\mathcal{R}_0$ in Models for Infectious-Diseases in Heterogeneous Populations. *Journal of Mathematical Biology*, 28, 365–382.
14. Dobson A. (2004) Population dynamics of pathogens with multiple host species. *American Naturalist*, 164, S64–S78.
15. Dobson A. & Foufopoulos J. (2001) Emerging infectious pathogens of wildlife. *Philosophical Transactions of the Royal Society of London Series B*, 356, 1001–1012.
16. Enserink M. (2004) Emerging infectious diseases - a global fire brigade responds to disease outbreaks. *Science*, 303, 1605–1606.
17. Ezenwa V.O., Godsey M.S., King R.J. & Guptill S.C. (2006) Avian diversity and West Nile virus: testing associations between biodiversity and infectious disease risk. *Proceedings of the Royal Society of London Series B*, 273, 109–117.
18. Gubler D.J. (2002) The global emergence/resurgence of arboviral diseases as public health problems. *Archives of Medical Research*, 33, 330–342.
19. Gubler D.J., Campbell G.L., Nasci R., Komar N., Petersen L. & Roehrig J.T. (2000) West Nile virus in the United States: guidelines for detection, prevention, and control. *Viral Immunology*, 13, 469–475.
20. Haefner J. (1996) *Modeling Biological Systems: Principles and Applications.* Chapman & Hall, New York.
21. Hayes C.G. (1988) West Nile fever. In: Monath TP (ed.) *The Arboviruses: Epidemiology and Ecology V. 5* , CRC Press, Boca Raton, FL, pp. 59–88.

22. Heesterbeek H. (2002) A brief history of $\mathcal{R}_0$ and a recipe for its calculation. *Acta Biotheoretica*, 50, 189–204.
23. Hethcote H.W. (2000) The mathematics of infectious diseases. *SIAM Review*, 42, 599–653.
24. Higgs S., Schneider B.S., Vanlandingham D.L., Klingler K.A. & Gould E.A. (2005) Nonviremic transmission of West Nile virus. *Proceedings of the National Academy of Sciences of the United States of America*, 102, 8871–8874.
25. Hilborn R. & Mangel M. (1997) *The Ecological Detective: Confronting Models with Data*. Princeton University Press, Princeton, NJ.
26. Kermack W. & McKendrick A. (1927) A contribution to the mathematical theory of epidemics. *Proceedings of the Royal Society of London Series A*, 115, 700–721.
27. Kot M. (2001) *Elements of Mathematical Ecology*. Cambridge University Press, Cambridge.
28. Lanciotti R., Roehrig J., Deubel V., Smith J., Parker M., Steele K., Crise B., Volpe K., Crabtree M., Scherret J., Hall R., MacKenzie J., Cropp C., Panigrahy B., Ostlund E., Schmitt B., Malkinson M., Banet C., Weissman J., Komar N., Savage H., Stone W., McNamara T. & Gubler D. (1999) Origin of the West Nile virus responsible for an outbreak of encephalitis in the Northeastern United States. *Science*, 286, 2333–2337.
29. Levins D. (1968) *Evolution in a Changing Environment*. Princeton University Press, Princeton.
30. Lewis M.A., Renclawowicz J. & van den Driessche P. (2006a) Traveling waves and spread rates for a West Nile virus model. *Bulletin of Mathematical Biology*, 68, 3–23.
31. Lewis M.A., Renclawowicz J., van den Driessche P. & Wonham M.J. (2006b) A comparison of continuous and discrete time West Nile virus models. *Bulletin of Mathematical Biology*, 68, 491–509.
32. Liu R., Shuai J., Wu J. & Zhu H. (2006) Modeling spatial spread of West Nile virus and impact of directional dispersal of birds. *Mathematical Biosciences and Engineering*, 3, 145–160.
33. Macdonald G. (1957) *The Epidemiology and Control of Malaria*. Oxford University Press, Oxford.
34. McCallum H., Barlow N. & Hone J. (2001) How should pathogen transmission be modelled? *Trends in Ecology & Evolution*, 16, 295–300.
35. Murray J. (2002) *Mathematical Biology I: An Introduction, 3rd ed.* Springer, Berlin Heidelberg New York.
36. Petersen L.R. & Roehrig J.T. (2001) West Nile virus: a reemerging global pathogen. *Emerging Infectious Diseases*, 7, 611–614.
37. Peterson A.T., Komar N., Komar O., Navarro-Siguenza A., Robbins M.B. & Martinez-Meyer E. (2004) West Nile virus in the New World: potential impacts on bird species. *Bird Conservation International*, 14, 215–232.
38. Rappole J., Derrickson S.R. & Hubalek Z. (2000) Migratory birds and spread of West Nile virus in the Western hemisphere. *Emerging Infectious Diseases*, 6, 1–16.
39. Ross R. (1911) *The Prevention of Malaria*. Murray, London.
40. Spielman A. (2001) Structure and seasonality of Nearctic *Culex pipiens* populations. *Annals of the New York Academy of Sciences*, 951, 220–234.
41. van den Driessche P. & Watmough J. (2002) Reproduction numbers and sub-threshold endemic equilibria for compartmental models of disease transmission. *Mathematical Biosciences*, 180, 29–48.
42. Wonham M.J., de-Camino-Beck T. & Lewis M.A. (2004) An epidemiological model for West Nile virus: invasion analysis and control applications. *Proceedings of the Royal Society of London Series B*, 271, 501–507.
43. Wonham M.J., Lewis M.A., Rencawowicz J. & van den Driessche P. (2006) Transmission assumptions generate conflicting predictions in host-vector disease models: a case study in West Nile virus. *Ecology Letters*, 9, 706–725.

Appendix

We include this appendix to illustrate a different approach to calculating the disease growth rate, λ, for the full West Nile model (14.9). Although this exercise is somewhat redundant with the earlier $\mathcal{R}_0$ calculations, it provides an alternative and pedagogically useful perspective.

We use linear analysis to calculate the disease growth rate for the two-level mosquito population shown in Fig. 14.6a. To simplify the ODE system (14.9), we non-dimensionalise by scaling time t with the quantity $1/\kappa$ by setting $\tau = \kappa t$, scaling all parameters to κ, and scaling the bird and mosquito densities by the initial bird density N_B^*. In the resulting dimensionless system (14.14), the two bird compartments s_b and i_b indicate the fraction of the initial bird density in susceptible and infected classes, with the total live bird density $0 \leq n_b = (s_b + i_b) \leq 1$. The four mosquito compartments l_m, s_m, e_m, and i_m, represent larval, susceptible, exposed, and infected females scaled to the initial bird density, with the total female mosquito population density $0 \leq n_m = (l_m + s_m + e_m + i_m)$. The rescaled system is:

$$\frac{ds_b}{d\tau} = -\alpha_b \beta_b i_m \frac{s_b}{n_b}$$
$$\frac{di_b}{d\tau} = \alpha_b \beta_b i_m \frac{s_b}{n_b} - \delta_b i_b$$

$$\frac{dl_m}{d\tau} = b_l (s_m + e_m + i_m) - m_l l_m - d_l l_m \qquad (14.14)$$
$$\frac{ds_m}{d\tau} = -\alpha_m \beta_b s_m \frac{i_b}{n_b} + m_l l_m - d_m s_m$$
$$\frac{de_m}{d\tau} = \alpha_m \beta_b s_m \frac{i_b}{n_b} - e_m - d_m e_m$$
$$\frac{di_m}{d\tau} = e_m - d_m i_m$$

where the subscripts b and m indicate the dimensionless versions of the dimensional variables and parameters defined in Tables 14.1 and 14.2. As in the dimensional system, we ensure a constant mosquito population density by setting $b_l = d_m(m_l + d_l)/m_l$.

For the DFE for this system, $(s_b, i_b, l_m, s_m, e_m, i_m) = (1, 0, b_l n_m^*/(m_l+d_l), n_m^*, 0, 0)$, we define small perturbations in each variable, $(\tilde{s}_b, \tilde{i}_b, \tilde{l}_m, \tilde{s}_m, \tilde{e}_m, \tilde{i}_m)$. The corresponding Jacobian matrix, J, describes the linearization with respect to $(\tilde{i}_b, \tilde{l}_m, \tilde{s}_m, \tilde{e}_m, \tilde{i}_m)$:

$$J = \begin{bmatrix} -\delta_b & 0 & 0 & 0 & \alpha_b \beta_b \\ 0 & -m_l - d_l & b_l & b_l & b_l \\ -\alpha_m \beta_b n_m^* & m_l & -d_m & 0 & 0 \\ \alpha_m \beta_b n_m^* & 0 & 0 & -d_m - 1 & 0 \\ 0 & 0 & 0 & 1 & -d_m \end{bmatrix}. \qquad (14.15)$$

(The term $\tilde{s}_b$ is not included because it decouples from the rest of the system; in other words, the 6×6 matrix that includes $\tilde{s}_b$ has an entire column of zeroes.) This yields the characteristic polynomial in λ:

$$0 = Det(J-\lambda I) = \lambda \left(\lambda + d_m + \frac{m_l b_l}{d_m} \right) (\lambda^3 + a_1\lambda^2 + a_2\lambda + a_3), \quad (14.16)$$

where I is the 5×5 identity matrix and $a_1 > 0$, $a_2 > 0$. The zero root of the 5th order polynomial comes from the steady-state condition $b_l = d_m(m_l + d_l)/m_l$ that means the disease-free mosquito population is constant. For $a_3 > 0$, and $a_1 a_2 > a_3$, by the Roth–Hurwitz conditions, all roots of the cubic polynomial in λ have negative real parts. Some algebra shows that $a_1 a_2 > a_3$, since

$$\begin{aligned}
a_1 &= 1 + \delta_b + 2d_m \\
a_2 &= d_m^2 + 2\delta_b d_m + \delta_b + d_m \\
a_3 &= \delta_b d_m^2 - \alpha_b \alpha_m \beta_b^2 n_m^* + \delta_b d_m
\end{aligned} \quad (14.17)$$

The disease outbreak threshold is thus when $a_3 = 0$ or equivalently, when zero is the largest eigenvalue of J. In biological terms this threshold may be thought of as a disease growth rate of zero, which corresponds directly to the reproduction number threshold, $\mathcal{R}_0 = 1$.

Table 14.1 Variables for West Nile virus model. Subscripts M and m refer to mosquitoes and B and b to birds; capital letters refer to dimensional forms and lower case to nondimensional forms, which are rescaled to N_B^*. Dashes indicate term not used

Meaning	Dimensional	Dimensionless
Mosquitoes		
Larval female mosquito density	L_M	l_m
Susceptible adult female mosquito density	S_M	s_m
Exposed adult female mosquito density	E_M	e_m
Infectious adult female mosquito density	I_M	i_m
Total female mosquito density,	N_M	n_m
$N_M = L_M + S_M + E_M + I_M$		
Total female mosquito density at the	N_M^*	n_m^*
disease-free equilibrium		
Threshold mosquito density for disease	$\hat{N}_M^*$	$\hat{n}_m^*$
outbreak, given constant population		
Average mosquito density across a season,	$-$	$\bar{n}_m$
given variable population		
Average mosquito density	$-$	$\bar{n}_m^*$
at the disease-free equilibrium		
Threshold average mosquito density for	$-$	$\hat{\bar{n}}_m^*$
disease outbreak, given variable population		
Birds		
Susceptible bird density	S_B	s_b
Infectious bird density	I_B	i_b
Total bird density $N_B = S_B + I_B$	N_B	n_b
Bird density at which frequency dependent and	$\tilde{N}_B$	$-$
mass action disease-transmission terms coincide,		
giving identical mosquito biting rates		
Total bird density at the disease-free equilibrium	N_B^*	1

Table 14.2 Parameters and estimated values for the West Nile virus models. Subscripts M and m refer to mosquitoes (parameterized primarily for *Culex pipiens* spp.), and B and b to birds (parameterized primarily for American crows, *Corvus brachyrhynchos*). Capital-letter subscripts refer to dimensional parameters and lower-case letters to dimensionless forms; numerical estimates given for dimensional forms only. All rates are *per capita* per day. For details and sources of parameter estimates, see [43] Tables 2–3 and sources therein. Dashes indicate term not used in numerical calculations

Meaning	Dimensional	Dimensionless	Mean	Range
Mosquitoes				
Birth rate for model with no larval class	$b_M = d_M$	b_m	–	–
Birth rate for model with larval class	$b_L = d_M(m_L + d_L)/m_L$	b_l	–	–
Natural death rate, adults	d_M	d_m	0.03	0.02–0.07
Natural death rate, larvae	d_L	d_l	0.02	0.01–0.06
Maturation rate	m_L	m_l	0.07	0.05–0.09
Biting rate under frequency-dependence	β_B	β_b	0.44	0.34–0.53
Biting parameter under mass action	$\beta'_B = \beta_B \bar{N}^*_B$	–	–	–
Probability of virus transmission to mosquito, per infectious bite	α_M	α_m	0.69	0.23–1.00
Virus incubation rate	κ_M	κ_m	0.10	0.09–0.12
Proportion surviving viral incubation period	$\phi_M = \kappa_M/(d_M + \kappa_M)$	ϕ_m	–	–
Birds				
Probability of virus transmission to bird, per infectious bite	α_B	α_b	0.74	0.27–1.00
Death rate from virus	δ_B	δ_b	0.20	0.17–0.25

Suggested Exercises and Projects

Exercises

1. A survey of Yale University freshmen in 1982 about an influenza outbreak reported that 91.1% were susceptible to influenza at the beginning of the year and 51.4% were susceptible at the end of the year. Assume that the mean infective period is approximately 3 days.

   (a) Estimate the basic reproduction number and decide whether there was an epidemic.
   (b) What fraction of Yale students in Exercise (a) would have had to be immunized to prevent an epidemic?
   (c) What was the maximum number of Yale students in Exercises (a) and (b) suffering from influenza at any time?

2. A disease is introduced by two visitors into a town with 1,200 inhabitants. An average infective is in contact with 0.4 inhabitant per day. The average duration of the infective period is 6 days, and recovered infectives are immune against reinfection. How many inhabitants would have to be immunized to avoid an epidemic?

3. A disease begins to spread in a population of 800. The infective period has an average duration of 14 days and an average infective is in contact with 0.1 person per day. What is the basic reproduction number? To what level must the average rate of contact be reduced so that the disease will die out?

4. An epidemic of a communicable disease that does not cause death but from which infectives do not recover may be modeled by the pair of differential equations

$$S' = -\beta SI, \quad I' = \beta SI.$$

Show that in a population of fixed size K such a disease will eventually spread to the entire population.

5. If a fraction λ of the population susceptible to a disease that provides immunity against reinfection moves out of the region of an epidemic, the situation may be modeled by a system

$$S' = -\beta SI - \lambda S, \quad I' = \beta SI - \alpha I.$$

Show that both S and I approach zero as $t \to \infty$.

6. Compare the qualitative behaviors of the models

$$S' = -\beta SI, \quad I' = \beta SI - \alpha I,$$

and

$$S' = -\beta SI, \quad E' = \beta SI - \kappa E, \quad I' = \kappa E - \alpha I,$$

with

$$\beta = 1/3,000, \quad \alpha = 1/6, \quad \kappa = 1/2, \quad S(0) = 999, \quad I(0) = 1.$$

These models represent an SIR epidemic model and an $SEIR$ epidemic model respectively with a mean infective period of 6 days and a mean exposed period of 2 days. Do numerical simulations to decide whether the exposed period noticeably affects the behavior of the model.

7. To the models in the previous exercise add a constant birth rate of $100/7$ births per year and a constant death rate of $1/70$ per year. Compare the behaviors of these models with each other and with the models of the previous exercise.

8. Consider the usual SEIR model

$$\begin{aligned}
S' &= \Pi - \mu S - \beta SI, \\
E' &= \beta SI - (\mu + \kappa)E, \\
I' &= \kappa E - (\mu + \alpha)I, \\
R' &= \alpha I - \mu R,
\end{aligned}$$

where individuals progress from compartment E to I at a rate κ and develop immunity at a rate α, natural mortality claims individuals at a rate μ, and there is a constant recruitment, Π, of susceptible individuals. The basic reproduction number $\mathcal{R}_0$ is calculated as $\mathcal{R}_0 = \frac{\kappa \beta \Pi / \mu}{(\mu + \kappa)(\mu + \alpha)}$, where $S_0 = \Pi / \mu$.

(a) Interpret the above formula for the basic reproduction number.
(b) Verify that the disease-free equilibrium is (locally asymptotically) stable for $\mathcal{R}_0 < 1$ and unstable for $\mathcal{R}_0 > 1$.

(c) Include disease induced death in the above model and compute the basic reproduction number.

9. Consider a model for a disease that confers only temporary immunity after recovery, so that recovered individuals lose their immunity at a per capita rate of c (per time unit). Formulate a model to describe such a disease and analyze its qualitative behavior. Is there a threshold condition for this SIRS model?

10. European fox rabies is estimated to have a transmission coefficient β of 80 km^2 years/fox and an average infective period of 5 days. There is a critical carrying capacity K_c measured in foxes per km^2, such that in regions with fox density less than K_c rabies tends to die out while in regions with fox density greater than K_c rabies tends to persist. Estimate K_c. [Remark: It has been suggested in Great Britain that hunting to reduce the density of foxes below the critical carrying capacity would be a way to control the spread of rabies.]

11. Consider a disease spread by carriers who transmit the disease without exhibiting symptoms themselves. Let $C(t)$ be the number of carriers and suppose that carriers are identified and isolated from contact with others at a constant per capita rate α, so that $C' = -\alpha C$. The rate at which susceptibles become infected is proportional to the number of carriers and to the number of susceptibles, so that $S' = -\beta SC$. Let C_0 and S_0 be the number of carriers and susceptibles, respectively, at time $t = 0$.

   (a) Determine the number of carriers at time t from the first equation.
   (b) Substitute the solution to part (a) into the second equation and determine the number of susceptibles at time t.
   (c) Find $\lim\limits_{t \to \infty} S(t)$, the number of members of the population who escape the disease.

12. In this exercise, the expected duration of an epidemic is calculated for different values of the population size N and the basic reproduction number $\mathcal{R}_0$ in the CTMC SIS epidemic model.

   (a) Let the population size $N = 25$, contact rate $\beta = 1$, and birth and recovery rates $b = 1/4 = \gamma$, so that $\mathcal{R}_0 = 2$ in the CTMC SIS epidemic model. Calculate the expected duration $\tau_k = E(T_k)$, $k = 1, \ldots, N$, i.e., $\tau = -D^{-1}\mathbf{1}$, where $\tau = (\tau_1, \ldots, \tau_N)^T$. Then sketch a graph of τ_k for $k = 1, \ldots, N$. Maple, MATLAB or other software may be useful in solving the linear system.
   (b) Use the mean τ computed in part (a) to find the second moment $\tau_k^2 = E(T_k^2)$, $k = 1, \ldots, N$ ($\tau^2 = -D^{-1}\tau$). Then compute the variance in the time to extinction, $\sigma_k^2 = \tau_k^2 - (\tau_k)^2$.

(c) Let $b = 1/2 = \gamma$ so that $\mathcal{R}_0 = \beta$. Calculate the expected duration $\tau_k = E(T_k)$ for different values of the contact rate β and the population size N. Suppose time units are expressed in terms of days. Give the value of τ_N in terms of months or years (whatever unit is appropriate). What happens to τ_N as β increases? as N increases?

13. In this exercise, the approximate quasistationary distribution for the infected population is computed for the CTMC SIS epidemic model and compared to the equilibrium solution of the deterministic model. Assume the approximate quasistationary distribution satisfies

$$p^1_{i+1} = \frac{b(i)}{d(i+1)} p^1_i,$$

where $\sum_{i=1}^{N} p^1_i = 1$, $b(i) = \beta i(N-i)/N$, and $d(i) = (b+\gamma)i$.

(a) Let the population size $N = 50$, contact rate $\beta = 1$, and birth and recovery rates $b = 1/4 = \gamma$. First find the equilibrium solution for I in the deterministic SIS epidemic model. Then find the approximate quasistationary distribution, p^1. Graph p^1 and compute its mean value. How does the mean value compare to the equilibrium solution?

(b) Let $b = 1/2 = \gamma$ so that $\mathcal{R}_0 = \beta$. Choose different values for N and β. For each choice of N and β compute the equilibrium solution for I in the deterministic SIS epidemic model and the approximate quasistationary distribution for the CTMC SIS epidemic model. How do the equilibrium solutions and the mean of the quasistationary distributions compare for different values of N and β?

14. In this exercise, sample paths for the Itô SDE SIS epidemic model are computed and compared to the equilibrium solution of the deterministic model.

(a) Write a computer program for the Itô SDE SIS epidemic model using Euler's method with $\Delta t = 0.01$, population size $N = 100$, contact rate $\beta = 2$, birth and recovery rates $b = 1/2 = \gamma$, and initial number of infected individuals $I(0) = 1$. Graph three sample paths of the Itô SDE for $t \in [0, 20]$. Then graph three sample paths for the same parameter values but for $I(0) = 5$.

(b) Graph the mean of 1,000 sample paths for the two different sets of parameter values in part (a). How do your results for the mean compare with the equilibrium solution of the deterministic model?

Projects

These suggested projects are taken in part from the recent book "A Course in Mathematical Biology: Quantitative Modeling with Mathematical and Computational Methods" by Gerda de Vries, Thomas Hillen, Mark Lewis, Johannes Müller, and Birgit Schönfisch, Mathematical Modeling and Computation 12, SIAM, Philadelphia (2006).

1 Cholera

The cholera virus, *Vibric cholerae*, is present in brackish water through algae blossom and through human faeces. Not every infection leads to sickness. Only 10–20% of infected individuals suffer from severe symptoms. Many do not show symptoms at all but their faeces are infectious. Cholera is a serious disease since the progress of symptoms can be very fast if not treated.

Large outbreaks are usually related to contaminated water. There are four major control mechanisms, which are recommended by the WHO: hygienic disposal of human faeces, adequate supply of safe drinking water, good food hygiene and cooking, washing hands after defecation and before meals. More information about this disease, control mechanisms and vaccination can be found at the WHO Web sites (www.who.int).

Develop a model for an outbreak of cholera:

1. Model the epidemic first without any control mechanism.
2. Extend your model to include the above control mechanisms and estimate which is most effective.

2 Ebola

The Ebola virus erupts occasionally in Africa. Ebola causes hemorrhaging and death in humans after about 10 days, and people in contact with infectives can be infected. Quarantine (isolation) of patients is an effective control procedure for Ebola. Develop a model for the spread of Ebola that includes quarantine of a fraction of the patients.

3 Gonorrhea

Gonorrhea is a sexually transmitted disease caused by a gonococcus bacteria. Assume that it is spread from women to men and from men to women.

Recovery from gonorrhea does not confer immunity. Formulate a model for gonorrhea with heterosexual transmission. How would you change your model to include consistent condom use by a fraction of the population?

4 HIV/AIDS

The human immunodeficiency virus (HIV), which is the etiological agent for acquired immunodeficiency syndrome (AIDS), emerged in 1981 and has become an important sexually-transmitted disease throughout the world. The two main components of transmission are needle sharing among injecting drug users, and prostitution.

1. In the absence of intervention methods or changes in social behavior, what is the expected size of the HIV epidemic, either as peak, final size or endemic level?
2. Some locations offer free needle exchange, so that injecting drug users can get clean needles. What effect would a needle-exchange program have on your model?

5 HIV in Cuba

In the article "A non-linear model for a sexually transmitted disease with contact tracing" by H. De Arazoza and R. Lounes in the IMA Journal of Mathematical Medicine and Biology, 19 (2002), pp. 221–234, we find the following data about HIV-positives, AIDS outbreak and death cases caused by AIDS from 1986 until 1997 in Cuba.

Year	HIV-cases	AIDS-cases	Death through AIDS
1986	99	5	2
1987	75	11	4
1988	93	14	6
1989	121	13	5
1990	140	28	23
1991	183	37	17
1992	175	71	32
1993	102	82	59
1994	122	102	62
1995	124	116	80
1996	234	99	92
1997	364	121	99
1998	362	150	98
1999	493	176	122
2000	545	251	142

Design a model which describes the epidemic spread of HIV after 1997 in Cuba and fit the above data. Which are the relevant parameters of your model? Try to introduce control mechanisms to lower the number of AIDS cases. Compare your control mechanism with the data of the given time period.

6 Human Papalonoma Virus

According to the Health Canada Web site (www.hc-sc.gc.ca), "HPV is likely one of the most common sexually transmitted infections (STIs) in Canada". Several types of HPV are know to circulate in the population. Some types lead to genital warts, while others lead to cancers. The virus is often asymptomatic and can switch between active and inactive states. Develop a model for HPV transmission. Include vaccination in you model, and use your model to estimate the fraction of the population that would need to be effectively vaccinated to control the disease.

7 Influenza

The recent rapid spread of avian influenza and the potential for the emergence of a pandemic strain of the virus are concerns of governments worldwide. Current influenza vaccines will most likely provide little protection against a shift in the virus, and the main control methods are antiviral treatments, quarantine and isolation.

1. Assuming a significant shift does occur, what is the expected size of an outbreak?
2. Would quarantine and isolation be as effective with influenza as they were with SARS?
3. How many doses of antivirals would be needed to control an epidemic in the Greater Toronto area?

8 Malaria

Now that mosquitoes are resistant to DDT, malaria has reemerged in many areas and is spreading into new regions as temperature changes occur. Malaria spreads from infected mosquitoes (the vector) to humans (the host) by biting, and susceptible mosquitoes can be infected when they bite an infected human. Humans can recover from malaria, but infected mosquitoes remain infected for their lifetime.

1. In the absence of intervention methods or change in social behavior, what is the expected size of the malaria burden, either as peak, final size or endemic level?
2. How effective could bednets be at reducing the cost of malaria?

9 Measles

Measles is no longer endemic in Canada, although small, isolated outbreaks can still occur among unvaccinated groups. Worldwide, an estimated 30 million infections occur each year with over 500,000 deaths in 2003 (www.hc-sc.gc.ca). Develop a model for measles transmission in Canada and estimate the fraction of the population that must remain vaccinated to maintain Canada's "herd immunity". Extend your model to a two patch model, with one patch vaccinated and one unvaccinated. How does travel between the two patches influence the dynamics in the vaccinated patch?

10 Poliomyelitis (Polio)

Polio is spread by a wild enteric coxackie virus and can cause paralysis in some people. The polio vaccine virus interferes with binding of the wild virus by filling the attachment site. Thus the vaccine virus interferes with the wild virus in the sense that a person cannot have both. Recovery from a polio infection gives immunity.

Most cases of polio are asymptomatic, but a small fraction of cases result in paralysis. In the 1950s in the United States, there were about 60,000 paralytic polio cases per year. In 1955 Jonas Salk developed an injectable polio vaccine from an inactivated polio virus. This vaccine provides protection for the person, but the person can still harbor live viruses in their intestines and can pass them to others. In 1961 Albert Sabin developed an oral polio vaccine from weakened strains of the polio virus. This vaccine provokes a powerful immune response, so the person cannot harbor the "wild-type" polio viruses, but a very small fraction (about one in 2 million) of those receiving the oral vaccine develop paralytic polio. The Salk vaccine interrupted polio transmission and the Sabin vaccine eliminated polio epidemics in the United States, so there have been no indigenous cases of naturally-occurring polio since 1979. In order to eliminate the few cases of vaccine-related paralytic polio each year, the United States now recommends the Salk injectable vaccine for the first four polio vaccinations, even though it is more expensive. In the Americas, the last case of paralytic polio caused by the wild virus was in Peru in 1991. In 1988 WHO set a goal of global polio eradication by the year 2000. Most countries are using the live-attenuated Sabin vaccine, because it is inexpensive (8 cents per dose) and can be easily administered into a mouth by an untrained volunteer. The WHO strategy includes routine vaccination, National Immunization Days (during which many people in a country or region are vaccinated in order to interrupt transmission), mopping-up vaccinations, and surveillance for acute flaccid paralysis. Polio has disappeared from many countries in the past 10 years, so that by 1999 it is concentrated in the Eastern Mediterranean region, South Asia, West Africa and Central Africa. It is likely that polio will be eradicated worldwide soon. WHO estimates that eradicating polio will save about $1.5 billion each year in immunization, treatment, and rehabilitation around the globe.

Formulate a model for polio with the wild and vaccine virus competing for the attachment site. How would your model be changed if the vaccine virus were transmissible?

11 Severe Acute Respiratory Syndrome (SARS)

Detailed data on the day-to-day probable and suspect cases for 2002–2003 Severe Acute Respiratory Syndrome (SARS) are given at
  http://www.hc-sc.gc.ca/pphb-dgspsp/sars-sras/cn-cc/index.html.
Go to this Web site, and see what data were recorded there.

Construct a model that can be used to determine the number of new SARS cases, SARS deaths, and SARS recoveries each day. Initially assume that there is no quarantining of SARS patients, and that there are no measures taken to reduce the likelihood of infection from one individual to another. One key disease control goal was to eradicate the outbreak of SARS through quarantining and preventative measures. Assess the effectiveness of these control measures on the disease dynamics.

12 Smallpox

An outbreak of smallpox in Abakaliki in southeastern Nigeria in 1967 has been reported by Bailey and Thomas. People living there belong to a religious group that is quite isolated and declines vaccination. Overall there were 30 cases of infection in a population of 120 individuals. The time (in days) between newly reported pox-cases is given in the following sequence:

$$13, 7, 2, 3, 0, 0, 1, 4, 5, 3, 2, 0, 2, 0, 5, 3, 1, 4, 0, 1, 1, 1, 2, 0, 1, 5, 0, 5, 5$$

Develop a model which describes these data and analyze the epidemic outbreak.

13 Tuberculosis

Worldwide, tuberculosis (TB) accounts for more deaths than all other diseases combined. The standard treatment for active tuberculosis is to give multiple drugs for at least 6 months. This therapy is effective if the person has drug-sensitive TB. Drug resistant strains of TB emerge when people do not complete the treatment.

1. Formulate a model for TB with drug-sensitive and drug-resistant strains of TB.
2. How would your model be changed to include improved compliance with drug therapy?

14 West Nile Virus

The West Nile virus is a vector born disease that has been found in over 150 bird species in North America. The virus is transmitted from bird to bird by mosquitoes. Bites by infected mosquitoes can also lead to infection in humans and other mammals. Develop a model of West Nile virus transmission between birds, mosquitoes and humans. What factors have the highest influence on the prevalence of the virus in the mosquito population? Currently, the West Nile virus season runs from April through September in Canada. Would you expect the prevalence of the virus in the mosquito population to increase as a result of global warming?

15 Yellow Fever in Senegal 2002

Yellow fever (YF) is a viral hoemorrhagic fever transmitted by infected mosquitoes. Yellow fever is spread into human populations in three stages:

1. *Sylvatic (or jungle).* YF occurs in tropical rain forests where mosquitoes, which feed on infected monkeys, pass the virus to humans who work in the forest.
2. *Intermediate.* YF occurs as infected individuals bring the disease into rural villages, where it is spread by mosquitoes amongst humans (and also monkeys).
3. *Urban.* YF occurs as soon as an infected individual enters urban areas. This can lead to an explosive epidemic in densely inhabited regions. Domestic mosquitoes carry the virus from person to person.

The epidemic can be controlled by vaccination. YF vaccine is safe and effective and provides immunity within 1 week in 95% of those vaccinated.

Below is a data set of YF cases and YF deaths of an outbreak in Senegal in 2002 collected from the internet archives of the World Health Organization (WHO). As soon as the virus was identified a vaccination program was started (Oct 1, 2002). On Oct 11, 2002 the disease was reported in Touba, a city of 800,000 residents. More information can be found on the WHO Web sites (www.who.int).

Report date	Cases (total)	Deaths (total)
Jan 18th	18	0
Oct 4th	12	0
Oct 11th	15	2
Oct 17th	18	2
Oct 24th	41	4
Oct 31st	45	4
Nov 20th	57	10
Nov 28th	60	11

1. Develop a model for the three stages of YF as outlined above.
2. Include a fourth stage that describes vaccination in urban areas.
3. Fit your model to the data.
4. What would have happened without vaccination?
5. Would you expect that the disease dies out, or that it becomes persistent?

Index

Lecture Notes in Mathematics

For information about earlier volumes
please contact your bookseller or Springer
LNM Online archive: springerlink.com

Vol. 1854: O. Saeki, Topology of Singular Fibers of Differential Maps (2004)

Vol. 1855: G. Da Prato, P.C. Kunstmann, I. Lasiecka, A. Lunardi, R. Schnaubelt, L. Weis, Functional Analytic Methods for Evolution Equations. Editors: M. Iannelli, R. Nagel, S. Piazzera (2004)

Vol. 1856: K. Back, T.R. Bielecki, C. Hipp, S. Peng, W. Schachermayer, Stochastic Methods in Finance, Bressanone/Brixen, Italy, 2003. Editors: M. Fritelli, W. Runggaldier (2004)

Vol. 1857: M. Émery, M. Ledoux, M. Yor (Eds.), Séminaire de Probabilités XXXVIII (2005)

Vol. 1858: A.S. Cherny, H.-J. Engelbert, Singular Stochastic Differential Equations (2005)

Vol. 1859: E. Letellier, Fourier Transforms of Invariant Functions on Finite Reductive Lie Algebras (2005)

Vol. 1860: A. Borisyuk, G.B. Ermentrout, A. Friedman, D. Terman, Tutorials in Mathematical Biosciences I. Mathematical Neurosciences (2005)

Vol. 1861: G. Benettin, J. Henrard, S. Kuksin, Hamiltonian Dynamics – Theory and Applications, Cetraro, Italy, 1999. Editor: A. Giorgilli (2005)

Vol. 1862: B. Helffer, F. Nier, Hypoelliptic Estimates and Spectral Theory for Fokker-Planck Operators and Witten Laplacians (2005)

Vol. 1863: H. Führ, Abstract Harmonic Analysis of Continuous Wavelet Transforms (2005)

Vol. 1864: K. Efstathiou, Metamorphoses of Hamiltonian Systems with Symmetries (2005)

Vol. 1865: D. Applebaum, B.V. R. Bhat, J. Kustermans, J. M. Lindsay, Quantum Independent Increment Processes I. From Classical Probability to Quantum Stochastic Calculus. Editors: M. Schürmann, U. Franz (2005)

Vol. 1866: O.E. Barndorff-Nielsen, U. Franz, R. Gohm, B. Kümmerer, S. Thorbjønsen, Quantum Independent Increment Processes II. Structure of Quantum Lévy Processes, Classical Probability, and Physics. Editors: M. Schürmann, U. Franz, (2005)

Vol. 1867: J. Sneyd (Ed.), Tutorials in Mathematical Biosciences II. Mathematical Modeling of Calcium Dynamics and Signal Transduction. (2005)

Vol. 1868: J. Jorgenson, S. Lang, $Pos_n(R)$ and Eisenstein Series. (2005)

Vol. 1869: A. Dembo, T. Funaki, Lectures on Probability Theory and Statistics. Ecole d'Eté de Probabilités de Saint-Flour XXXIII-2003. Editor: J. Picard (2005)

Vol. 1870: V.I. Gurariy, W. Lusky, Geometry of Müntz Spaces and Related Questions. (2005)

Vol. 1871: P. Constantin, G. Gallavotti, A.V. Kazhikhov, Y. Meyer, S. Ukai, Mathematical Foundation of Turbulent Viscous Flows, Martina Franca, Italy, 2003. Editors: M. Cannone, T. Miyakawa (2006)

Vol. 1872: A. Friedman (Ed.), Tutorials in Mathematical Biosciences III. Cell Cycle, Proliferation, and Cancer (2006)

Vol. 1873: R. Mansuy, M. Yor, Random Times and Enlargements of Filtrations in a Brownian Setting (2006)

Vol. 1874: M. Yor, M. Émery (Eds.), In Memoriam Paul-André Meyer - Séminaire de Probabilités XXXIX (2006)

Vol. 1875: J. Pitman, Combinatorial Stochastic Processes. Ecole d'Eté de Probabilités de Saint-Flour XXXII-2002. Editor: J. Picard (2006)

Vol. 1876: H. Herrlich, Axiom of Choice (2006)

Vol. 1877: J. Steuding, Value Distributions of L-Functions (2007)

Vol. 1878: R. Cerf, The Wulff Crystal in Ising and Percolation Models, Ecole d'Eté de Probabilités de Saint-Flour XXXIV-2004. Editor: Jean Picard (2006)

Vol. 1879: G. Slade, The Lace Expansion and its Applications, Ecole d'Eté de Probabilités de Saint-Flour XXXIV-2004. Editor: Jean Picard (2006)

Vol. 1880: S. Attal, A. Joye, C.-A. Pillet, Open Quantum Systems I, The Hamiltonian Approach (2006)

Vol. 1881: S. Attal, A. Joye, C.-A. Pillet, Open Quantum Systems II, The Markovian Approach (2006)

Vol. 1882: S. Attal, A. Joye, C.-A. Pillet, Open Quantum Systems III, Recent Developments (2006)

Vol. 1883: W. Van Assche, F. Marcellàn (Eds.), Orthogonal Polynomials and Special Functions, Computation and Application (2006)

Vol. 1884: N. Hayashi, E.I. Kaikina, P.I. Naumkin, I.A. Shishmarev, Asymptotics for Dissipative Nonlinear Equations (2006)

Vol. 1885: A. Telcs, The Art of Random Walks (2006)

Vol. 1886: S. Takamura, Splitting Deformations of Degenerations of Complex Curves (2006)

Vol. 1887: K. Habermann, L. Habermann, Introduction to Symplectic Dirac Operators (2006)

Vol. 1888: J. van der Hoeven, Transseries and Real Differential Algebra (2006)

Vol. 1889: G. Osipenko, Dynamical Systems, Graphs, and Algorithms (2006)

Vol. 1890: M. Bunge, J. Funk, Singular Coverings of Toposes (2006)

Vol. 1891: J.B. Friedlander, D.R. Heath-Brown, H. Iwaniec, J. Kaczorowski, Analytic Number Theory, Cetraro, Italy, 2002. Editors: A. Perelli, C. Viola (2006)

Vol. 1892: A. Baddeley, I. Bárány, R. Schneider, W. Weil, Stochastic Geometry, Martina Franca, Italy, 2004. Editor: W. Weil (2007)

Vol. 1893: H. Hanßmann, Local and Semi-Local Bifurcations in Hamiltonian Dynamical Systems, Results and Examples (2007)

Vol. 1894: C.W. Groetsch, Stable Approximate Evaluation of Unbounded Operators (2007)

Vol. 1895: L. Molnár, Selected Preserver Problems on Algebraic Structures of Linear Operators and on Function Spaces (2007)

Vol. 1896: P. Massart, Concentration Inequalities and Model Selection, Ecole d'Été de Probabilités de Saint-Flour XXXIII-2003. Editor: J. Picard (2007)

Vol. 1897: R. Doney, Fluctuation Theory for Lévy Processes, Ecole d'Été de Probabilités de Saint-Flour XXXV-2005. Editor: J. Picard (2007)

Vol. 1898: H.R. Beyer, Beyond Partial Differential Equations, On linear and Quasi-Linear Abstract Hyperbolic Evolution Equations (2007)

Vol. 1899: Séminaire de Probabilités XL. Editors: C. Donati-Martin, M. Émery, A. Rouault, C. Stricker (2007)

Vol. 1900: E. Bolthausen, A. Bovier (Eds.), Spin Glasses (2007)

Vol. 1901: O. Wittenberg, Intersections de deux quadriques et pinceaux de courbes de genre 1, Intersections of Two Quadrics and Pencils of Curves of Genus 1 (2007)

Vol. 1902: A. Isaev, Lectures on the Automorphism Groups of Kobayashi-Hyperbolic Manifolds (2007)

Vol. 1903: G. Kresin, V. Maz'ya, Sharp Real-Part Theorems (2007)

Vol. 1904: P. Giesl, Construction of Global Lyapunov Functions Using Radial Basis Functions (2007)

Recent Reprints and New Editions

4. Manuscripts should in general be submitted in English. Final manuscripts should contain at least 100 pages of mathematical text and should always include

   – a general table of contents;

   – an informative introduction, with adequate motivation and perhaps some historical remarks: it should be accessible to a reader not intimately familiar with the topic treated;

   – a global subject index: as a rule this is genuinely helpful for the reader.

   Lecture Notes volumes are, as a rule, printed digitally from the authors' files. We strongly recommend that all contributions in a volume be written in the same LaTeX version, preferably LaTeX2e. To ensure best results, authors are asked to use the LaTeX2e style files available from Springer's web-server at

   ftp://ftp.springer.de/pub/tex/latex/svmonot1/ (for monographs) and

   ftp://ftp.springer.de/pub/tex/latex/svmultt1/ (for summer schools/tutorials).

   Additional technical instructions, if necessary, are available on request from:

   lnm@springer.com.

5. Careful preparation of the manuscripts will help keep production time short besides ensuring satisfactory appearance of the finished book in print and online. After acceptance of the manuscript authors will be asked to prepare the final LaTeX source files (and also the corresponding dvi-, pdf- or zipped ps-file) together with the final printout made from these files. The LaTeX source files are essential for producing the full-text online version of the book. For the existing online volumes of LNM see:

   http://www.springerlink.com/openurl.asp?genre=journal&issn=0075-8434.

   The actual production of a Lecture Notes volume takes approximately 8 weeks.

6. Volume editors receive a total of 50 free copies of their volume to be shared with the authors, but no royalties. They and the authors are entitled to a discount of 33.3 % on the price of Springer books purchased for their personal use, if ordering directly from Springer.

7. Commitment to publish is made by letter of intent rather than by signing a formal contract. Springer-Verlag secures the copyright for each volume. Authors are free to reuse material contained in their LNM volumes in later publications: a brief written (or e-mail) request for formal permission is sufficient.

Addresses:

Professor J.-M. Morel, CMLA,
École Normale Supérieure de Cachan,
61 Avenue du Président Wilson, 94235 Cachan Cedex, France
E-mail: Jean-Michel.Morel@cmla.ens-cachan.fr

Professor F. Takens, Mathematisch Instituut,
Rijksuniversiteit Groningen, Postbus 800,
9700 AV Groningen, The Netherlands
E-mail: F.Takens@math.rug.nl

Professor B. Teissier, Institut Mathématique de Jussieu,
UMR 7586 du CNRS, Équipe "Géométrie et Dynamique",
175 rue du Chevaleret, 75013 Paris, France
E-mail: teissier@math.jussieu.fr

For the "Mathematical Biosciences Subseries" of LNM :

Professor P. K. Maini, Center for Mathematical Biology,
Mathematical Institute, 24-29 St Giles,
Oxford OX1 3LP, UK
E-mail : maini@maths.ox.ac.uk

Springer, Mathematics Editorial I, Tiergartenstr. 17,
69121 Heidelberg, Germany,
Tel.: +49 (6221) 487-8410
Fax: +49 (6221) 487-8355
E-mail: lnm@springer.com